THE LIVES OF DESERT ANIMALS IN JOSHUA TREE NATIONAL MONUMENT

Frontispiece: SCOTT ORIOLE

GMC

THE LIVES OF DESERT ANIMALS IN JOSHUA TREE NATIONAL MONUMENT

Alden H. Miller

Robert C. Stebbins

ILLUSTRATED BY

Gene M. Christman

UNIVERSITY OF CALIFORNIA PRESS • BERKELEY AND LOS ANGELES

1964

UNIVERSITY OF CALIFORNIA PRESS
BERKELEY AND LOS ANGELES, CALIFORNIA

CAMBRIDGE UNIVERSITY PRESS
LONDON, ENGLAND

LIBRARY OF CONGRESS CATALOG CARD NUMBER: 64-18643
DESIGNED BY WILLIAM SNYDER
PRINTED IN THE UNITED STATES OF AMERICA

CONTENTS

INTRODUCTION

Deserts hold the interest of modern man for many reasons. Foremost is their radical departure from the environment in which he has grown up and to which he is comfortably accustomed. Thus he is moved to awe. In some people this verges on fear, and in others it is a challenge to adventure. When deserts are thus sought, understanding and appreciation grow, and an inquiring approach emerges.

But there are other compelling attractions. People have many conflicting interests. One of these is that by preference and economic necessity they crowd into great metropolitan areas, which condition is only partly relieved by resort to well-settled suburbs. Yet they wish to escape from this periodically, if not permanently, in order to have a measure of isolation and quiet which deserts so well afford. The open desert vistas, smog-free and superbly beautiful in their changing lights, the dramatic changes of the seasons, and the occasional lavish burst of bloom when rains do strike serve as aesthetic attractions, refreshingly relieving tensions and the turmoil of everyday living.

When civilized man comes to the desert, as he does in such large numbers now in southeastern California, he is not prepared to meet desert wilds on their own terms. He wants to live, in reality, close to his usual pattern. This is partly dictated by his innate physiology, which has no special adaptations to resist the water shortage of these dry areas. But even more he does not readily change the pattern of life he has learned. He tends to arise late and be abroad in the most severe part of the desert day. Instead of relaxing and slowing down his pace in the desert heat, he fights it, frets about it, buys air conditioners, seeks frequent shower baths, and sees to it that he has refrigerated drinks. Within the fairly broad limits of his innate physiologic capacities, he could do better than this, and he can learn much from the way desert animals live and meet the daily problems of existence. The sympathetic desert adventurer thus will inquire into these matters and find them fascinating and not a little useful.

It is our purpose in this book to describe the vertebrate animals of the desert area centered in Joshua Tree National Monument and to tell what we have learned of their manner of desert living. Fortunately many kinds of desert animals have by now been carefully studied in the laboratory, especially with respect to their water requirements, metabolism, and temperature tolerances. From these foundation studies we can to a degree extrapolate the physiologic requirements of related species. In all instances, however, we find extensive behavioral adjustments in desert animals which allow them to live safely within the limits of their physiologic tolerances. We as naturalists then seek to tell the story of the way these animals have ordered their lives, balancing their activities with their physiologic requirements, and coming through successfully, or, if not, dying out to permit the better adapted individuals to populate these arid lands.

The mammals, birds, and reptiles of the desert are in many instances numerous, the populations of small mammals and of lizards being especially dense. Yet these terrestrial animals are often inconspicuous to the visitor because of his own habits and limited ability to observe. In the first place he will miss the teeming night life of the desert rodents because he does not usually venture abroad after dark except on roads. Actually the desert is usually bright enough at night so that kangaroo rats and pocket mice can be watched as they move about on the surface. And the visitor, although seeing a few common species of lizards readily enough, will miss the concealingly colored lizard that crouches against rock or sand or in a bush or is a mere blurred streak as it flashes across the ground to bury itself out of sight. The amphibians, another terrestrial group with which we deal, are very much water-dependent, and the few kinds present live only in the vicinity of springs. The birds are common, indeed conspicuous, to the casual observer, chiefly about oases, where they tend to concentrate, and by reason of their daytime activity. But even these should be sought in early morning and in late evening when they are most active, for in midday in summer they usually remain quietly in the shade as do rabbits and squirrels.

To enjoy the desert animals and learn about them, the visitor should then adjust his schedule to theirs. He should be abroad at night, looking and listening. He can watch, or where permitted, capture small mammals at night. He may patrol roads on warm nights for interesting snakes that crawl onto the pavement to take up its warmth and thus maintain their activity. He can watch bats in their evening flights, especially concentrated about water, and listen for calls of coyotes and owls. In the early morning all diurnal species, although on different individual schedules, will be emerging, active in search of food and readily watched, and then too the trackways of the nighttime may be read in the sandy trails. In midday he should find a shady spot among the rocks or under a Joshua tree or beside a juniper, or even better at an oasis, and sit quietly, conserving water and keeping his temperature down through inactivity. Soon he will see the daytime species doing likewise, near about, often at close range, because not disturbed by the watcher's motion. Thus with patience and the right approach, he may observe many kinds of animals. Our work in the Monument, though still not uncovering every rarity and vagrant presumed to occur, has over the years and around the seasons brought us in touch with 42 species of mammals, 167 birds, 36 reptiles, and 5 amphibians. Thus there are very many animals in these regions that on casual meeting seem rather desolate.

The desert is characterized above all by limited rainfall and low atmospheric humidity (Leopold, 1961). These are relative matters, as Buxton (1923) has well stressed. An annual rainfall below 5 inches a year always produces a desert, by any ecologist's concept. More important than low rainfall itself is the speed with which moisture is lost through evaporation. Here temperature, atmospheric humidity, and wind velocity are complexly interrelated, and direct measures of the net effect shown in evaporative rate need to be taken. Few data of this kind have been systematically recorded, but for North American deserts, the index or ratio of rainfall to evaporation given by Livingston and Shreve (1921: table 15) shows that all desert areas of the continent have percentages in the order of 20 or less and usually under 10 per cent. The actual evaporation values in the Mohave and Colorado desert basins are in the order of 90 to 100 inches a year. The severity of this water loss is accentuated if the rainfall is in most years focused in a single winter rainy season and with a long summer drought following. This is generally true of the Joshua Tree area and unlike the distinctly double rainy period of the well-studied Tucson district. Above all, the sudden brisk rains, widely spaced, which are typical of deserts, mean rapid runoff and loss of water, so that total annual rainfall is less effective than it would otherwise appear to be.

Although we do not have evaporation data for the Joshua Tree Monument, the information on rainfall and temperature compiled by the National Park Service (table 1; fig. 1) from the local weather station in Twentynine Palms gives a readily grasped picture of conditions in the lower levels of the Monument. Here it is seen that annual average precipitation over a recent 24-year period has been 4.19 inches, with April to June being the driest period and December and January the time of most likely rainfall. A weak and very erratic tendency to some rain from July to October has been noted and is reflected in these averages, but such summer rain, because of rapid evaporation, contributes little moisture that is usable.

Table 1 and figure 1 also show that the period of high daily maximum temperature, low humidity, and rare or ineffective rain—the heat stress period for animals—runs from May to September. Actually this was noted in our own field experience to run generally from late May to the first week of September, with its greatest manifestation in June, July, and August. Conversely the nighttime low-temperature periods critical to many desert animals, and especially at elevations of 4,000 to 5,000 feet in the mountains, is November to March, inclusive. With the low humidity and rarity of cloud cover, the daily range in temperature is great. One may regularly expect 30 degrees of difference over a twenty-four hour span in both summer and winter, and at times this may be as much as 40 to 50 degrees. This daily fluctuation and

TABLE 1

CLIMATIC DATA FOR TWENTYNINE PALMS, 1970 FEET ALTITUDE, FOR THE YEARS 1936 TO 1959 INCLUSIVE

Inches Precipitation

Fall and Winter Months		*Spring and Summer Months*		*Annual*
October	0.48	April	0.12	
November	.31	May	.03	
December	.58	June	.02	
January	.60	July	.54	
February	.28	August	.56	
March	.30	September	.37	4.19

Average Daily Temperatures

	Min.	*Max.*		*Min.*	*Max.*	
October	52.9	84.5	April	49.9	81.1	
November	41.4	71.6	May	56.9	89.5	
December	36.9	63.5	June	64.2	98.4	
January	35.0	61.6	July	72.0	104.5	
February	37.9	66.1	August	70.5	102.7	
March	42.0	72.4	September	63.9	97.3	67.4

Average Relative Humidity

Fall and Winter Months	9:00 A.M.	1:00 P.M.	5:00 P.M.	*Spring and Summer Months*	9:00 A.M.	1:00 P.M.	5:00 P.M.	*Annual*	
October	29.4	21.4	24.0	April	27.6	18.4	20.1		
November	33.7	24.6	31.2	May	25.0	16.3	15.5		
December	40.8	30.0	35.2	June	20.6	12.6	12.6		
January	43.0	32.1	38.1	July	23.6	18.1	17.6		
February	37.8	26.7	28.8	August	27.7	19.9	18.9		
March	32.2	22.1	24.2	September	25.4	16.9	16.7	9 A.M.	30.6
								1 P.M.	21.6
								5 P.M.	23.6

Degree of Cloudiness

	Clear	*Partly Cloudy*	*Cloudy*		*Clear*	*Partly Cloudy*	*Cloudy*
October	24	4	3	April	21	6	3
November	22	5	3	May	25	5	1
December	19	7	5	June	27	2	1
January	20	7	4	July	21	8	2
February	19	6	3	August	21	7	3
March	23	6	2	September	25	4	1

the invariably cool nights are factors to which desert animals must adjust but from which they also derive real benefit by using the part of the 24-hour cycle most beneficial to them.

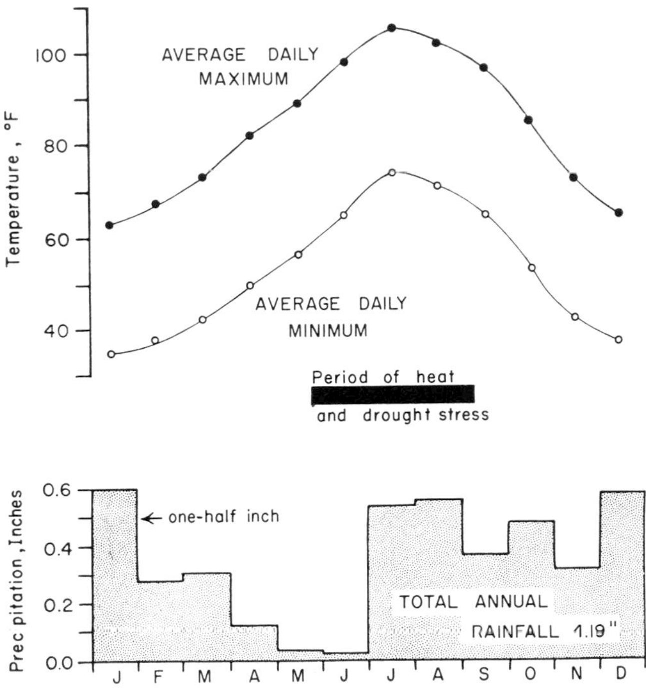

Fig. 1. Diagram showing seasonal features of temperature and rainfall at Twentynine Palms on the northern border of Joshua Tree National Monument. The conditions of climate shown are representative of the lower elevations of this section of the desert.

Temperatures in the sun of course range very high and through much of the year would if long experienced elevate the body temperature of a vertebrate animal far above a lethal point, which can never be higher than 115° F. and in most animals is significantly lower. The heights of the sun and substrate temperatures are then not so important for their actual extreme values, which may range to 150 or 160° F., but for the fact that they are reached for part of each day in the warm season. They must be avoided for any long periods of time, but they can be effectively and critically used in appropriate small doses to elevate the body temperature of reptiles to optimum functioning level, and in warm-bloods or endothermic mammals and birds they can partly compensate for the internal production of body heat which is expensive in calories of food.

The broad problems of desert existence thus fall into five categories. These are: (1) Water shortage, in that every animal and plant must have and must appropriately conserve water for its essential metabolism. (2) Coping with both high and low temperatures, which in deserts run to extremes, seasonally and daily, and have real impact on the efficient operational temperature of the species and which can affect adversely the water balance; high temperatures in our area of study are more often than low temperatures the important problem. (3) Great irregularity in water supply, locally and over a span of years, leading to fluctuating conditions of moisture, which demand a large safety factor in the operating plan of a species in order to meet frequent crises. (4) Limited concealment owing to the sparseness of desert vegetation, so that predators are difficult to avoid and protection from the sun, wind, and torrential rain is poor. (5) The difficulties presented by prevalent dust and by dust and sandstorms, leading to the need of protecting the respiratory system and of maintaining burrows and retreats in the sand; related to this is the requirement of locomotion on and digging into loose alluvium or sand.

ONE

THE SOLUTION
OF PROBLEMS OF DESERT LIFE

In order that an over-all view may be gained of desert adaptation, we draw attention to examples of ways in which species of the Joshua Tree area are meeting the several problems of desert existence. These are summarized from our detailed studies and speculations on the nature of these adaptations recorded in the accounts of the various kinds of animals.

Water. Each species must maintain water balance so that intake offsets loss, although the balance may temporarily be upset to varying degrees. Large species in general have more reserve capacity in withstanding temporary shortages. The key aspects of water balance are (1) intake sources and (2) conservation.

A simple, direct way for a species to meet the problem is to live only near supplies of free water where it can drink extensively and freely. Such species are in a sense not true desert dwellers, and modern man himself falls in this category. Other examples of this kind of animal are mourning doves, goldfinches, house finches, mountain quail, and the amphibians, the latter absorbing water through their skin and taking it in whenever necessary during their active seasons. These groups of animals use water copiously and with this mode of attack on the problem develop no special adaptations for water saving. Among the strong-flying birds, several of which we have just cited, long trips to springs are feasible and permit living as far away as five or six miles.

A second, very important solution of the water problem by vertebrates

Fig. 2. A hole dug in sand to a depth of about 40 inches by coyotes seeking water in August, 1950, in the canyon below Upper Covington Flat. The water exposed by the coyotes here was also used by quail and goldfinches.

is reliance on moist plant food and moist animal matter, chiefly insects. This is then a form of dependency on the water-storing and conserving adaptations of plants primarily and of insects secondarily. Desert plants, because they are in fixed and exposed position at all times, in order to exist have evolved both extensive root systems for obtaining underground moisture and devices for restricting water loss by transpiration from their stems and leaves. Insects through means of a controlling hard and dry exoskeleton can hold and conserve moisture derived from plant food in highly efficient adaptation to desert conditions. The vertebrate species that are exclusively insect-eaters then have no problem of adequate intake of water, as the water content of their food may be 60 to 85 per cent, as estimated, for example, for the food of the grasshopper mouse. Bats, flycatchers, warblers, and vireos, all primarily insectivores, then have their water problems solved for them by the adaptation of their prey species. The same of course applies to carnivores such as snakes, coyotes, foxes, and bobcats. These animals cannot be wasteful of water, and some even supplement their moist diet with drinking water when feasible (fig. 2), but they are never in difficulty so long as their food sources are adequate.

Other groups depend on succulent vegetation, including fruits. Thus house finches can resort to cactus fruits in season, antelope squirrels select moist green yucca pods and flower parts, the vegetarian tortoises and the desert iguana eat moist green plants, and rabbits eat juicy cactus pads when other green food fails. A man can likewise use cactus, both the fruits and the green storage stems.

A basic source of water unappreciated by the untrained observer is the metabolic process itself. All animal activity in which nutrients are used up involves of course an oxidative reaction in which carbohydrates and the carbon and hydrogen radicals of proteins and fats are converted to energy and carbon dioxide and water. A very considerable source of water is here involved. Thus for every calorie of a dry food such as pearl barley that is used in this way a little over 13 grams of water is formed within the animal's body. Or put differently, one gram of dry starch food will on oxidation yield 0.6 gram of water; and one gram of fat will provide a little over one gram of water as a by-product of the energy that this burning affords.

The real issue remaining, then, is whether this is enough water to carry on all the functions of the water-dependent chemistry of the animal body and the inner fluid-transport and flushing systems. Can this metabolic water be conserved sufficiently for the assured success of these operations? The metabolic water is naturally equally available to animals of all types proportional roughly to their food utilization. We are struck with the importance of this source, however, when conservation of water in a species is such that it can get along with almost nothing but this basic water supply. This is the extraordinary situation in the kangaroo rats and some of their close relatives among desert rodents, such as pocket mice. Metabolic water can of course be increased by using more food, but no sure gain is realized by eating dry food to get more metabolic water, because this elevated energy use probably means proportionately greater expenditure of water.

Water may be conserved in several different ways. Broadly these fall in three categories. (1) use of as little water as possible in passing feces and urinary products; (2) reduction of evaporation from the skin and lung surfaces; (3) and adjustment of behavior to minimize evaporative losses.

Concerning the first, the simplest adaptation is that of drying the feces before discharge. Of the groups here considered, only the mammals discharge the feces separately from the urinary fluid. Resorption of water from the undigested food residue can be carried on in the large intestine to a high degree. For example, in Merriam kangaroo rats the feces are dried to the point of having only about $\frac{1}{5}$ the water content of those of non-desert rodents. Similarly one may note relatively dry feces of mountain sheep, rabbits, and pocket mice.

In the birds and reptiles the fecal mass is variously intermixed with the urinary discharge, and the latter in all cases has a somewhat greater, though variable, water content because of the essential need of carrying urea and various salts, by-products of metabolism, from the body. Thus the critical problem of concentrating the urinary discharge is the governing one in these groups and masks somewhat the conservation of the fluids of the feces, although the latter saving is clearly also a factor.

In reptiles the concentration of urea in the urine as it leaves the kidney is not greater than that of the blood stream from which it is derived, but the urine is then held in the cloaca, where urea to varying degrees is changed to uric acid and thus precipitates. This means that the fluid mass then has less concentration of urea than the blood, and the osmotic relations are such that water is resorbed. The ability to precipitate uric acid from the urine and flush or extrude the precipitate with little water loss is then a basic water-conserving mechanism most important to desert reptiles. The white urinary extrusion of snakes and lizards, almost a solid, though moist, mass, represents a great saving in water. The varying amounts of dark fecal material passed results from mixture in the common cloacal chamber. In desert tortoises water is held in the bladder and this perhaps serves as a storage measure. But at the same time uric acid is precipitated here to some degree, and thereafter only a small amount of water is needed to facilitate the normal extrusion of the precipitate.

In birds and mammals the urine is concentrated in the kidney and, with respect to urea, to a point above the urea level of the blood. This is done through the kidney tubules. The fluid passed may then be conspicuously concentrated, or hypertonic, in relation to the blood, and the nitrogenous wastes are moved out of the body with relatively low water loss. In birds part of the concentrated waste is precipitated as uric acid and appears as the white material of the droppings. This, as in reptiles, is a highly efficient method of saving water. In desert-dwelling birds, the relatively solid, though moist, combined urinary and fecal discharge loses very little water. In desert mammals the urine, concentrated with urea primarily, may be a semifluid pasty discharge. Again in kangaroo rats, as an example, this concentration is such that it is four times that in man. But even this is a much less effective conservation of excretory water than in birds and reptiles.

A second approach to water conservation is to reduce its evaporation from the body. The outer surface of the body of amphibians can in no instance completely prevent such loss, although a partial reduction of loss is achieved by the skin through less active mucous outpouring from the skin glands. Thus desert-dwelling amphibians must avoid exposure to low atmospheric humidity for any substantial number of hours or they must, if they are exposed, be so situated that they can quickly enter water and regain their losses. Such is the situation in the treefrogs and the red-spotted toad of the oases of Joshua Tree Monument.

All the reptiles, birds, and mammals possess dry skins in which horny dead cell layers at the surface resist any very significant evaporative loss. But in some lizards, such as the nocturnal banded gecko, some drying out through the skin seems possible, and even in a bird or in a mammal without sweat glands a small water loss from the surface takes place in dry atmosphere.

But the surface losses just described in the truly terrestrial animals are but a fraction of the total water loss by evaporation, most of which occurs through the surface of the lungs and respiratory passages. In the lung the loss is an inevitable consequence of the system of exchange of oxygen and carbon dioxide which must be carried on there. To reduce this loss there are two solutions. (*a*) The metabolism may be held at a low level whenever possible, thus reducing the total of respiratory activity and the consequent loss of water vapor from the lungs. In the reptiles and in a very few hibernating mammals and birds, periods of quiescence with lowered body temperature, either daily or seasonally, decrease respiration. Such periods occur in bats, pocket mice, poor-wills, and swifts of our desert area. Even among the warm-bloods, or homeotherms, the birds and mammals, the pace of activity may be reduced while their temperatures are up to normal, so that metabolism and respiration are less rapid. This is a physiologic adaptation of the cactus mouse, for example, compared with its non-desert relatives of the same genus. Reduction in birds by this means is more difficult because of their especially high constant body temperature, of the order of 106° F., which means that even at rest, and with metabolism at as low or economical level as possible consistent with this temperature, respiratory loss will be high. It is especially high in the small species, those weighing 60 grams or less. (*b*) The second method of reducing water loss from respiration is to recapture it by condensing water vapor as it passes out through the nasal passages. In kangaroo rats the hot air expelled from the lungs passes through nasal passages with temperature perhaps 20 degrees lower. Thus some water must be condensed and recaptured by cooling. The complex nasal passages of some lizards, such as the fringe-toed lizard, may similarly recapture some moisture.

The third broad approach to water saving is behavioral. Much is gained by avoiding exposure. Thus many hours of each 24-hour period are spent by desert amphibians, reptiles, and small mammals underground or in tight crevices where the microclimate is more humid and where drying winds do not sweep the animals' skin. The humidity in a desert rodent burrow is two to five times that on the exposed surface of the ground. Trips to the surface can be limited to those hours at night or at other times when relative humidity is greatest. This higher humidity would be effective in holding down evaporation from the skin and from the lungs of those species other than homeotherms.

The behavioral cycle also is such as to avoid activity during high temperatures, for if the temperature of the body approaches a lethal level, panting and other emergency water-evaporating mechanisms are brought into play which are very expensive to the water supply even though for the moment they are vital to the animal. And lastly, among migratory birds, the whole schedule of the year may be adjusted so that the desert is used only in the less arid and cool seasons; these species thus merely flee from the most difficult aspects of the desert. So it is with Audubon warblers and ruby-crowned kinglets, which spend only the winter in the Monument.

Finally, in a few species there have evolved special capacities to store water against emergency periods, as in the bladder of the desert tortoise, or to sustain water loss without serious effect and eventually to recoup the supply. The latter is possible especially in animals of somewhat larger body mass, but perhaps scarcely or not at all in small species whose reserve tissue fluids are relatively less compared to their surface and to their metabolic rate. For example, the mourning dove, a moderate-sized bird, can go without drinking water for four or five days in an emergency, but then it will in ten minutes drink 17 per cent of its own weight in water to catch up on the supply. Storage has another aspect, used especially by lizards and by migratory birds passing through the desert. Fat deposited in the abdomens and tails of lizards like geckos and under the skin surface in birds is a food type which we have seen yields an especially high amount of metabolic water when it is utilized. Thus an emergency water reserve is involved here as well as a reserve of food used for energy.

Temperature. Truly terrestrial vertebrates, the reptiles, birds, and mammals, function most efficiently, and effectively, in their active periods when their body temperatures are high. These high values lie just a few degrees below the lethal point for certain of their tissues, so that refined adjustments are necessary in order that they may operate within these narrow limits. In the birds and mammals internal heat production, as necessary, sustains a relatively constant level, of the order of 97 to 99° F. in mammals and 104 to 108° in most birds. External heat sources readily available in the desert through much of the year reduce the amount of energy that must be used within the body to sustain the high body temperature. In the reptiles, which must gain their high temperature from outside sources, or in other words are ectotherms, optimum operating temperatures at levels often of 94° to 100° are achieved by carefully scheduling the times and kind of exposure to the sun and to warm substrates. Thus for these animals even more than for the stable temperature types, or endotherms, the high temperatures of the deserts afford rich opportunities and advantages for the conduct of their lives.

But desert temperatures run to extremes, the more important of which, or at least the more impressive to humans in our area, are the high extremes. When the environmental temperature exceeds that of the body temperature and even in endotherms equals or closely approaches it, the heat becomes dangerous. In reptiles this matter is simply adjusted by going into the shade, or if this is not sufficiently cool, going underground to their retreats, where at depths of only a few inches temperatures never reach a lethal point for them. But should they be caught out too long in the sun, they quickly die. There is no simpler way of killing a snake, such as a sidewinder, than to force it to stay out in hot sun for a few minutes.

Most small desert mammals, because they are nocturnal, avoid the high temperature problem. Their daytime retreats for the most part are underground or certainly in the shade. When occasionally the shade temperature in such places reaches levels higher than that of their bodies, emergency cooling mechanisms are used, such as panting or the spreading of saliva about the face, as may be noted in white-footed mice. But these evaporative cooling mechanisms, as already stressed, are highly wasteful of water and cannot be afforded for long.

In birds the extraordinarily high body temperature means that there are only short periods of each day even in summer when shade temperature exceeds this level (fig. 3) and when dangerous heat flow from the environment

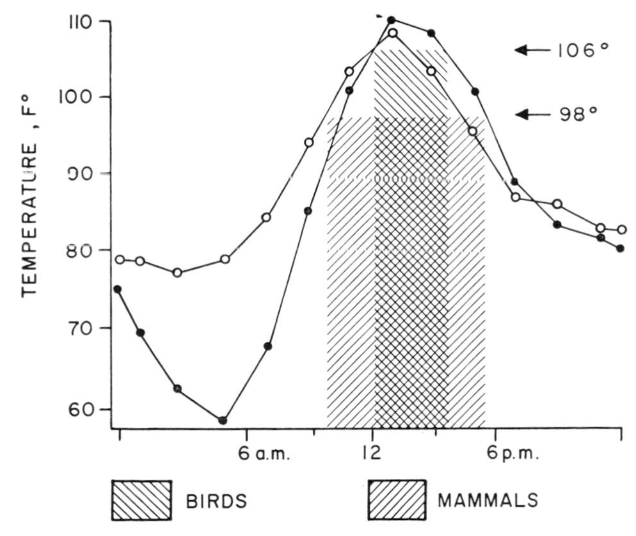

Fig. 3. Diagram showing the parts of representative summer days when air shade temperatures exceed the body temperatures characteristic of birds (106° F.) and mammals (98° F.). The lines connecting solid dots show hourly air temperatures on a representative day in June; open dots, a day in July (data from Dawson, 1954). The breadth of the column for mammals in contrast to the narrower column for birds reflects the longer daily periods when mammals on the surface encounter temperatures that exceed their normal body temperature.

into the body occurs. Of course warm-up from radiation in exposed situations can be avoided largely by getting into the shade. Thus birds, by having a high metabolic rate and a high temperature developed and perfected for other reasons, are to a degree preadapted to desert life. On the other hand, when birds do encounter too high an environmental temperature, they cannot by their nature go underground. They must use emergency and moisture-wasteful cooling mechanisms which consist essentially of more rapid breathing and panting. Their mobility in flying to water sources, their extensive use of moist insect food, and their conservation of fluids in the urinary system help them in providing a margin of safety for their water resources when high temperatures become acute. By and large, birds, compared with reptiles and small mammals of comparable size, seem extraordinarily able to withstand high temperatures and even to stay exposed in the hot sunlight for several hours. Thus black-throated sparrows spend much time daily in hot sun, singing on exposed bush tops with surrounding temperatures well over 120°. Some such desert species doubtless have a special tolerance of elevated body temperatures a few degrees above normal, but we need experimental verification of this. This trend in adaptive tolerance is suggested by the finding that cactus mice can stand higher body temperature before putting their emergency cooling mechanism into operation than can their relatives the deer mice, which are less adapted to deserts. Emergency cooling in birds can be seen in any hot midday period as species such as horned larks sit in the shade of a post, with mouth open, panting, or scrub jays gape as they rest in the shade of a piñon tree. Their inactivity except for breathing means of course holding to a minimum the heat production of muscular activity.

But again, as in the case of the water problem, many species avoid the critical high temperatures by absenting themselves part of the year. This may be a period of underground quiescence or estivation in the summer, in a species such as the Beechey ground squirrel, or in birds the use of the area only in transit in spring and fall or merely for winter residence. Migrants that must pass across our western deserts in moving to and from our coastal areas and mainland México, in the fall passage particularly, encounter high temperatures to which they are poorly adjusted by physiologic mechanisms, behavior, or learned familiarity with retreats. The water and high-temperature problems at times overcome them, especially those with poor fat reserves, with resultant high mortality. Warblers and small flycatchers have demonstrated this particularly and are discussed in detail later.

Low temperatures in winter present more of a problem than first meets the eye. For the ectothermous amphibians and reptiles, a few hibernating birds such as poorwills, and mammals such as bats and little pocket mice, body temperatures simply fall and the animals may stay below the surface where it does not freeze. The endotherms may maintain just enough heat production to offset any chance of freezing. For others that cannot do this, the bitterly cold midwinter nights can be withstood, if the species are of nocturnal habit, by restricting the time of activity on the surface to short periods when muscular heat production will be high. Such would seem to be true for kangaroo rats and deer mice. In small diurnal mammals and birds, protected retreats for sleeping, where cooling is less extreme and chilling winds are avoided, ameliorate the low-temperature problem. Covered roost nests of cactus wrens and verdins and roosting holes of plain titmice and ladder-backed woodpeckers are examples of ways of avoiding low temperatures, and among the rabbits, holes and forms and heavy insulating fur are the answers. Antelope squirrels especially, and Merriam chipmunks to lesser degree, change from their summer pelage to a heavier winter fur and thus are better able to check heat loss.

Irregular Extremes. Not only do deserts present extremes of temperature and rainfall, seasonally at these latitudes, and even daily, but—characteristically of arid areas—the rainfall is most erratic from year to year and is locally spotty. Rainfall fluctuates violently over a 10- or 20-year period in a given area and so accordingly do the vegetation and animal food supply. Thus one section of the Joshua Tree Monument may receive rain several years in succession preceding the growing season such that a rich temporary plant growth ensues.

The desert vertebrates in their long-range evolution have had to develop ways as species of coping with this problem, else long ago they would have become extinct. Vagrancy is a notorious line of solution in deserts, and it is effective in the more mobile kinds of animals such as birds and large mammals. For example the dry span of several years in Pinto Basin in the late 1940's apparently meant the departure of Le Conte thrashers and a quick resettlement locally when more favorable conditions returned (fig. 4). Horned larks colonized here and in Pleasant Valley only so far as we know in the favorable year of 1960, finding these places through vagrancy.

Fig. 4. Desert bloom in the floor of Pinto Basin in early April of 1960 following good local rains. The abundant caterpillars of sphinx moths shown among the sand verbenas form a rich source of moist food for birds such as Le Conte thrashers.

But for less mobile kinds that cannot cruise and find the favorable areas, most of the population dies off. The species of the desert must then be adapted to this mortality. Some individuals must be able to persist and tolerate extremes and then build up the numbers from small, wide-scattered remainders that are left in the most favored local refuges; they must have a high reproductive rate. Such capacity seems well exhibited in the little pocket mice. Contrarily, wasteful and hopeless reproductive effort and its energy expenditure is advantageously avoided. Thus in poor years Gambel quail breed not at all, as may well be true for desert tortoises and certain snakes.

A corollary of the necessary capacity to repopulate is an ability to reinvade areas of complete depopulation. Innate dispersal tendencies on the part of desert species, especially among young animals seeking establishment, is favorable.

Concealment. The open vegetation of the desert and the bright lighting of it, even at night, make concealment of animals a special issue. Every animal, except the large carnivore, when abroad and active is in danger of falling prey to its natural predators, and concealment, broadly speaking, is the line of solution to this danger at some time in the activity period of almost all such species. Even the master predator may benefit from its own concealment or inconspicuousness as it approaches its prey. The pressures then for evolution of concealing coloration in desert animals is especially acute because of constant predation and openness of habitat.

We see extraordinary examples of protective resemblance in several of

Fig. 5. Desert horned lizard crouched on its normal background of desert sand and gravel, showing the concealing effect of pattern and color which disrupt the animal's outline and make it blend with the background.

the lizard species. Thus the fringe-toed lizards in the dunes of Pinto Basin are sand-colored and finely mottled. When at rest on the surface, they are extremely difficult to see. Similarly remarkable is the concealment of the long-tailed brush lizard as it rests with body and tail parallel to the nearly vertical branches of bushes such as creosote, its dimensions and color blending with the twigs. Le Conte thrashers with their light sand-colored backs, at a little distance when quiet, match their background so as to be seen with utmost difficulty, and so to a greater or less degree do rock wrens, gray vireos, and pocket gophers.

At night any appreciable contrast with the background can be detected especially when coupled with motion. And thus we find that most of the nocturnal rodents have evolved coloration which approximates that of their background.

Sand and Wind. The open desert is subject to heavy and sometimes prolonged wind, which coupled with the dry uncompacted sand and ground surfaces means an unstable and moving substrate in parts of the desert. Many desert species have expanded foot structures enabling them better to run on soft surfaces or dig in them. Such are the scale fringes of the lizards of the genus *Uma,* the stiff brush of hair on the feet of kit foxes and of desert kangaroo rats, the elongated toes of roadrunners, and the large hind feet of jackrabbits.

In burrowing types especially, and in those living low to the ground in sandy habitat, special protection of ears and nostrils may be noted. Thus the nasal passages of lizards of the abundant iguanid family have a sharp bend and valves to prevent inspiration of sand grains. In the extreme case of the fringe-toed lizard, special valvular ear flaps, overlapping eyelids, and a countersunk lower jaw guard against entry of sand grains to body passages and sense organs. Kit foxes have heavy hair brushes within the ear which guard the ear tube from sand.

In the snakes and certain lizards the eyes are at all times protected from sand by the hard, dry transparent spectacle that covers them. Thus all snakes are preadapted in this way to sand and to burrowing in it.

Locomotion of snakes in soft sand reaches its greatest specialization in the sidewinder's procedure, in which the body is thrown into a succession of lateral loops, thus minimizing slippage. Each transverse or diagonal section presses down on the sand, somewhat like the transverse or diagonal ridge treads of a tractor wheel or belt.

Burrowing within the sand or gravel is relatively easy, but constant reopening of burrows or maintenance is necessary because of cave-ins. Many species use situations in or about roots of bushes or alongside rocks to reduce this trouble. Striking are the locations of the large burrows of the giant desert kangaroo rats among the stems or roots of creosote bushes where danger of collapse of the burrows and drifting of loose sand is reduced.

TWO

THE ENVIRONMENTS
OF JOSHUA TREE MONUMENT

The Joshua Tree Monument encompasses a mountain system that extends southeastward from the Morongo Valley at the eastern face of the high San Bernardino Mountains of southern California. The main part of this desert upland is known as the Little San Bernardino Mountains. On their southwestern flank the land drops abruptly to the low desert trough of the Coachella Valley, the bottom of which is below sea level. To the east the Little San Bernardino Mountains give way to somewhat isolated mountain masses such as the Cottonwood and Eagle mountains. On the north side is a series of plateaus with a northern border of low mountains beyond which in turn are the lower flats of the valley centered on Twentynine Palms at 1900 feet elevation. Eastwardly the intermountain plateaus drop off to the low Pinto Basin, of less than 2000 feet elevation, which is continuous with fairly low desert to the northeast and southeast through passes north and south of the Coxcomb Mountains.

This complex highland of the Joshua Tree Monument forms a divide between the low-lying environments of the Colorado Desert and the moderately high basins and sinks of the Mohave Desert on the north (fig. 6). The uplift also constitutes an extension eastward of the coastal ranges, for the Morongo Valley is a fairly narrow and not very deep cleft between the coastal mountains and the Monument. The connecting divide across its northeast end is 3500 feet in elevation, whereas the crests of the Little San Bernardino Mountains are generally only about 4500 feet with the highest points reaching 5500 feet.

The area which we have studied includes all the Monument as formerly constituted. The present boundaries were established in 1950 and encompass 872 square miles. The original area, which was larger, was modified because of a combination of considerations, among which were a need to exclude certain mining areas, the impracticability of surveillance of some remote sections by Monument personnel, and retention of the most significant areas for desert animal and plant life. The latter consideration was influenced by some of our findings in our field trips of 1945 and 1946. The only area of particular biologic interest that was removed was Little Morongo Canyon, which originally was only partly within the Monument and which because of this and its location could not be effectively administered as a preserve.

The biologic investigations we report have not been confined rigorously to the Monument area, old or new. When useful data have been gathered a few miles outside the boundaries, we have included them. We regard the portrayal of the pertinent biologic conditions as more important than formal adherence to a precise area of record. In similar vein we are most concerned with the dominant and important examples of the desert vertebrate animals of this region in contrast to casual or accidental occurrences in the study area.

By including all sections formerly in the Monument and certain bordering areas, we offer observations made in Little Morongo Canyon and some from Morongo Valley to the west of it. The Virginia Dale mining area in the northeastern section and Twentynine Palms and the roads east and west into it just north of the boundary likewise are involved. We have given essentially no attention to the Pinto Mountains and have not visited the Coxcomb Mountains, neither of which seems likely to afford much new biologic information. The total spread of the region we have surveyed is from west to east about 50 miles and north to south, at the longitude of Cottonwood Spring, about 30 miles.

Plant Belts. The general physiographic features just described dictate differences in the desert ecology governed by elevation. A coastal group of plants and animals has been carried eastward, so to speak, in a belt along the highlands. The low desert environments are continuous around the east end of the mountain system but are separated into northern and southern arms toward the west. These relations may be expressed by mapping (fig. 6) a succession of three altitudinal plant belts each marked by a particular kind of dominant or conspicuous vegetation. The three are, from low to high, the creosote bush, yucca, and piñon belts. There is interdigitation of the dominant or marker species of these belts depending on slope exposure, drainage, and soil conditions, but there are also transitional zones between them. In the main the creosote bush belt extends up to 3000 feet elevation, the yucca belt from 3000 to 4200 feet, and the piñon belt from 4200 feet to the summits at 5500 feet.

Even though the borders of the belts vary in position some 300 feet because of local conditions, the boundaries are generally 500 feet (300–700 feet) lower than the same belts in the Providence Mountain section of the eastern part of the Mohave Desert, where a similar survey was conducted (Johnson, Bryant, and Miller, 1948). The difference probably is related to the location of the Providence Mountains, which are farther removed from coastal climatic influences than are the Little San Bernardino Mountains. There may be associated, but as yet undetermined, differences in rainfall and temperature that influence the position of the plant zones. That the differences in the two areas are real is attested by comparison of a locality in the creosote bush belt in the Ivanpah Valley of the Providence Mountain region at 3700 feet (*op. cit.:* fig. 6) with the development of yuccas in the flats near Quail Springs in the Joshua Tree Monument at 3700 feet. Similarly, the Joshua tree "forest" at 4200 feet 4 miles north of Cima (*op. cit.:* fig. 7) may be compared with the piñon belt at 4200 feet at Black Rock Spring in the Monument.

Went (1948:243) has briefly indicated the altitudinal occurrence of creosote bush (*Larrea tridentata*), Joshua trees (*Yucca brevifolia*), and piñon (*Pinus monophylla*) in the Monument. He reports correctly the occurrence of *Larrea* up to 4000 feet, but at such levels we have not found it to be more than local in occurrence and less prominent there than the species of yuccas; other shrubs such as blackbrush (*Coleogyne ramosissima*), paper-bag bush (*Salazaria mexicana*), and cactuses are here common. Thus, the creosote bush belt must be regarded as terminating at a point substantially lower than that of the maximum altitudinal occurrence of the plant species itself. Went states that piñons characterize the plant cover of rocky slopes at about 5000 feet. Actually on north slopes above Indian Cove piñons are common at 3500 feet, but in general they do not become dominant until 4200 feet. But at that level they are certainly prominent and may form a woodland, even on moderate slopes, as at Black Rock Spring and Pinyon Wells. The junipers (*Juniperus californica*) that are usually associated with piñons in the piñon belt are not restricted to rocky or well-drained slopes, and in some high alluvial basins, such as Upper Covington Flat at 5000 feet, piñons are replaced by an open woodland of junipers and Joshua trees, the latter here far exceeding its normal altitudinal limits for the area.

Habitats. Environmental settings characterized by relatively circumscribed features of vegetation, soil, and drainage occur within the broader plant belts and in some instances are not confined to a single belt. These habitats are of extreme importance to the vertebrate fauna, governing the occurrence and manner of existence of the animal species.

SAND DUNE. Loose sand, either on flat ground or mounded into large dunes, occurs in the lower parts of the Monument within the creosote bush belt. Creosote bushes are very widely spaced on such sandy areas or are lacking

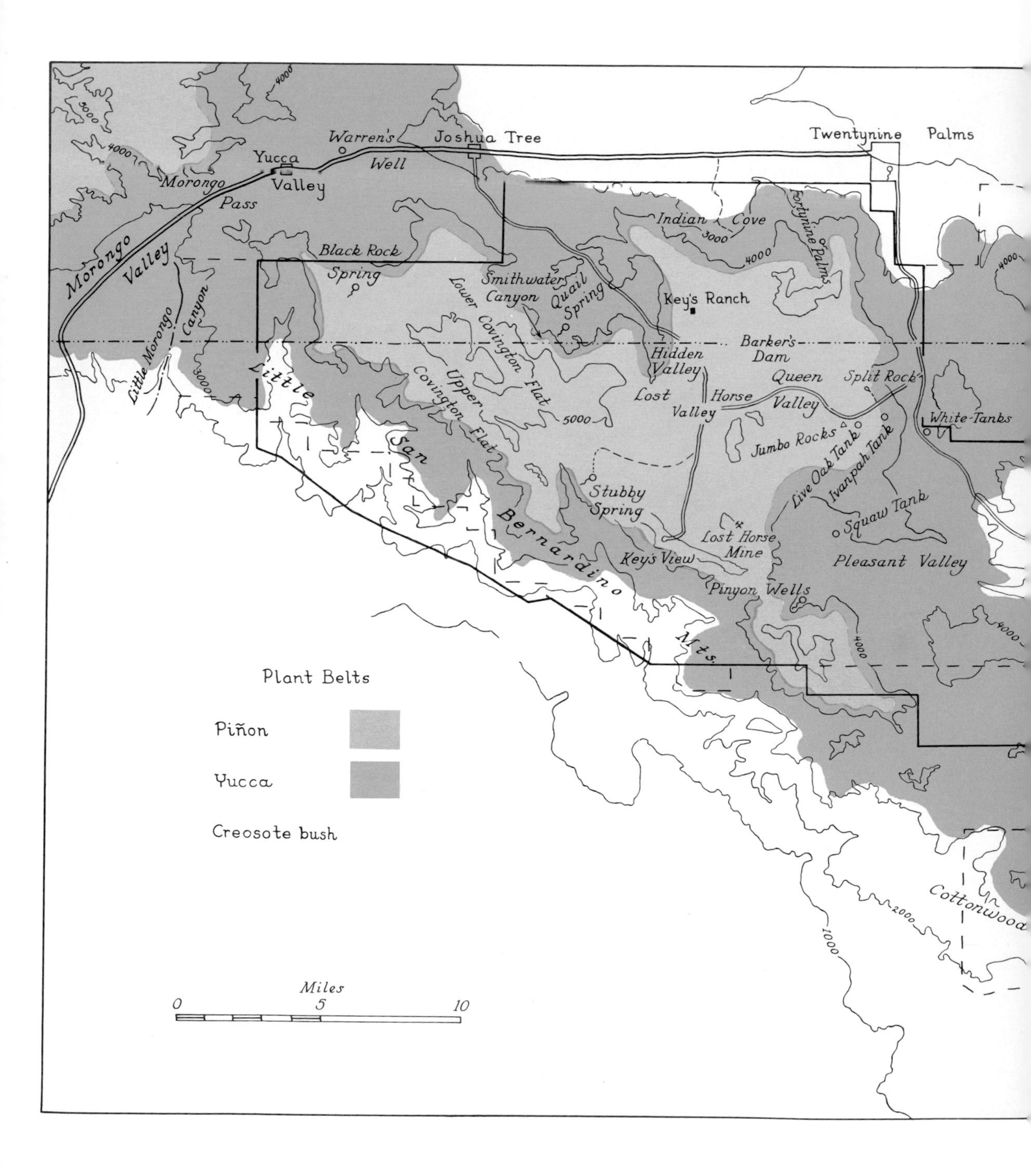

Fig. 6. Map of the Joshua Tree area, showing old and new boundaries of the Monument, physiographic features, contours, localities, and the three altitudinal plant belts.

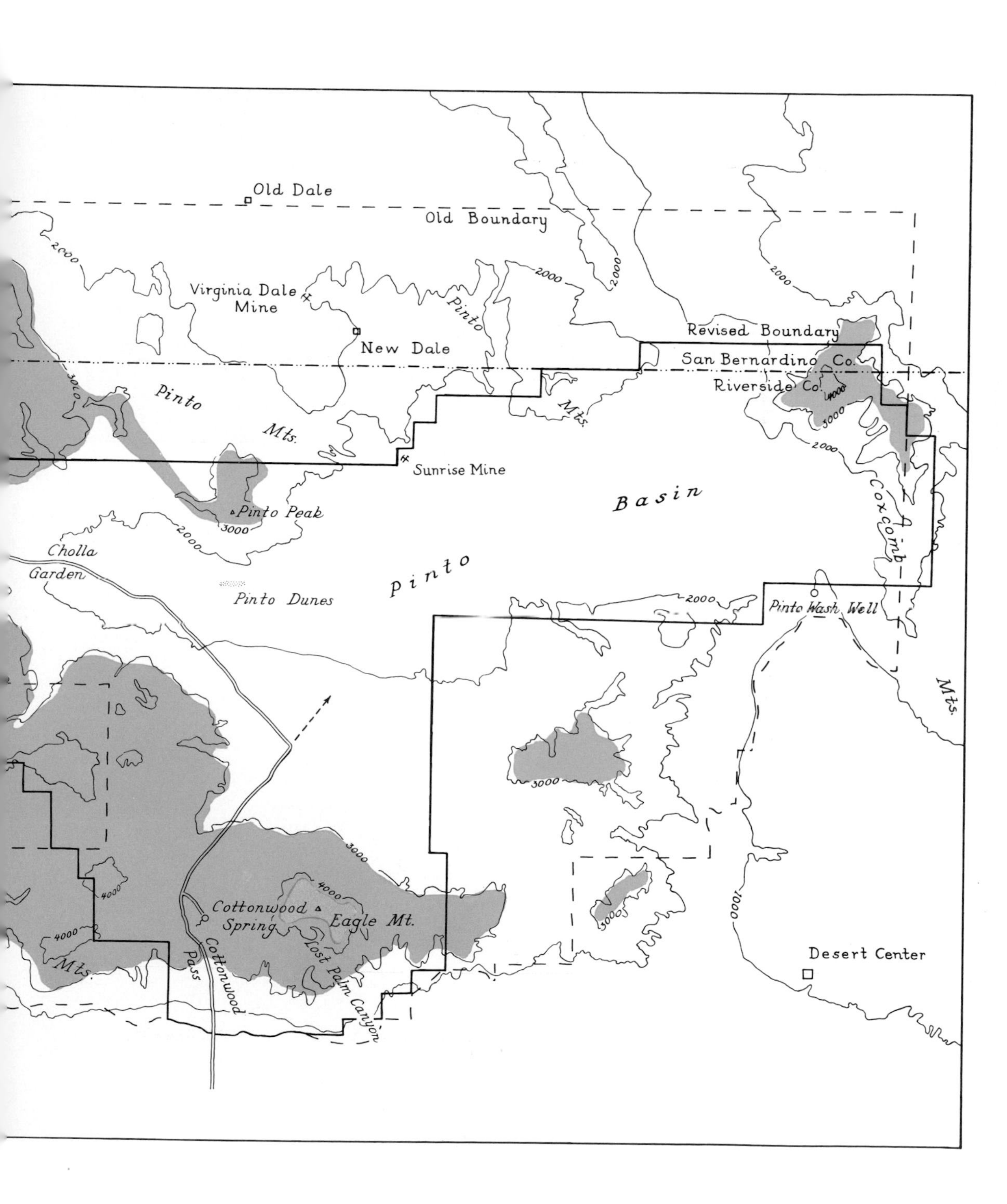

Old Dale
Old Boundary
Virginia Dale
Mine
New Dale
Pinto
Pinto
Mts.
Mts.
Revised Boundary
San Bernardino Co.
Riverside Co.
Sunrise Mine
Basin
Coxcomb
Pinto Peak
Cholla
Garden
Pinto Dunes
Pinto
Pinto Wash Well
Mts.
Cottonwood
Spring
Eagle Mt.
Lost Palm Canyon
Cottonwood
Pass
Mts.
Desert Center
2000
3000
4000
1000

Fig. 7. Sand dune habitat in Pinto Basin, showing dune grass against the background of a flat dominated by creosote bush.

Fig. 8. Grasses and primroses on the Pinto Basin sand dune in a year of desert bloom, April 11, 1960.

entirely on the dunes themselves. Dune grasses and, in some years of suitable rains, annual plants grow on the dunes (figs. 7, 8). Sandy tracts and small dunes are to be found east of Twentynine Palms, especially near Virginia Dale Mine, but the most conspicuous dunes are in Pinto Basin. There the largest dune, some 1600 yards long, is situated near the drainage channel at the south base of Pinto Peak. The sand habitat is essential to fringe-toed lizards and desert kangaroo rats and it is favored by desert tortoises, Le Conte thrashers, and kit foxes.

DESERT WASH. Extending through low hills and out over alluvial fans and through desert basins are drainage channels which in periods of rare heavy rainfall carry flood waters but which in the main have underground moisture sources tapped by certain large plants. The soil is loose, sandy or gravelly, with pockets of fine alluvium and much cross-bedding of deposits. As a consequence, shrubs and annual plants are diverse here and the large bushes and small trees characteristically present form irregular borders or strips. The larger plants consist of smoke trees (*Parosela spinosa*) in the drier places (fig. 55), and desert willows (*Chilopsis linearis*) and mesquite (*Prosopis* sp.) in those with better underground moisture sources (fig. 9). Catclaw (*Acacia greggii*) is an element in these washes also, but it tends to occur higher up in the mountains than do the other trees. There is an especially large spread of it in the wash below Quail Spring. In the wash below Cottonwood Spring palo verde (*Parkinsonia microphylla*) is conspicuous in the border vegeta-

Fig. 9. Desert wash below Indian Cove in which desert willow predominates.

Fig. 10. A creosote bush in Pinto Basin, showing the sand-holding capacity of the clump it forms. Typical spacing of this plant may be judged from the dark-appearing bushes seen in the background.

tion. Vertebrate animals regularly present in this habitat are the verdin, phainopepla, black-tailed gnatcatcher, roadrunner, Costa hummingbird, mockingbird, western whiptail, and zebra-tailed lizard.

CREOSOTE BUSH HABITAT. Great expanses of the desert valleys and alluvial fans are dominated by creosote bush, whether the soil be sandy, gravelly, or stony. The bushes (fig. 10) are widely spaced with extensive root systems, competing for the limited moisture. Lesser bushes, sparse grasses, and at times substantial growths of annual plants may be present. We know of no species of vertebrate that finds this habitat exclusively to its advantage. Many species occupy it but in numbers usually lower than those in other, adjoining habitats. One may regularly find the Merriam kangaroo rat, desert tortoise, antelope ground squirrel, and black-throated sparrow in unmixed creosote bush habitat.

CHOLLA CACTUS. Cholla cactuses occur sparsely in the creosote habitat and more commonly in the yucca habitat. In these places they provide important facilities for a number of bird and mammal species. In a few areas, however, cholla cactuses dominate the vegetation. This is true probably only of the Bigelow cholla (*Opuntia bigelovii*), which may form large patches on alluvial slopes. The most conspicuous of these in the Monument is the Cactus Garden area (fig. 11), on the west slope of Pinto Basin at 2200 feet in the creosote bush belt. Cactus patches are regularly used by desert wood rats and black-throated sparrows, and house finches may be expected in them if there are water sources in the vicinity.

OASIS. The springs which provide water for animals also support plant growth of special advantage to a number of vertebrate species, some of which are entirely dependent on it. In addition to small plants growing in the water or at its edge, sedges, chrysothamnus bushes, *Zauschneria latifolia,* scrub willows (*Salix*), and larger water-dependent trees occur. The trees (fig. 12) are cottonwood (*Populus fremontii*), mesquite (*Prosopis* sp.), willow, and California fan palm (*Washingtonia filifera*). Not all oases have all these tree species, and the trees at others may be so few in number as to form an inadequate amount of habitat for birds dependent on them. The best developments of trees are at the oases at Lost Palm Canyon, Cottonwood Spring, Twentynine Palms, Fortynine Palms, Smithwater Canyon, and Little Morongo Canyon. At Twentynine Palms a particularly large thicket of mesquite occurs. The higher water sources, such as Black Rock Spring, Quail Spring, Stubby Spring, and Pinyon Wells, have little or no associated tree growth. Originally there were ponds of open water at Twentynine Palms, and in Little Morongo Canyon a permanent small stream flows for half a mile, in a few places spreading into ponds and small sedge marshes. Elsewhere a few artificial ponds have been created by low dams. These provide temporary spreads of shallow water and mud margins. Such are White Tanks, Ivanpah Tank, and Barker's Dam.

Fig. 11. Cholla cactus habitat formed by the Bigelow cholla at the Cactus Garden area on the west slope of Pinto Basin.

Fig. 12. California fan palms and cottonwoods at the oasis of Cottonwood Spring. The spring is situated at the base of the cliff at the left, among the trees.

Fig. 13. Ocotillos on a gravelly slope on the west side of Pinto Basin. This spectacular tall plant occurs on well-drained alluvial fans at middle elevations but forms only a sparse vegetative cover of limited use to vertebrate animals.

Many kinds of birds and mammals come to the water sources and also some snakes such as the gopher snake and speckled rattlesnake, although the snakes may be drawn to water by the presence of prey rather than by an urge to drink. The species especially dependent on the tree and bush growth at the oases are the hooded oriole, Bullock oriole, brown towhee, and song sparrow, whereas species seeking the oases primarily to drink are the mourning dove, mountain quail, Gambel quail, house finch, lesser and Lawrence goldfinches, bats of several species, coyote, mule deer, and mountain sheep.

YUCCA. This habitat is rather diverse but has one general feature throughout, namely the presence of fairly tall, well-spaced plants, chiefly yuccas, but also junipers. Thus are provided elevated lookout posts and nest sites for several types of birds, protecting spines and retreats for birds and lizards, and certain fruiting bodies and wood sources. The Joshua tree (*Yucca brevifolia*) is of course the most conspicuous element (fig. 14) and with the juniper (*Juniperus californica*) forms an open woodland. Joshua trees cover great tracts at middle elevations of the western half of the Monument, but not east of the longitude of Twentynine Palms. Other, smaller yucca species such as Spanish dagger (*Yucca mohavensis*) and nolina (*Nolina parryi*) mark the yucca belt beyond the western limit of the Joshua tree. On the mountain

Fig. 14. A clump of Joshua trees, rising to a height of about 20 feet and showing flowering stalks on the tips of the branches. These trees are the dominant features of the yucca habitat and plant belt in the western part of the Monument.

slopes in the yucca belt a diversity of shrubs and cactuses is found. Frequently present are antelope brush (*Purshia glandulosa*), blackbrush (*Coleogyne ramosissima*), paper-bag bush (*Salazaria mexicana*), ephedra, buckwheats (*Eriogonum* sp.), and, in the drainage channels especially, species of *Chrysothamnus*.

Animals that show particular attachment to the yucca habitat are the night lizard, spiny lizard, ladder-backed woodpecker, Scott oriole, cactus wren, loggerhead shrike, sparrow hawk, and desert wood rat. Also common are antelope ground squirrels, San Diego pocket mice, and cactus mice.

DESERT GRASSLAND. Some desert sinks fill with water sufficiently after heavy rains as to preclude bush growth and favor, in certain years, the development of short grass and annuals of variable density. Such places are to be found in Queen Valley, but the best habitat of this kind is on the floor of Pleasant Valley. There in April of 1960 (fig. 15) was a dense growth of filaree (*Erodium* sp.) and pepper grass (*Lepidium*), one foot high in the bottom of the sink, and grass two inches tall on slightly higher ground. Western meadowlarks and horned larks favor this plant environment.

ROCKY CANYON. The essential features of this habitat are steep slopes consisting of cliffs or rock-dominated surfaces which have sparse interspersed vegetation and provide retreats among the rocks and local patches of shade during parts of the day (figs. 16, 17). Although developed especially in the

Fig. 15. Floor of the desert playa in Pleasant Valley on April 10, 1960, when rains had stimulated a growth of grasses, filaree, and other annual plants.

Fig. 16. Smithwater Canyon showing rocky canyon habitat and an oasis with willow clumps and cottonwoods, below which is a water seep.

canyons of the mountains, this habitat also involves steep rocky mountain slopes apart from significant drainage channels. Because the steep mountains of the Monument are largely above the 3,000 foot level, the rocky canyon environment occurs principally within the yucca belt and to some extent within the piñon belt. Characteristic inhabitants are the chuckwalla, collared lizard, canyon wren, rock wren, white-throated swift, western pipistrelle, Beechey ground squirrel, canyon mouse, Audubon cottontail, and mountain sheep.

PIÑON. Well-spaced piñons (*Pinus monophylla*) form open woodland (fig. 18) on the crests of the Little San Bernardino Mountains and a sparse scattering of small trees on rocky slopes, especially on north exposures. On the woodland floor or among the rocks, needle duff accumulates, and understory or intermixed shrubs of mountain mahogany (*Cercocarpus* sp.), scrub oak (*Quercus dumosa*), and antelope brush (*Purshia glandulosa*) are common. Interspersed as trees are junipers (*Juniperus californica*) and even Joshua trees in small numbers. Piñon habitat is extensive from the west end of the Monument southeast along the crests to the heights back of Pinyon Wells.

Fig. 17. A flowering nolina, member of the yucca group, growing on a rocky wall in the narrow part of Smithwater Canyon, April 8, 1960.

Fig. 18. Piñon habitat along the base of a hill slope at Lower Covington Flat.

Fig. 19. Chaparral composed chiefly of scrub oak and manzanita on the crest of the Little San Bernardino Mountains, 5000 feet, west of Upper Covington Flat.

On the lesser mountains on the north edge of the plateau and on the north side of Eagle Mountain it occurs chiefly as scattered trees and clumps on rocky slopes. Animals either common in piñons or closely confined to them and the associated oaks are the gray fox, Merriam chipmunk, piñon mouse, mule deer, mountain quail, screech owl, scrub jay, piñon jay, plain titmouse, bushtit, gray vireo, Gilbert skink, and California striped racer.

CHAPARRAL. Within the piñon belt are some limited areas of chaparral growth (fig. 19), seldom as dense as on coastal slopes, and usually consisting of patches of manzanita (*Arctostaphylos*) and scrub oak (*Quercus dumosa*) between which are alleyways of bare ground. In the area between Black Rock Spring and Upper Covington Flat are frequent patches of chaparral, and there are some others on the Little San Bernardino Mountains south of Pinyon Wells. Species which find this habitat especially favorable are the California thrasher, mountain quail, rufous-sided towhee, Bewick wren, gray fox, and mule deer.

TABLE 2
PRINCIPAL OCCURRENCE OF HABITATS IN TERMS OF PLANT BELTS AND LIFE-ZONES

Habitat	Lower Sonoran Zone		Upper Sonoran Zone
	Creosote bush belt	Yucca belt	Piñon belt
Sand dune	x		
Desert wash	x	x	
Creosote bush	x		
Cholla cactus	x		
Oasis	x	x	x
Yucca		x	
Desert grassland		x	
Rocky canyon		x	x
Piñon			x
Chaparral			x

THREE

PLAN AND SCOPE
OF FIELD STUDY

The need for a study of the vertebrate animal life of the Joshua Tree National Monument was felt keenly by James E. Cole during his period of service as superintendent of this unit of the national park system. Following informal discussions between Cole and Joseph S. Dixon and E. Lowell Sumner, Jr., of the research biology group of the National Park Service, the Museum of Vertebrate Zoology of the University of California, Berkeley, was urged in 1945 to undertake this task. Its purpose emerged as twofold: to conduct an original investigation of the adaptations, ecology, and distribution of the animals of this part of the Californian deserts; and to provide a source book on natural history for the park naturalists, useful to them and to the increasing number of visitors to the desert parks and monuments who as part of their intellectual recreation seriously inquire about the lives of desert animals and the problems they entail.

In May, 1945, a reconnaissance trip to the Monument was led by the senior author, and much time was spent in the field with Cole discussing the undertaking. As the plan developed, a series of trips was inaugurated by parties from the Museum directed toward becoming acquainted with all segments of biologic interest in the Monument, and scheduled to represent different seasons of the year. By spreading the field work over several years, chiefly in the period from 1945 to 1951, the annually variable conditions typical of deserts could in some measure be assessed. Basic to such a survey was extensive collecting of specimens* to determine with precision the

* Marked by asterisks in accounts of occurrences; see also p. 36.

species and subspecies represented. These documentary collections and the copious field notes on the occurrence and activities of the animals form the substance on which our report rests. Our experience in other desert areas is drawn upon in some instances to supplement our information obtained in the Monument.

The principal field trips and the personnel involved were as follows.

May 12–21, 1945: Alden H. Miller, Loye Miller, Virginia D. Miller, James E. Cole

October 12–24, 1945: A. H. Miller, Robert C. Stebbins, Ward C. Russell, Monroe D. Bryant

June 30–July 16, 1946: R. C. Stebbins, Henry G. Weston, Wade Fox, A. H. Miller

April 21–29, 1947: R. C. Stebbins, George W. Salt

August 22–September 15, 1950: A. H. Miller, W. C. Russell, Robert E. Bailey, John R. Hendrickson, Robert A. Norris, Jerry Russell

March 30–April 11, 1951: A. H. Miller, W. C. Russell

October 11–16, 1953: R. C. Stebbins, Robert Cogan

April 7–12, 1960: A. H. Miller, James D. Anderson, Gene M. Christman

The winter season, which is not represented by the Museum's field trips, was covered in very helpful manner by the continuous observations on birds by Frances Carter (1937), who was in residence at Twentynine Palms from December 30, 1933, to May 17, 1934, and from October 17, 1934, to May 30, 1935; and by Charles Sibley's work at Quail Spring from January 17 to 27 and from February 9 to 18, in 1946. Additionally a number of shorter winter trips were made by Loye Miller, and his notes with the specimens taken by him and by A. J. van Rossem, who often participated, have been made available to us.

Prior to 1945 Annie M. Alexander and Louise Kellogg had collected for the Museum in or near the Monument in February of 1935 and in April and May of 1941. Miss Alexander's direct acquaintance with the region and her enthusiasm for a concerted study of the area led to her support of the Museum's program of work begun a few years later.

The principal localities for field work are shown on the map (fig. 6). Many of these are near springs which were a convenience to the camping parties and of critical interest in observing their importance to desert animals. But at each camp we ranged widely in the vicinity. Even so, the field program could be only one of extensive sampling, as time and access did not permit us to visit every mountain sector and desert flat. Inevitably, then, our accounts must be incomplete in terms of rare and local occurrences. Certain occurrences in the Monument have been reported to us, often indirectly, without sufficient detail or substantiation to enable us to evaluate or incorporate the information in the book. Some of these items are part of continuing studies by other investigators, which are much to be desired, and which will in due course be presented in proper form in their reports. We believe that such additional data that may be forthcoming on rare or casual occurrences of animals will not materially alter the over-all picture of distribution of desert faunas which this book attempts to provide, even though they will undoubtedly improve it.

The faunal survey in the Joshua Tree Monument is a further unit in the regional surveys conducted over the years by the Museum of Vertebrate Zoology. It complements particularly those made before in nearby areas of California, namely the studies of the San Bernardino and San Jacinto mountains and, in the desert environments, the Death Valley, Providence Mountain, and lower Colorado River Valley regions (see map, fig. 20).

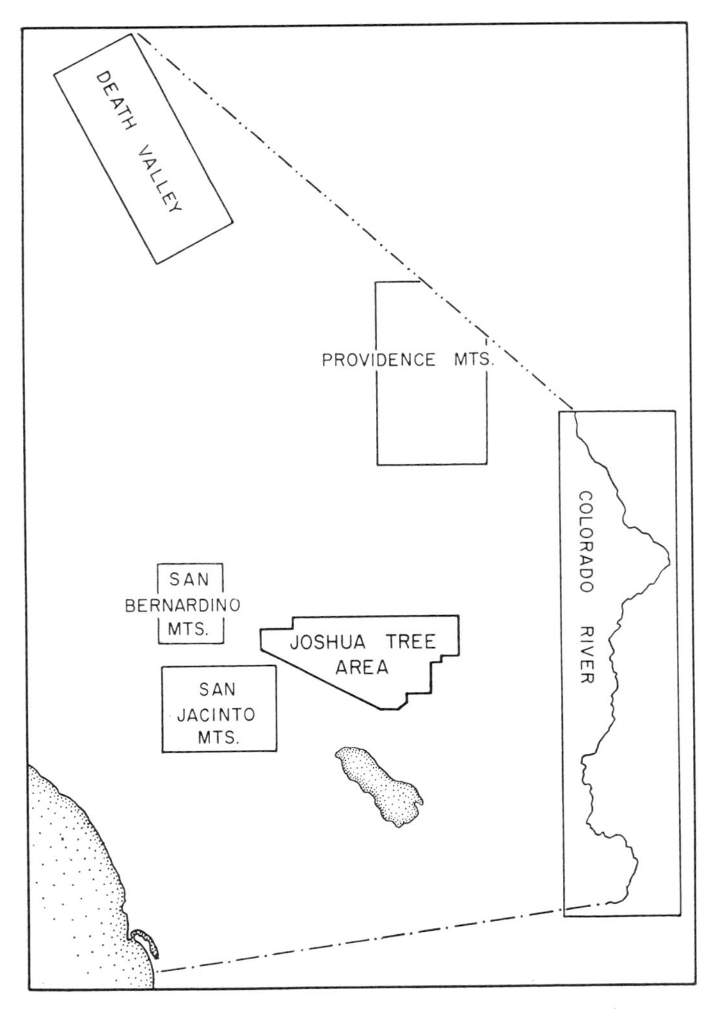

Fig. 20. Map showing position of the Joshua Tree area in relation to other sectors of southern California where faunal surveys of vertebrate animals have been made by the Museum of Vertebrate Zoology.

ACKNOWLEDGMENTS

The persons mentioned in the preceding account of field work have contributed essentially and critically to our undertaking. These some twenty people have represented a wide range of interests and special abilities. Additionally collections or notes at our disposal have been of substantial assistance and have derived from the field work of the following persons: Lowell Adams, Seth B. Benson, John Davis, Dorothy Groner, C. A. Harwell, David H. Johnson, Ernest L. Karlstrom, Carl B. Koford, Robert D. Moore, Chandler P. North, and Gerson M. Rosenthal.

Dennis G. Rainey and Richard B. Loomis have kindly permitted us to record certain specimens[1] taken in the Monument by investigators from Long Beach State College when this has not anticipated special publications which they are developing.

The personnel of the National Park Service at Joshua Tree Monument have been most helpful in arranging for our field work and in supplying us with significant information. Beside James E. Cole, who was so vitally involved, we record with appreciation the assistance and interest of Frank R. Givens, Samuel King, Hesmel L. Earenfight, William R. Supernaugh, James R. Youse, and Bruce Black of the National Park Service.

The illustrations have in large measure been planned and executed by Gene M. Christman. His color and wash drawings reflect his wide acquaintance with Californian deserts, and many of the photographs were taken by him for the particular purposes of this report. Other photographs and outline drawings are by the authors or are drawn from their previously published works. Ten photographs of reptiles and amphibians have been generously contributed by Nathan W. Cohen and are acknowledged in the legends to the figures. Similarly, nine photographs of reptiles have been made available by Robert G. Crippen.

Our thanks are also due to the Lida Scott Brown Fund for a liberal grant that aided us in publishing the color plates.

Robert I. Bowman, George F. Fisler, and Larry L. Wolf assisted over a period of several years in assembling and checking data. William Z. Lidicker, Jr., has critically read the section on mammals, Ned K. Johnson the material on flycatchers, and Don R. Medina the section on water balance and conservation.

[1] Marked by superior numeral 1 in accounts of occurrences.

FOUR

FAUNAL ANALYSIS

As we have said, the Joshua Tree Monument lies at the junction of the Mohave and Colorado deserts, and a first consideration in analyzing its fauna of vertebrate animals is to see which of the two desert areas contributes most to the complex of species present in the Monument. A second problem is the degree to which the fauna of the coastal mountains and slopes has spread eastward into the desert highlands of the Monument, for early in our study of the area it became evident that there was a strong influence from that direction and that the highlands were a peninsula which in one respect supported a modified coastal fauna.

Faunistic relations are reflected in several ways. These are: (1) the presence of species in the area concerned which occur in the adjoining faunas; (2) the racial or subspecific affinities shown in the species that are present; and (3) the apparent parental stocks and their geographic location from which local endemic races seem to have been derived. In addition there is the related question of the degree to which racial features are conspicuously intergradient or steeply clinal in the populations in the Monument area.

Table 3 summarizes the data necessary to these analyses. In entering the facts in the table, a degree of arbitrariness is inevitable. A species present in the Monument may occur only marginally or very locally in an adjoining faunal area. The definitions of these areas and faunas can never be precise, and one must make judgments then that attempt to represent the general truth of occurrence. Thus the Scott oriole, *Icterus parisorum,* is not considered to be a member of the Colorado Desert fauna, even though it occurs variously about the borders of this desert, for the species does not nest in

TABLE 3
OCCURRENCE OF SPECIES AND RACES OF THE MONUMENT IN ADJACENT FAUNAS AND IN PLANT BELTS

Species name	Species present in adjacent faunal area			Races representative of faunal areas				Intergradation involving faunal representatives			Endemic race	Plant belt occupied in Monument			Names of races present
	Mohave	Colorado	San Diego	Mohave and Colorado	Mohave	Colorado	San Diego	Mohave and Colorado	Mohave and San Diego	Colorado and San Diego		Creosote	Yucca	Piñon	
AMPHIBIANS															
Bufo boreas			x				x						x		halophilus
Bufo punctatus	x	x											x	x	
Hyla californiae			x										x	x	
Hyla regilla			x										x		
Rana catesbeiana	[Introduced]														
REPTILES															
Gopherus agassizi	x	x										x	x		
Coleonyx variegatus	x	x	x	x								x	x	x	variegatus
Dipsosaurus dorsalis	x	x		x								x			dorsalis
Crotaphytus collaris	x	x		x								x	x	x	baileyi
Crotaphytus wislizenii	x	x		x								x	x	x	wislizenii
Sauromalus obesus	x	x		x								x	x		obesus
Callisaurus draconoides	x	x		x								x	x		rhodostictus
Uma scoparia	x											x			
Sceloporus magister	x	x		x								x	x	x	uniformis
Sceloporus occidentalis			x				x							x	longipes
Uta stansburiana	x	x	x	x			x		x	x		x	x	x	hesperis; stejnegeri
Urosaurus graciosus	x	x										x			
Phrynosoma coronatum			x				x							x	blainvillei
Phrynosoma platyrhinos	x	x		x								x	x		calidiarum
Xantusia vigilis	x	x	x										x	x	vigilis
Cnemidophorus tigris	x	x	x	x								x	x	x	tigris
Eumeces gilberti			x				x							x	rubricaudatus
Leptotyphlops humilis	x	x	x		x		x					x			humilis

Species name	Species present in adjacent faunal area			Races representative of faunal areas				Intergradation involving faunal representatives			Endemic race	Plant belt occupied in Monument			Names of races present
	Mohave	Colorado	San Diego	Mohave and Colorado	Mohave	Colorado	San Diego	Mohave and Colorado	Mohave and San Diego	Colorado and San Diego		Creosote	Yucca	Piñon	
Lichanura trivirgata	x	x	x						x	x		x	x		
Masticophis flagellum	x	x	x									x	x	x	piceus
Masticophis lateralis			x				x							x	lateralis
Salvadora hexalepis	x	x	x			x		x	x	x		x	x	x	hexalepis
Phyllorhynchus decurtatus	x	x		x								x	x		perkinsi
Arizona elegans	x	x	x			x		x	x	x		x	x	x	eburnata
Pituophis melanoleucus	x	x	x		x			x	x	x		x	x	x	deserticola
Lampropeltis getulus	x	x	x									x	x	x	californiae
Rhinocheilus lecontei	x	x	x	x								x	x		clarus
Chionactis occipitalis	x	x			x							x	x		occipitalis
Hypsiglena torquata	x	x	x	x					x	x		x	x	x	deserticola
Trimorphodon vandenburghi			x										x		
Tantilla eiseni	x		x											x	[race uncertain]
Crotalus viridis			x				x							x	helleri
Crotalus scutulatus	x												x	x	
Crotalus atrox	x	x										x			
Crotalus mitchelli	x	x	x			x	x					x	x	x	pyrrhus
Crotalus cerastes	x	x			x			x				x	x		cerastes
BIRDS															
Cathartes aura	x	x	x									x	x	x	teter
Buteo jamaicensis	x	x	x									x	x	x	calurus
Aquila chrysaëtos	x		x									x	x	x	canadensis
Falco mexicanus	x	x	x									x	x	x	
Falco sparverius	x	x	x										x	x	sparverius
Lophortyx gambelii	x	x		x								x	x	x	gambelii
Oreortyx pictus			x								x		x	x	russelli
Zenaida asiatica		x				x						x	x		mearnsi
Zenaidura macroura	x	x	x									x	x	x	marginella
Geococcyx californianus	x	x	x									x	x	x	
Tyto alba	x	x	x									x	x	x	pratincola
Otus asio		x	x			x				x		x	x	x	yumanensis

TABLE 3 (*Continued*)

Species name	Species present in adjacent faunal area			Races representative of faunal areas				Intergradation involving faunal representatives			Endemic race	Plant belt occupied in Monument			Names of races present
	Mohave	Colorado	San Diego	Mohave and Colorado	Mohave	Colorado	San Diego	Mohave and Colorado	Mohave and San Diego	Colorado and San Diego		Creosote	Yucca	Piñon	
BIRDS (*Continued*)															
Bubo virginianus	x	x	x	x								x	x	x	pallescens
Micrathene whitneyi		x				x							x		whitneyi
Speotyto cunicularia	x	x	x									x			hypugaea
Asio otus			x											x	wilsonianus
Phalaenoptilus nuttallii	x	x	x		x			x				x	x	x	nuttalli
Chordeiles acutipennis	x	x	x									x	x		texensis
Aëronautes saxatalis	x	x	x									x	x	x	saxatalis
Calypte costae	x	x	x									x	x	x	
Dendrocopos scalaris	x	x		x									x	x	cactophilus
Tyrannus verticalis	x	x	x									x	x	x	
Tyrannus vociferans	x		x									x	x	x	vociferans
Myiarchus cinerascens	x	x	x									x	x	x	cinerascens
Sayornis nigricans			x										x		semiatra
Sayornis saya	x	x	x									x	x	x	saya
Eremophila alpestris	x	x	x		x							x	x		ammophila
Petrochelidon pyrrhonota	x	x	x										x		[race uncertain]
Aphelocoma coerulescens	x		x				x				x			x	cana; obscura
Corvus corax	x	x	x									x	x	x	sinuatus
Gymnorhinus cyanocephalus	x				x									x	rostratus
Parus inornatus	x		x								x			x	mohavensis
Auriparus flaviceps	x	x		x								x	x		acaciarum
Psaltriparus minimus	x		x								x			x	sociabilis
Thryomanes bewickii	x		x				x		x				x	x	correctus
Campylorynchus brunneicapillus	x	x	x									x	x	x	couesi
Catherpes mexicanus	x		x										x	x	conspersus
Salpinctes obsoletus	x	x	x									x	x	x	obsoletus
Mimus polyglottos	x	x	x									x	x	x	leucopterus

Species name	Species present in adjacent faunal area			Races representative of faunal areas				Intergradation involving faunal representatives			Endemic race	Plant belt occupied in Monument			Names of races present
	Mohave	Colorado	San Diego	Mohave and Colorado	Mohave	Colorado	San Diego	Mohave and Colorado	Mohave and San Diego	Colorado and San Diego		Creosote	Yucca	Piñon	
Toxostoma bendirei	x												x		bendirei
Toxostoma redivivum			x				x							x	redivivum
Toxostoma lecontei	x	x		x								x	x		lecontei
Polioptila caerulea	x		x										x	x	amoenissima
Polioptila melanura	x	x	x	x								x	x		lucida
Phainopepla nitens	x	x	x									x	x		lepida
Lanius ludovicianus	x	x	x		x	x		x				x	x	x	nevadensis; sonoriensis
Vireo vicinior	x		x											x	
Sturnella neglecta	x	x	x										x		[race uncertain]
Icterus cucullatus	x	x	x				x			x		x	x		californicus
Icterus parisorum	x												x	x	
Icterus bullockii	x	x	x			x	x					x	x		parvus
Molothrus ater	x	x	x									x	x	x	obscurus
Passer domesticus	[non-native]														
Carpodacus mexicanus	x	x	x									x	x	x	frontalis
Spinus psaltria	x	x	x										x	x	hesperophilus
Spinus lawrencei			x										x	x	
Pipilo erythrophthalmus	x		x				x							x	megalonyx
Pipilo fuscus			x				x						x		senicula
Amphispiza bilineata	x	x		x								x	x	x	deserticola
Spizella atrogularis	x		x		x									x	evura
Melospiza melodia		x	x				x						x		cooperi
MAMMALS															
Macrotus californicus		x	x									x			
Myotis thysanodes	x	x	x											x	thysanodes
Myotis volans	x	x	x											x	interior
Myotis californicus	x	x	x	x									x	x	stephensi
Pipistrellus hesperus	x	x	x	x								x	x	x	hesperus
Eptesicus fuscus	x	x	x	x								x	x	x	pallidus
Lasiurus cinereus	[Migrant only]														

TABLE 3 (*Continued*)

Species name	Species present in adjacent faunal area			Races representative of faunal areas				Intergradation involving faunal representatives			Endemic race	Plant belt occupied in Monument			Names of races present
	Mohave	Colorado	San Diego	Mohave and Colarado	Mohave	Colorado	San Diego	Mohave and Colorado	Mohave and San Diego	Colorado and San Diego		Creosote	Yucca	Piñon	
MAMMALS (*Continued*)															
Euderma maculatum	x	x	x									x			
Antrozous pallidus	x	x	x	x								x	x		pallidus
Sylvilagus audubonii	x	x	x	x								x	x	x	arizonae
Lepus californicus	x	x	x	x								x	x	x	deserticola
Eutamias merriami			x				x							x	merriami
Ammospermophilus leucurus	x	x										x	x	x	leucurus
Citellus beecheyi	x		x		x								x	x	parvulus
Citellus tereticaudus	x	x		x								x			tereticaudus
Thomomys bottae	x	x	x		x	x	x					x	x	x	cabezonae; mohavensis; rupestris
Perognathus longimembris	x	x	x		x							x	x	x	longimembris
Perognathus formosus	x	x			x			x				x	x		mohavensis
Perognathus spinatus		x				x						x			spinatus
Perognathus fallax	x	x	x	x								x	x	x	pallidus
Dipodomys merriami	x	x			x			x				x	x	x	merriami
Dipodomys microps	x				x									x	occidentalis
Dipodomys deserti	x	x		x								x			deserti
Reithrodontomys megalotis	x	x	x	x									x		megalotis
Peromyscus crinitus	x	x		x								x	x	x	stephensi
Peromyscus eremicus	x	x	x	x								x	x	x	eremicus
Peromyscus maniculatus	x	x	x	x								x	x	x	sonoriensis
Peromyscus boylii			x				x							x	rowleyi
Peromyscus truei	x		x				x							x	chlorus
Onychomys torridus	x	x	x	x								x	x	x	pulcher
Neotoma lepida	x	x	x	x								x	x	x	lepida
Neotoma fuscipes			x				x						x	x	simplex
Microtus californicus	x		x										x		sanctidiegi
Canis latrans	x	x	x	x								x	x	x	mearnsi

Species name	Species present in adjacent faunal area			Races representative of faunal areas				Intergradation involving faunal representatives			Endemic race	Plant belt occupied in Monument			Names of races present
	Mohave	Colorado	San Diego	Mohave and Colorado	Mohave	Colorado	San Diego	Mohave and Colorado	Mohave and San Diego	Colorado and San Diego		Creosote	Yucca	Piñon	
Vulpes macrotis	x	x	x	x									x	x	arsipus
Urocyon cinereoargenteus	x	x	x				x		x	x			x	x	californicus
Procyon lotor	x	x	x										x		[race uncertain]
Taxidea taxus	x	x	x	x								x	x	x	berlandieri
Felis concolor	x	x	x											x	[race uncertain]
Lynx rufus	x	x	x										x	x	[race uncertain]
Odocoileus hemionus		x	x				x							x	fuliginatus
Ovis canadensis	x	x	x									x	x	x	nelsoni
Subtotals															
Reptiles and Amphibians (40)	31	28	24	13	4	3	9	4	6	6	0	27	29	23	
Birds (60)	50	40	50	7	5	5	8	2	1	2	4	35	50	43	
Mammals (41)	35	34	33	18	6	2	7	2	1	1	0	23	28	31	
GRAND TOTAL (141)	116	102	107	38	15	10	24	8	8	9	4	85	107	97	

the very openly vegetated, low-lying desert environments typical of the Colorado basin, whereas it is widespread in several habitats of the Mohave Desert. Also in tabulating we have had to adapt the complex information on racial affinities and intergradation laid out in each species account to a rather few, simplified categories.

Species Composition. One hundred and forty-one species are considered in the analysis. Excluded are species of birds and bats that may be presumed to be only transients or nonbreeding visitants to the area and also non-native species. The number of species common to both the Monument and to the Mohave Desert stands at 116 (82 per cent), whereas those common to the Monument and the Colorado Desert and to the Monument and coastal southern California (San Diegan district) are only moderately lower at 102 (72 per cent) and 107 (76 per cent), respectively. The contrast thus shown in affinity with the Mohave and Colorado deserts is probably a true reflection of difference. The figure for the San Diegan district is somewhat misleading, for clearly the fauna of the Monument as a whole is desertlike and the similarity to the Mohave and Colorado desert faunas is substantial. A somewhat more accurate picture is presented by contrasting the combined desert total of 123 (87 per cent) with the coastal total of 107 (76 per cent).

The foregoing totals for faunal elements in common show some meaningful differences when broken down by animal groups. Thus the mammals show essentially equal relations to the three adjoining faunas. The birds display a marked disparity (Mohave and San Diego, each 80 per cent; Colorado, 66 per cent) in which the relations to the Colorado fauna are relatively weak. The herpetofauna on the other hand shows the weakest link with the San Diegan (60 per cent), and this would be even weaker if one excluded the four amphibian species and considered the reptiles alone (58 per cent). In each animal group, however, the species composition shows the derivation from the Mohave Desert to be as strong or stronger than that from either of the others.

Another way of reflecting faunal affinities at the species level might be to focus attention on those species which are related to only one of the three faunal areas being compared, in as much as such species by their restrictions are particularly meaningful when they appear among the elements in the Joshua Tree Monument. But to do this legitimately the three faunas should differ from one another in approximately the same degree, and we stress again the obvious fact that they do not and that the two desert faunas are most alike and stand in contrast with the coastal fauna. Of the total species of the Monument listed for the Mohave Desert only there are 6, and for the Colorado Desert there are 2. But species listed for both are 25 as contrasted with 19 for the coastal district; this latter contrast is regarded as valid.

Racial Affinities. Many of the species in the Monument are represented by races or subspecies peculiar to a fairly restricted region and in many instances with characteristics in some way related to the environment of that region. In other instances the races present are those widespread over the western part of the continent and they tell us nothing of the affinities of the species in the Monument with respect to adjoining areas and faunas. Accordingly in table 3, we have tallied only those races which are representative of one or more of the adjoining faunal areas. Among the desert races a single race may be present in both desert areas concerned (38 instances) and more or less similarly representative of each. Only three cases occur involving a race of both the San Diegan district and one or the other desert areas.

The total instances of racial occurrences relating to the Mohave Desert are 53, for the Colorado Desert 48, and for the San Diegan coastal district 24. These values are believed to reflect in a very meaningful way the faunal affinities and the resemblances of the environments of the Monument to those of the three areas adjoining. These differentials, then, supplement the data on species composition and stress even further the contrasts drawn from that consideration.

In the different animal groups involved, the greatest contrast in number of races of the deserts and the coastal area is shown by the mammals and the herpetofauna. In these groups the two deserts have contributed similar numbers, the totals slightly favoring the Mohave Desert, and the coastal area is only a half or a third as well represented as either of the deserts. The races of birds are much fewer, absolutely and relatively, in the total of those typical of the adjoining areas concerned, and they tend to be more equally divided among these areas, but even so the coastal group is but two-thirds that of either of the others.

For the whole vertebrate fauna there are 60 races representative of the deserts combined or collectively (26 mammals, 18 reptiles and amphibians, 16 birds) and only 20 of the coastal fauna (7 mammals, 6 reptiles and amphibians, 7 birds).

Endemic Races. The development of endemic races of vertebrates limited to relatively small areas is almost always dependent on a high degree of isolation of the areas. The Joshua Tree Monument is not isolated from either the Mohave or Colorado deserts and, as indicated by the record of intergradations of races within its limits, is a region of contact; certainly it is not one of desert environments isolated significantly from both the adjoining desert districts.

However, the uplands of the Monument provide an isolated area of environment that is ecologically related to the coastal districts but modified in significant degree from them. The isolation at the westward end is of varying degrees of effectiveness for different species.

In these upland environments are four endemic races of western faunal affinity. All of these are birds typical of the piñon belt. Two, the mountain quail (*Oreortyx pictus russelli*) and the plain titmouse (*Parus inornatus mohavensis*) are primarily modifications of a coastal type from which they must have been derived in the main. The bushtit (*Psaltriparus minimus sociabilis*) and the scrub jay (*Aphelocoma coerulescens cana*)—the latter restricted to the highly isolated Eagle Mountain—predominate in coastal affinities but, more than the first two, show some evidence of deriving some of their attributes from eastward sources.

The fact that all the endemic races are birds may reflect the greater influence on birds of the vegetative cover on which they depend and the isolation of these vegetational formations in the peninsular uplands of the Monument. It is possible that one or two other endemic forms may become evident on further study of better series of certain mammals and reptiles of the piñon belt.

The existence of these endemic races of mainly coastal derivation of course augments the evidence of faunal tie in that direction in one sense. But this tie is already represented in the listing of the species concerned from this fauna. On the contrary, the fact that there are differentiated races, with characteristics of pale coloration reflecting the desert environment of these uplands, strengthens the distinction from the coast and adds to the totality of desert racial relations in the Monument. The endemics in their apparent history show coastal affinities, but their direction of change stresses the desert environment of the present and its strong contrast with that of the coast.

Intergradation. Races of vertebrates, somewhat arbitrarily delimited for practical purposes, are in fact usually parts of gradient or clinal systems of characters in the species concerned. These clines tend to show very steep or abrupt changes along borders of strongly contrasted ecologic regions. Such is often true at desert borders. In dealing with large numbers of species we have not been in a position to refine the data on clinal systems as we would hope to do in monographing a particular species or genus. However, we have noted instances where gradation from one racial "norm" to another was especially evident in the populations in the Monument. Such evident gradations are tallied in table 3; probably still others will be revealed by further study.

The tabulation does show 8 or 9 cases each of considerable intergradation, steep clinal change, or intermixture of features involving pairs of contrasting faunal areas. It is generally true that intergradation in the Monument proper is considerable, perhaps more than shown, between populations of the Mohave and Colorado deserts and between these deserts and the coastal province. These intergrading phenomena are quite predictable in terms of the intermediate location of the Monument and its physiography.

Faunal Elements in Relation to Vegetation Belts. The occurrence of the 141 resident or summer resident vertebrates in the three vegetation belts of the Monument (fig. 6) is summarized in table 3. Both because the belts are somewhat arbitrarily delimited, as successive altitudinal bands, and because any one species can be classed as related to certain belts only approximately and by ignoring numerous local exceptions and special circumstances, the results of the tabulation should not be given undue weight and should not be subjected to refined quantitative analysis. It should be evident from general inspection, however, that the species that are part of the Mohave or Colorado faunas are rarely absent (2 instances out of 31) from the creosote and yucca belts and that the San Diegan faunal elements are usually present (13 of 18) in the piñon belt and in no instance reach down to the creosote belt. These altitudinal and ecologic distributions are then much in accord with the occurrence and adjustments of the species in the adjoining and contrasting faunal districts.

In reviewing the number of species present in each of the three vegetation belts it is noteworthy (table 3) that the yucca belt has the most species and the piñon and creosote belts successively fewer. This is apparently a reflection of the relative uniformity and openness of vegetation of the creosote belt and of the diversity of plant types, soil types, and physiography in the yucca belt. Even with the addition of winter visitant species of birds, the relative numbers of species in the three belts would remain about the same.

Examination of the total species for each group of vertebrates in the vegetation zones shows some significant contrasts. The reptile species are fewer in the piñon belt than in the creosote belt, a reflection generally of the heat requirements of this group and the fact that they are less dependent than birds on the protective cover of large plants. The mammals show the greatest number of species in the piñon belt, but this is probably not a fully reliable difference. The count is unduly influenced by the fact that several rarely detected species have been found there and perhaps largely by chance were not encountered in the yucca belt, where they might be expected.

TABLE 4

NUMBER OF SPECIES OCCURRING IN ONE, TWO, OR THREE PLANT BELTS

	1		2		3	
	Number	Per Cent	Number	Per Cent	Number	Per Cent
Reptiles and amphibians	14	35	13	33	13	33
Birds	17	28	18	30	25	42
Mammals	16	39	9	22	16	39
All groups	47	33	40	28	54	38

An examination of the degree to which species are limited to one belt rather than spread through two or three is shown in table 4. The interesting fact emerges that more species occur in one belt than in two but that even more occur in all three. However, this results from a rather large number of birds (42 per cent) that range through all three, whereas considerably fewer species of reptiles and amphibians (33 per cent) do so. The mammals show an intermediate situation. Birds as a whole then tend to be less narrowly limited than reptiles and mammals in a zonal sense, or in terms of altitudinal plant belts, in this area.

Summary. The vertebrate fauna of the Joshua Tree Monument is one predominantly of desert affinities and includes in greatest proportions those species that occur in the Mohave Desert to the north. This faunal affinity is even more strongly indicated by the high proportion of races characteristic of the Mohave Desert that range into the Monument. The faunal contribution of the Colorado Desert to the south is also high, especially in terms of characteristic races present, but it is not so great as that of the Mohave Desert. This is true, it would seem, because the Monument is a high-altitude desert in the main and ecologic conditions normal to the Mohave Desert extend with little or no interruption south into it. A number of instances of intergradation and clinal change occur in the Monument between forms of the Mohave and Colorado deserts.

The uplands of the Monument, particularly the piñon belt, form a peninsula of ecologic conditions which at the west tend to divide rather sharply the low-elevation elements of the Colorado Desert from those of the Mohave Desert. This peninsula also permits eastward extension and isolation of faunal elements, species and races, of the coastal districts of southern California. These coastal intrusions tend to remain confined to the piñon belt and show intergradation with their desert counterparts in a number of instances. The isolation of the peninsula is such that several coastal species have evolved endemic races reflecting the desert conditions to which they have been subjected.

FIVE

BIRDS

WATER BIRDS

A miscellaneous group of water birds has been recorded as migrants and vagrants in the Monument. Only among the shorebirds do we have records of several species of one order (see p. 71). The other water birds reported are loons, grebes, herons, ducks, and rails, none of which use the Monument much in migration. Rather these kinds fly routes that avoid it or they cross over it without stopping, some traveling at night, in part at least, when heat problems are not encountered. Those species that do stop at the few watering places do so primarily to rest, and they are usually species that can alight and take off from land, such as herons and teal. The diving birds require considerable water surfaces even for a rest stop, and they are not to be found in the Monument except on a very limited seasonal basis.

Most of the water birds have been recorded from Twentynine Palms in years past when exposed water surfaces or marshy spots were available. Reëstablishment of such facilities would probably soon draw occasional visitants in special need of rest in their desert flights. Ponds and marshes are easily detected from the air, and areas adjoining the monument in various desert sinks, as near Dale, have quickly proved their attracting power for water birds in transit, bringing in species which have not yet actually been recorded in the Monument.

Rest stops, even without foraging opportunities, may be important for individual birds for purposes of water intake and for metabolism of fat reserves which may not keep pace with energy use while they are actually engaged in flight.

Loons and grebes are much more likely to tire in the course of migratory flights than are ducks or shorebirds, inasmuch as they are relatively heavy bodied for their wing equipment. The ponds and tanks in the Monument are, however, inadequate for these swimming and diving types and at best must provide only temporary stopping points without significant food resources for them. The grebes and loons must land on water surfaces, as they can do no more than hobble or shuffle along on land and probably could not take off from it. The diving birds that occur in the Monument are usually exhausted birds; few of them would be able to survive and continue.

Common Loon. *Gavia immer*

Description. A large diving bird, the size of a small goose, with straight, stout, pointed bill and white breast and belly; back dark gray or (breeding plumage) checkered black and white; head gray or (breeding plumage) black with partial white collar and white thoat mark. Feet webbed and set far posteriorly. Short tail.
Range. Breeds in arctic and subarctic America from Aleutian Islands to Greenland; also Iceland; south to northeastern California, North Dakota, northern Iowa, the Great Lakes, and Connecticut. Winters in North America from southern Alaska, the Great Lakes, and Newfoundland south to Baja California, Sonora, and Gulf coast of United States.
Occurrence in Monument. Occasional migrant. Recorded from: vicinity of Twentynine Palms, Nov. 11; Pinto Wash Well, Apr. 29 [1].

In the fall of 1934 a common loon was found exhausted by the roadside nearly 5 miles from water. Miss Carter reports (1937:212) that "it was brought to the oasis [at Twentynine Palms] and placed in the pasture pool. On the following day it seemed quite revived, although still permitting close approach. It swam freely and dived, uttering a low chuckling sound and also a high quavering 'laugh.' [It] was not seen again."

On April 29, 1962, a loon was taken by a party from Long Beach State College as the bird came in to alight on a temporarily flooded area near Pinto Wash Well in the eastern part of the Monument.

A possible additional occurrence, with some uncertainty as to species, was that of a loon killed on the road in about 1946, as reported to one of our parties by Mr. and Mrs. W. D. Robinson.

Eared Grebe. *Podiceps caspicus*

Description. A small diving bird, somewhat smaller than a coot, with straight, thin, pointed bill, and slender neck; back dark gray or blackish; breast and belly white; crown and neck gray; throat and postauricular area white; postauricular white patch not sharply defined nor extended to hind neck; in breeding plumage, head and neck black, with buffy feather tufts on sides of head. Toes with lateral flaps. Tail absent.
Range. Breeds in North America from central British Columbia and southern Manitoba south to northern Baja California and southern Texas. In Eurasia, from British Isles, Denmark, Russia, and Manchuria south to Mediterranean area and Asia Minor; also southern and eastern Africa. In North America winters from Washington to southern Baja California and through México to Guatemala; also to Colombia.
Occurrence in Monument. One record of a migrant: Twentynine Palms, Apr. 16.

In 1935 Miss Carter saw one in full breeding plumage in the "pasture pool" at Twentynine Palms. Concerning the occurrence of migrant diving birds, see p. 49.

Green Heron. *Butorides virescens*

Description. A small, relatively short-legged heron with crown, back, and wings dull greenish or blue-green; neck reddish, except for midventral part which is white with longitudinal streaks of dark gray; belly streaked with gray; wing coverts edged or spotted with buff. Legs greenish yellow. Bill dark or, in immatures, yellow ventrally.
Range. Breeds from Washington, Nebraska, and southeastern Canada south to the West Indies and Panamá. Winters from central California (rarely), southern Arizona, southern Texas, the Gulf coast, and South Carolina to northern South America.
Occurrence in Monument. Occasional spring migrant. One winter record. Recorded from: Twentynine Palms, Feb. 8, May 1, 5, 6, 9–14.

Miss Carter observed single individuals in two different years in May in the "pasture pool" at Twentynine Palms. The February record she reports may represent a winter resident or merely a vagrant. In the spring what was apparently the same individual tarried as long as five days.

Unlike water birds that require ponds on which to settle or swim, the herons can in transit utilize any sort of marshy or moist ground as a stopping place. The green heron probably would not stay long at a rest point unless there was vegetation about the water in which to hide. Herons fly easily, although slowly compared with shorebirds, and like them may seldom settle for rest periods when crossing desert terrain.

The green herons occurring in the area probably belong to the race *Butorides virescens anthonyi* of the Pacific coast and Great Basin areas.

Common Egret. *Casmerodius albus*

Description. A large, entirely white heron with yellow bill and black legs and feet.
Range. Breeds from southern Oregon, southern Minnesota, and New Jersey south through southern United States, México, Central America, the West Indies, and South America to the Straits of Magellan; in Old World from southeastern Europe and southern Asia south to southern Africa, Ceylon, the East Indies, Australia, and New Zealand. In post-breeding season may wander farther north. Partly migratory in northern part of breeding range.
Occurrence in Monument. Recorded only from Twentynine Palms: Apr. 23, 24, July 13.

Single individuals have been seen in spring and summer as migrants or wanderers, stopping in at the oasis and "swamp" at Twentynine Palms.

The American members of this species belong to the race *Casmerodius albus egretta.*

American Bittern. *Botaurus lentiginosus*

Description. A heron intermediate in size between the common egret and the green heron, rich brown mottled with ochraceous above; yellowish brown stripes on white of throat, neck, and breast; sides of neck with broad, black, longitudinal stripe.
Range. Breeds from central British Columbia, southern Mackenzie, northern Manitoba, northern Ontario, and Newfoundland south to southern California, central Arizona, Louisiana, and Florida. Winters from southwestern British Columbia, southern Arizona, Oklahoma, the Ohio Valley, and Delaware south through México to Panamá and the West Indies.
Occurrence in Monument. Noted only at Black Rock Spring in fall migration: Sept. 4.

Early on a morning in September an American bittern flew southeast through the piñon-covered basin south of Black Rock Spring. It was visible for a long time as it flew in a straight line over the tops of the piñons and yuccas on a southeast course in the general direction of the Salton Sea. Late in the evening of the same day what must have been another individual was seen flying south in this area against a rather strong wind. Although these bitterns may at times come to rest at oases, they apparently for the most part cross the area in direct flight, possibly moving from lakes or marshes in the San Bernardino Mountain area along the crests of the Little San Bernardino Mountains, as at Black Rock Spring, into the marshes of the Salton Sea and the Colorado River delta.

Green-winged Teal. *Anas carolinensis*

Description. A small duck with bend of wing gray and with green speculum bordered by buff and white anteriorly and black medially and laterally. Male: head chestnut, the postocular area green, the chin and nape black; back and flanks finely vermiculated gray and white; white patch in front of wing; breast white, spotted with black; belly white; rump and crissum tan bordered with black. Female: gray and buff mottled throughout with dusky, lighter on under parts.
Range. Breeds from north-central Alaska, northern Mackenzie, northern Manitoba, and James Bay south to central California, northern Utah, northern New Mexico, southern Minnesota, northern Ohio, and western New York. Winters from southern British Columbia, northern Nebraska, southern Illinois, and Nova Scotia south through southern United States, México, and the West Indies to Honduras.
Occurrence in Monument. Occasional transient. Recorded from: Twentynine Palms, Jan. 25, Feb. 10, 17.

Miss Carter reported a single individual in late winter on the dates indicated. Such occurrences, although too early for spring migration, should not be regarded as showing winter residence in the Monument. They probably reflect irregular movements during the winter season between more favorable foraging and resting grounds elsewhere.

Cinnamon Teal. *Anas cyanoptera*

Description. A small duck with large light blue areas on anterior wing surface from bend of wing inward. Male dark cinnamon over most of body; female buff-brown, mottled with dusky, lighter on under parts than on back. Female distinguishable from female blue-winged teal only on comparison of bill width in specimens—bill broader, especially terminally, in cinnamon teal.
Range. Breeds from central British Columbia east to eastern Wyoming, southwestern Kansas, and west-central Texas, south to southern California, and in México to Jalisco and Tamaulipas; also in Colombia and from southern Perú and Bolivia to Chile, Patagonia, and the Falkland Islands. In winter, North American form ranges chiefly from southwestern United States south to Panamá, Colombia, and Ecuador.
Occurrence in Monument. Fairly frequent spring and fall migrant at suitable watering places. Recorded from: Ivanpah Tank, Sept. 12; Twentynine Palms, Mar. 22–25, Apr. 10–16.

This teal is the species of duck most likely to settle at the limited areas of water in the Monument in the course of migration through the desert. Even a muddy irrigation ditch will attract it. It does not require actual water surface on which to land as do grebes and loons and most of the diving ducks. At Twentynine Palms Miss Carter saw as many as nine in one flock, but at other times only single individuals were present. Unless males are seen, as they were at Twentynine Palms, the species identification of this type of teal by sight remains in doubt. The blue-winged teal may occur in the Monument, but thus far it has not been positively identified.

The race of cinnamon teal in North America is now regarded as distinct from those in South America and has been designated *Anas cyanoptera septentrionalium.*

HAWKS AND ALLIES

Turkey Vulture. *Cathartes aura*

Description. A large, soaring bird with wingspread of about 6 feet. Plumage black; silvery gray in certain lights on under wing surface; head unfeathered and red; wings typically held above horizontal and tips upturned and spread as a result of air pressure.
Range. Breeds from southern Canada south through Central America, the Greater Antilles, and South America to the Straits of Magellan and the Falkland Islands. Winters from California, Nebraska, and the Ohio Valley southward.
Occurrence in Monument. Year-round resident in small numbers; also a spring migrant. Recorded from: Morongo Pass, Apr. 28; Little Morongo Canyon, Apr. 3; Black Rock Spring, July 6; Quail Spring, Apr. 22, May 19–21, July 3; Key's View, July 3, 4; Twentynine Palms, Apr. 3, July 13; 5 mi. S. Twentynine Palms, Apr. 10; Live Oak Tank, Apr. 6–8; Pinto Basin, Apr. 10; Cottonwood Spring, Apr. 26, 29, May 14.

Turkey vultures have been seen occasionally through the spring and summer, and it is probable that a few nest in the Monument, although no nest site has been found. Frances Carter states that she observed them intermittently through the periods of her residence in the area, which included two winter seasons. They are seen most often in the spring, and some of the observations then, especially in April, doubtless represent northward migrants passing through. More specifically, on April 3 at Twentynine Palms a flock of 29 was observed which was almost certainly a migratory aggregation. A food source on the desert roadways at other seasons usually attracts no more than four birds.

With the elimination of cattle grazing on government land in the Monument area, sources of carrion for these vultures has probably diminished. However, road kills of jack rabbits, antelope squirrels, and snakes on the desert highways generally provide food for a small number of vultures. The food is usually moist when they find it, and consequently there is little or no dependence of this species on sources of drinking water.

Turkey vultures of this general section of the continent belong to the race *Cathartes aura teter.*

Sharp-shinned Hawk. *Accipiter striatus*

Description. A small (90—160 gram), long-tailed, and round-winged hawk, the tail square cut. Adults: dark blue-gray above, the crown black; sides of head light gray; under parts reddish brown, incompletely cross-barred with white; under tail coverts white; tail with 4—5 cross bars of black. Immatures: brown above; under parts white with longitudinal streaks of dark brown.
Range. Breeds from northern Alaska and northern Canada in the transcontinental forests south in the mountains to southern California, the highlands of México as far as Guerrero, Texas, Louisiana, and the Greater Antilles. (Forms breeding in mountains of Central and South America probably are conspecific.) Winters from southern British Columbia, western Montana, southern Nebraska, southern Minnesota, southern Michigan, southern Ontario, New York, and southern Maine south through México and Central America to Panamá.
Occurrence in Monument. Sparse winter resident. Recorded from: Smithwater Canyon, Nov. 19[1]; Quail Spring, Feb. 10; Stubby Spring, Feb. 10; Twentynine Palms, Feb. 16; Pinyon Wells, Oct. 9 *, 17; Cottonwood Spring, Apr. 30.

Single individuals, totaling eight, have been reported in the winter period from October 9 to April 30. All have been seen in the piñon belt or else about the larger trees at oases. Wintering flocks of sparrows should provide a good food source for this small hawk. On May 14 a hawk, probably of this species rather than the closely similar Cooper hawk, took three newly hatched Say phoebes from under the eaves of the Inn at Twentynine Palms. This date is not unreasonably late for a wintering or migrating Sharp-shinned Hawk. Cooper hawks should be on their breeding grounds at this time.

The specimens taken are adults and immatures of the race *Accipiter striatus velox.*

Cooper Hawk. *Accipiter cooperii*

Description. Coloration essentially as in sharp-shinned hawk (see above); size larger (275—445 grams) and tail somewhat more rounded.
Range. Breeds from southern British Columbia, central Alberta, southern Ontario and Quebec, and Nova Scotia south to northwestern Baja California, Sinaloa, Chihuahua, Nuevo León, the Gulf states, and central Florida. Winters from Oregon, Colorado, Nebraska, Iowa, southern Ontario, New York, and Maine south throughout the United States and México to Guatemala and Costa Rica.
Occurrence in Monument. Sparse winter visitant. Recorded from: Black Rock Spring, Sept. 2; Lower Covington Flat, Aug. 27 *, 31[1]; Quail Spring, Oct. 15; Queen Valley, Oct. 14; Fortynine Palms, Oct. 12; Twentynine Palms, Apr. 29, Sept. 21[1], Nov. 4; Pinyon Wells, Oct. 9, 10; Cottonwood Spring, Sept. 15 *, 17[1], Oct. 14.

Cooper hawks probably spend the entire winter in the Monument, although specific midwinter records do not happen to have been made. The dates extend from August 27 to April 29. It is even possible that they breed in the piñon belt, but we have no evidence of their doing so in the Monument area.

At Twentynine Palms on November 4, one swooped to the ground and caught a fox sparrow. On September 15 at Cottonwood Spring, an immature made an unsuccessful strike at the antelope squirrels that were feeding in the campground, and on October 14 one was plucking a small bird it had captured. At Quail Spring on October 15 one swooped at a Gambel quail but did not catch it, but on August 31 at Lower Covington Flat a Cooper hawk had succeeded in capturing one of these quail. Prey of the types indicated by these observations afford moist food, and in the winter season ample small bird populations are present for the predations of this species.

All occurrences have been in the piñon belt or about the larger trees or thickets at oases.

All birds taken have been first-year individuals.

Red-tailed Hawk. *Buteo jamaicensis*

Description. A large broad-winged, short-tailed, soaring hawk with wingspread about 5 feet. Upper surface brown; lower surface variable, black or dark brown to white, but usually with some band of dark streakings on belly; tail rufous in adults, gray, banded inconspicuously with dusky in immatures.

Range. Breeds from central Alaska and central Canada south through United States, México, the West Indies, and Central America to Panamá. Winters from southwestern British Columbia, southern Minnesota, southern Michigan, central New York, and southern Maine southward.

Occurrence in Monument. Resident in small numbers. Recorded from: Little Morongo Canyon, Apr. 3; 1 mi. S Warren's Well, Oct. 16; Black Rock Spring, Aug. 29, Oct. 16; Upper Covington Flat, Apr. 9; Lower Covington Flat, Apr. 8, 9, June 30[1], July 1[1], Aug. 24, 27; Quail Spring, Jan. 23, Oct. 15; Hidden Valley, Oct. 14, 15; Indian Cove, Oct. 16; Fortynine Palms, Oct. 12; Twentynine Palms, Jan. 12; 8.5 mi. S, 1 mi. W Twentynine Palms, Oct. 11 *; Live Oak Tank, Jan. 24, Apr. 6–8; Pinto Basin, Oct. 14; Cottonwood Spring, May 1.

Occasional red-tailed hawks are seen the year around, principally about the mountain faces and canyons and in the piñon and Joshua tree belts. On April 3 about cliffs in the head of a side canyon draining into Little Morongo Canyon, two birds were watched as they called frequently. Their behavior suggested there was a nest site in the cliffs. Similarly two birds seen along the canyon draining Lower Covington Flat acted as though stationed for nesting on April 8.

On April 9 in Upper Covington Flat a red-tail was flushed from a nest in a large Joshua tree standing in a grove of these trees. The nest was 25 feet up in the crown branches and was constructed of juniper sticks, old flowering stalks of yuccas, and strips of tree yucca fibers. Outside dimensions of the nest were about 30 by 18 inches deep. The cup was shallow, about 12 inches across, and had a few bits of down on the rim. The nest held one egg, the set as yet evidently incomplete. There were no food remains on or under the nest. The adult left silently and stayed away as we inspected the nest.

An adult female taken on October 11 on the road leading to Pinyon Wells was perched in a Joshua tree at about 7:30 A.M. It had eaten a long-nosed snake (*Rhinocheilus lecontei*), six inches of the tail of which was still intact in the stomach. Probably this snake had been taken early on this same morning. One mile south of Warren's Well on October 16 a red-tail was feeding

Fig. 21. Adult red-tailed hawk in flight.

on a road-killed jack rabbit. Rabbits, ground squirrels, pocket gophers, and larger species of snakes are the principal categories of food used by these hawks and of course they provide food with adequate water content.

The specimens taken belong to the western race of red-tailed hawk, *Buteo jamaicensis calurus.*

Swainson Hawk. *Buteo swainsoni*

Description. A fairly large soaring hawk, with wingspread of about 4 feet. Breast usually with dark band, unlike most red-tailed hawks; occasionally nearly entirely white or entirely black below; tail gray with narrow black bars; light under wing coverts contrast with darker primaries. In migration seasons usually seen in flocks.
Range. Breeds from Alaska, northwestern Mackenzie, Saskatchewan, Manitoba, Minnesota, and Illinois south to Baja California, Sonora, Durango, south-central Texas, and Missouri. Winters in central Argentina, mainly in provinces of Córdoba and Buenos Aires.
Occurrence in Monument. One report of spring migration: Live Oak Tank, Apr. 6–8.

This hawk is not known to nest in the deserts near Joshua Tree Monument, although a nest has been found in the northern Mohave Desert near Cima. Apparently it is only a migrant, therefore, in the Monument. The occurrence in April at Live Oak Tank involved a flock of 20 birds, a typical situation in the migration of this species. Such groups often are seen in spring in the Coachella Valley just south of the Monument, where in years of lush annual vegetation they may stop to feed on large insects, particularly the caterpillars of sphinx moths.

Golden Eagle. *Aquila chrysaëtos*

Description. A large, dark, soaring bird, with wingspread of about 7 feet. Plumage black; head fully feathered; wing tips not strongly upturned; white at base of primaries and at base of tail in immatures; hind neck golden brown in adults.
Range. Northern Eurasia south to Spain, mountains of northern Africa, central Asia, Korea, and Japan. In North America, breeds locally from northern Alaska and Mackenzie to Labrador and south to northern Baja California, Sinaloa, Durango, Guanajuato, and Nuevo León; in the east, south to Maine and Nova Scotia. In fall and winter wanders south to western and central Texas, Mississippi, and northern Florida.
Occurrence in Monument. Resident in small numbers. Recorded from: Morongo Pass, Mar. 28; Quail Spring, Sept. 8; Key's View, Mar. 30; Barker's Dam, Apr. 27; Pinyon Wells, Oct. 16; Pinto Basin, Apr. 11; Eagle Mountain, Oct. 20.

Golden eagles have not been seen at all periods of the year, but their occurrence during the spring when the species should be occupied with nesting indicates residence in the area. We have not located any nest sites. On many of the dates of observation two eagles have been seen soaring in the same vicinity, possible representing a pair. Thus, two flew past our camp at the north base of Eagle Mountain on October 20. Two were soaring on April 27 at Barker Dam. On September 8, three were soaring over a granite outcrop near the road leading to Quail Spring. Possibly such a group at this date involved an immature, but the birds were too far away to determine this with certainty. Rabbits are probably the principal type of food utilized by golden eagles in the area, although we have no specific records of their food habits in the Monument.

The golden eagles of North America belong to the race *Aquila chrysaëtos canadensis*.

Marsh Hawk. *Circus cyaneus*

Description. A slim, narrow-winged, long-tailed hawk or harrier with wingspread of about 4 feet. Conspicuous white patch at base of tail; body rich brown in females and subadult males; light gray in adult males; facial disc composed of specialized feathers. The wings seldom are held for long in a typical soaring position; they are either beating or are tilted upward.
Range. Northern Hemisphere, nesting from borders of tundra areas south to Mediterranean region and to southwestern United States, northern Baja California, Missouri, Ohio, and Virginia. In winter south to northern Africa, northern India, southern China, Colombia, and the West Indies.

Occurrence in Monument. Rare migrant. Recorded from: Lower Covington Flat, Nov. 6 [1]; Pinto Basin, Apr. 10.

A marsh hawk taken on Lower Covington Flat by a party from Long Beach State College was carrying a mountain quail. Another marsh hawk was seen beating along the base of the large sand dune in Pinto Basin in 1960 over an area which in that year was teeming with large sphinx moth caterpillars in lush spring vegetation. These hawks may be expected to traverse the Monument occasionally in migration, although they generally avoid deserts or pass through them quickly.

Osprey. *Pandion haliaetus*

Description. A large hawk, with angular and fairly narrow wings; wingspread of 5 to 6 feet. Under surface of body white; dorsal surface black; head white with black line running through and behind the eye and variable amount of black on crown; under surface of wings generally light, but with black patch at bend of wing.
Range. In Eastern Hemisphere from Lapland and Kamchatka south to Spain, northern Africa, southern Arabia, southern China, East Indian archipelago, Australia, and Tasmania. In Western Hemisphere breeds from northwestern Alaska, northwestern Mackenzie, southern Labrador, and Newfoundland south along the Pacific coast to Baja California and from Sonora to Guatemala, including the Tres Marías Islands; also south to New Mexico, the Gulf states, and Florida. Winters from central California, southern Arizona, and the Gulf states south to Perú and Brazil.
Occurrence in Monument. Recorded from: Key's View, Mar. 23, 30.

Ospreys, which are partly migratory, move in small numbers through San Gorgonio Pass and the Coachella Valley, where they have been seen on a number of occasions. It is not surprising, therefore, that they should occasionally appear in the adjoining uplands of the Monument, as has been noted near Key's View on the escarpment of the Coachella Valley.

This species is strongly tied to bodies of water, as it is almost exclusively a fish eater. If the birds forage at all while crossing the deserts, they might conceivably resort to snakes and large lizards for food.

The ospreys of North America belong to the race *Pandion haliaetus carolinensis*.

Prairie Falcon. *Falco mexicanus*

Description. A moderate-sized hawk, with pointed wings and rapid, powerful wing beat; wingspread about 3 feet and tail long. Upper surface grayish brown; under parts pale buff and white, variably streaked with dark brown; malar and subauricular stripe of brown.
Range. Breeds from southern British Columbia, southern Alberta, and southern Saskatchewan south to southern Baja California, southern Arizona, southern New Mexico, and northern Texas; east to North Dakota and Kansas. Winters from the northern part of breeding range south to Oaxaca and Hidalgo.
Occurrence in Monument. Sparse resident. Recorded from: Indian Cove, May 14; Split Rock Tank, Oct. 29–Nov. 2; White Tanks, Feb. 18.

Prairie falcons are scarce residents in the Monument, but there evidently are a few nest sites. Harwell reported visiting a nesting pair on April 5 and

Fig. 22. Sparrow hawks, female (left) and male, at entrance to a nesting cavity in a Joshua tree.

subsequently at an unspecified locality in the Monument. The pair was using an old raven nest, presumably on a cliff. At Indian Cove in May birds were apparently located for nesting somewhere in the cliffs of the nearby canyons.

Sparrow Hawk. *Falco sparverius*

Description. A small (85—140 gram) hawk, with long pointed wings and long tail. Male: dorsal surface and base of tail cinnamon; forehead, margins of crown, and most of upper wing surfaces blue-gray; primaries and bars on secondaries and lower back, black; tail black subterminally, with white tips; under parts pale cinnamon, spotted with black on belly and flanks; face marked with black malar bar and black postauricular bar. Female: upper surface generally dusky, barred on wings and back with cinnamon; tail cinnamon barred with black; under parts whitish, streaked with brown; facial markings as in male.
Range. Breeds from northern Alaska, Mackenzie, Alberta, northern Manitoba, southern Ontario, southern Quebec, and Nova Scotia south through United States and northern México, locally to Guatemala; also West Indies and much of South America, south as far as Tierra del Fuego. Winters from southern British Columbia, Utah, Colorado, Nebraska, Illinois, southern Ontario, and New Hampshire southward.
Occurrence in Monument. Fairly common resident. Recorded from: 1 mi. S Warren's Well, Oct. 16; Black Rock Spring, Sept. 3; Upper Covington Flat, June 11 [1], 12 [1]; Lower Covington Flat, Mar. 27 [1], Apr. 9, May 19, Aug. 26, 27, Dec. 5 [1]; Joshua Tree P.O., Aug. 22; Quail Spring, Jan. 20, 22, 23, 25, 27, Feb. 10, 11, May 19—21, July 1, Sept. 10; Stubby Spring, Apr. 10; Queen Valley, Apr. 10; Twentynine Palms, Apr. 4, Oct. 7, 29; Pinyon Wells, Oct. 8, 9; Pleasant Valley, Apr. 10; Live Oak Tank, Apr. 6—8, Dec. 4—6; Pinto Basin, Apr. 4.

Sparrow hawks may be seen in all sections of the Monument, but they especially favor the Joshua trees where the wide-open spaces provide ground surface they can easily scan and hunt and the tree tops afford lookout posts. Nesting probably takes place in cavities in the Joshua trees and piñons as well as in small cavities in cliffs. However, we have not actually examined any nest sites in the Monument, although dependent juveniles have been taken in June.

Food consists of large, ground-dwelling insects, lizards, and probably occasionally small rodents and small birds. These are seen at a distance from lookout perches or from the air as these small falcons hover in fixed location to watch for prey on the ground. In Pleasant Valley on April 10 two of these hawks were working on the abundant grasshoppers present there at that season. In winter at Quail Spring, one was seen frightening winter flocks of Oregon juncos. Food of the kinds indicated has adequate water content, and sparrow hawks have not been seen resorting to water at springs.

Sparrow hawks are usually seen singly, although occasionally pairs are noted. A pair was reported roosting in an old cabin near Quail Spring in winter. At this season there may be some influx of winter visitant individuals from the north. In the course of a day's observation, one seldom records more than one bird or a pair in any one locality, but in winter the days on which birds are seen are a little more numerous than in midsummer.

The sparrow hawks of the Monument belong to the race *Falco sparverius sparverius*.

QUAIL

Gambel Quail. *Lophortyx gambelii*

Description. A gray-backed quail with erect, black, recurved plumes on head. Male with chestnut crown, bordered by white anteriorly; forehead gray; throat black, bordered with white; breast plumbeous; belly whitish with central black area; flanks streaked with dark brown and white. Female without conspicuous patterning of the gray head; lacks black throat and belly patches of male; otherwise similar to male.
Range. Resident from southern Nevada, southwestern Utah, and western Colorado south to northeastern Baja California, central Sonora, northwestern Chihuahua, and western Texas.
Occurrence in Monument. Common resident. Recorded from: Black Rock Spring, July 6, Aug. 28; Lower Covington Flat, Apr. 7, 29 [1], 30 [1], May 15 [1], 17–21, June 11 [1], Aug. 22, 31; Quail Spring, Jan. 23 *, 26, 27, Feb. 11, 17, 18, Apr. 10, 22, 25, May 14, June 3, July 1, Sept. 9 *, 10 *, Oct. 15, Dec. 21 *; Key's Ranch, Mar.; Indian Cove, Apr. 7, 8 *, 27, May 13, 14, Oct. 12, 15; Twentynine Palms, Mar. 16, 21, May 15, June 2, July 13; Pinyon Wells, Oct. 9 *, 10, 16; Split Rock Tank, Oct. 29–31; White Tanks, Apr. 26; Eagle Mountain, May 16, Oct. 19; Cottonwood Spring, Feb. 7–9, Apr. 5, 26, 29, May 14, 18, June 22, July 11 [1], 12 [1], 14, Sept. 13, 14, 15 *, Oct. 13; Lost Palm Canyon, Oct. 23 *; Pinto Wash Well, Mar. 18 [1].

The Gambel or desert quail occurs chiefly in the vicinity of springs, for, although it is less rigidly tied to water supplies than the mountain quail, it cannot maintain its population indefinitely without water or succulent vegetation. When the latter fails to grow, as it does at times for long periods in this part of the desert, the residual populations of Gambel quail become dependent on the water sources. The dry seed food they then utilize demands water supplements. The daily water loss in quail, because of their size and the consequent favorable relation of surface to body mass, is not high (see Bartholomew and Cade, 1956; 1 per cent daily loss at 68° to 73° F. environmental temperature), so that daily use of water during moderate warmth would not be essential; but it would be required rather frequently if not daily in connection with use of seasonally dry food.

Another factor tending to concentrate quail populations is the growth of food plants suitable for them, chiefly in canyons and on alluvial fans near temporary or permanent water sources. Gullion (1956:40) has shown that areas with catclaw, desert willow, indigo bush, chollas, yuccas, and other normally associated plants of desert bajadas are those with suitable foods for this quail.

Probably when water is needed, a covey stays within a mile's radius of it, according to Gullion's estimates. We have always found the species in the Monument in the vicinity of springs or of water which, from other evidence, must have existed in some hidden location, even if we did not find it. The greatest distance from possible water at which we have recorded the species was one and a half miles, but this was in May, below Indian Cove, and succulent vegetation was then widespread.

Coveys are most conspicuous and large close to springs during dry periods.

Thus on September 15 at Cottonwood Spring, a group of about 30 was within 100 yards of water, and a similar-sized group was at Black Rock Spring on July 6; 50 were seen at Quail Spring on January 26, and on March 21, 1942, at least 100 were seen at the oasis at Twentynine Palms. Coveys otherwise, except obvious family groups, have been estimated by us as follows: 6, 6, 6, 8, 8, 12, 15, 15, 15, 20, 20, 25, 25, 25.

In winter at Quail Spring a Gambel quail was collected on January 26 in which the crop held seeds of catclaw (*Acacia greggii*) and desert willow (*Chilopsis linearis*) and green tips of leaves of filaree (*Erodium*). There also was a beetle with bronze-green elytra. A few days earlier, crops of two birds from this area contained seeds and green plant material but no insects.

In the spring seasons when there has been sufficient rainfall to start growth of succulent annuals, Gambel quail separate out in pairs and males without mates isolate themselves to give the single *cow* note or cock call. But on April 5, 1951, at Cottonwood Spring a covey of 20—probably the remnant of the covey of 30 of the preceding September—showed no signs of pairing, and no pairs were scattered in the area. At this time in this dry year there was scarcely any small annual vegetation; the country locally seemed to show the cummulative effects of several seasons of severe drought. In other years and other places cock-calling and pairing have been noted on April 7, 8, 10, 22, 25, and May 16, and young appear by mid-April. The population at Cottonwood Spring therefore may not have bred in 1951.

Young were noted on April 26, 1947, at Cottonwood Spring, when a pair had a brood of at least 5 chicks. On May 18, 1945, a covey of 12 half-grown young was seen here, and on July 1, 1950, at Quail Spring, 6 half-grown young were noted. We judge that eggs are laid, varying with the year, chiefly in the period from the first of April to the end of May.

Gambel quail range in the Monument from the lowest levels, as at Twentynine Palms and Pinto Wash Well, up to the crests of the Little San Bernardino Mountains at 4500 to 5000 feet. Over much of this range, except at the lowest points, they occur within the area occupied by the mountain quail. The only upper-level spring where we have not recorded Gambel quail is Stubby Spring, and this may reflect a fortuitous gap in our information. In these extensive areas of common range the two species of quail may form mixed coveys in the nonbreeding season, as was specifically noted on October 16 at Pinyon Wells, on October 19 on Eagle Mountain, and on October 15 at Quail Spring. In the first two instances one and two individuals were running and flushing with a larger group of mountain quail. In the last case groups of 15 of each species were distinct, but the Gambel quail were following at the rear of the mountain quail in moving to water. Gambel quail are much more likely to form compact coveys than are mountain quail, and in flight they are less inclined to scatter widely.

On October 15 at Quail Spring a Cooper hawk swooped unsuccessfully at

Fig. 23. A family of Gambel quail in a desert wash in August; male, female, and three well-grown young.

a flock of Gambel quail at 7:10 A.M., and on August 31 at Lower Covington Flat a Cooper hawk had captured a Gambel quail.

Adult quail taken on July 11 and 12 were in molt, and those of September 9 and 10 and October 9 were in late stages of the annual molt in which the outer two or three primary feathers were not yet fully grown.

A hybrid between *Lophortyx gambelii* and *Lophortyx californicus* was taken on November 20, 1903, in Morongo Pass by H. E. Wilder and is now in the collection of the Museum of Vertebrate Zoology. It was first recorded by Hachisuka (1928), who we may suppose saw it in Wilder's collection (then at Fortuna, Calif.) or had some rather specific report on it, for he gave measurements of the specimen and stated that a female "taken from the same covey is also a hybrid." Allan Brooks prepared a painting (Hachisuka, *op. cit.*) of the male, but whether from direct comparison or not we do not know. We have not been able to verify the female, but there is no doubt of the hybrid background of the male. We now find that this male has wing feathers that are clipped and infer that it must have been a captive at one time. Wilder was known to be careful in keeping records, and we believe that the bird must have come from the place indicated. Morongo Valley was little settled in 1903, and yet the bird may have been caught and held captive locally. It is less likely that it was pen-reared there or released from a crossing in a game farm elsewhere. If Hachisuka obtained his information from Wilder correctly and expressed it accurately, especially the notation "taken from the same covey," the inference is that we are dealing with a wild-produced hybrid, but there will always remain some doubt that, contrary to an earlier supposition (see A. H. Miller, 1955), it is that. The area in Morongo Pass is of course one where, at the east base of the San Bernardino Mountains, the ranges of the California quail and the Gambel quail adjoin, and other hybrids of these species are known from nearby San Gorgonio Pass that were clearly of wild origin.

The Gambel quail of the Monument are of the race *Lophortyx gambelii gambelii.*

Mountain Quail. *Oreortyx pictus*

Description. A large quail with long, straight black topknot. Head, neck, and breast blue-gray, the throat chestnut bordered by a white line around base of bill and from eye along sides of neck; back, wings, and tail olive; abdomen and sides chestnut, with wide white and black bars on flank; under tail coverts black. Male and female alike.

Range. Resident from southern Washington and southwestern Idaho south through western Nevada and California to northern Baja California.

Occurrence in Monument. Common resident. Recorded from: Little Morongo Canyon, Apr. 2, 3, 4; Black Rock Spring, Apr. 3, 4, Aug. 28, 29, 30, 31 *, Sept. 1, 2, 4, Dec. 20; Lower Covington Flat, Apr. 8 *, Aug. 23, 24 *, 25 *, 26, 27, 28 *; Smithwater Canyon, Apr. 10 *, May 21, June 28 [1]; Quail Spring, Jan. 22, Feb. 11, 21 *, 28 *, Apr. 23, May 19–21, July 1, 3, 4, Sept. 9, Oct. 15, Dec. 20 *; Stubby Spring, Sept. 5, 7, Nov. 1–2, Dec. 4–6; Key's View, July 15; Indian Cove, Apr. 8, May 13 *, 14; Fortynine Palms, Apr. 6, 7 *; Pinyon Wells, Apr. 10, Oct. 8, 9, 12 *, 13, 14 *, 15, 16 *, 17 *; Live Oak Tank, Apr. 6–9, May 9 *; Eagle Mountain, May 16 *, 20 *, Oct. 19 *.

Mountain quail have proved to be surprisingly common in the mountains of the Monument at the upper levels, chiefly in the scrub oaks and chaparral. They occur even on Eagle Mountain across the gap of some twenty miles of unsuitable habitat between this mountain and the east end of the Little San Bernardino Mountains. Although able to flourish in these desert uplands, they seem to be strictly dependent on sources of drinking water. We have never seen them in places where they could not find such water within about a mile, although in two instances we ourselves have not been able to locate the actual water source. Apparently they are adept at using water that does not support an oasis but which may be available in catchment tanks or crevices in the rocky canyons. For example, on the crest of Eagle Mountain on October 19 mountain quail were collected at the head of a rocky defile. They had drunk so recently that the water ran in a stream from their mouths when we picked them up, and yet a half hour's search of this rocky canyon system failed to reveal the water that must have existed, water that we were eager to find ourselves after scrambling up the dry mountain face. At Key's View this species was heard in the distance. We know of no water at this point, but there is a spring no more than two miles away, and the birds, which were in that direction, may have been no more than one mile from this water; indeed, there may have been nearer sources in the rocky escarpments. The most flourishing coveys are about the larger, more obvious water sources where water is available throughout the year and where it is apparently used by them daily. Such is true of Little Morongo Canyon, Quail Spring, Smithwater Canyon, Indian Cove, and Stubby Spring. Although it is possible that at times, especially in winter, mountain quail forego daily water, or can survive well without it, we have seen no sign of nesting success except close by springs.

Mountain quail come to water on foot, usually single file in groups of 6 to 20, except when they are separated in pairs in the spring. They approach water cautiously, one individual taking the lead and the initiative. Typical of such activity in the afternoon watering period are the following events, watched from concealment at a water hole in the canyon below Lower Covington Flat on August 24. At 4:15, 7 mountain quail flew across canyon from a steep slope and began giving the assembly call, *cle cle cle cle cle cle*. They remained in view a long time and then started walking down toward the water, single file, with extreme caution. The assembly notes were given repeatedly, along with various softer notes, these apparently most often by the lead bird. The lead seemed to be maintained by one bird. At the edge of the large rock above the water hole, the first bird stopped, peered over the edge, cocking its head from side to side and giving the soft notes. When it turned around to look at the other quail, it called softly and they responded with the assembly call. Evidently the quail sensed some danger because the group started back up the hill calling. At 4:35, 7 quail again moved down to within 30 feet of the water. A Merriam chipmunk moved through the group to the rock above the water, and in a few minutes the quail arrived there too, calling continuously and scanning the area nervously. Three more quail came in and mixed with the original 7. Others arrived, until the number was 13. At 4:50 one quail slid and slipped down one end of the large rock and started to drink. The others waited a few minutes, peering down, and then one or two at a time scrambled down to drink. Only one or two flew from the rock to the ground 9 feet below. At 4:53, 8 more quail flew into the area; 19 were in view at once. At 5:00, 4 more quail arrived, but some were now leaving. The observer then moved to leave, and only then was the alarm *scree* and sputter sounded. At once all quail began running rapidly up the canyon slopes, giving assembly calls. This section of the canyon seemed deserted by mountain quail in April of 1960 when the springs had dried up.

Covey sizes estimated in late summer through winter are as follows: 3, 3, 5, 5, 5, 6, 6, 6, 7, 7, 8, 8, 10, 10, 10, 12, 15, 15, 15, 15, 20. Thus the average is low, at 9.1. We have excluded higher figures that quite clearly represent composites of several coveys, as is true of the three groups described above. Other estimates of 30, 100, and 150 in a given area, as at Quail Spring, represent crude population estimates for the population centering about the springs and not coveys as such. We have never seen compact, unified coveys of 25 and 30 as noted in Gambel quail, and in general even the small coveys of mountain quail, except when they are moving to water, are looser than those of *Lophortyx,* each bird often being 10 feet from its nearest neighbor. Mountain quail when alarmed usually run, although in the brush of the Monument, which is more open than in most parts of the species' range, flight is not rare. A covey, when it does take wing, tends to scatter in many directions in an open, spread pattern. Use of flight when not pressed is very limited and is chiefly seen along steep canyon walls, when birds will cross the canyon by this means. Flight is, however, strong. Once one flushed when surprised 40 yards away and flew 150 yards upslope at a 25° angle.

The food of mountain quail in the autumn consists in major degree of acorns of the scrub oak. At Pinyon Wells in October birds we collected had many shelled acorns in the crop and one manzanita seed. At this season the acorns were soft and green at the base, and the bases could be chipped with one's fingernail. Under oaks where the quail had been feeding, acorns were found with cuts in the green base which fitted the cutting edge of the tip and sides of the upper and lower mandibles of the quail (see fig. 25). Evidently they cut into the shell at the base, open a hole, and split or cut away the remainder of the covering. Even later in the year acorns may still be gleaned in the leaf duff. The quail have been seen above ground in the oak scrub, and it is possible that they pull off green acorns before they are fully ripe.

Fig. 24. A family of mountain quail among juniper trees.

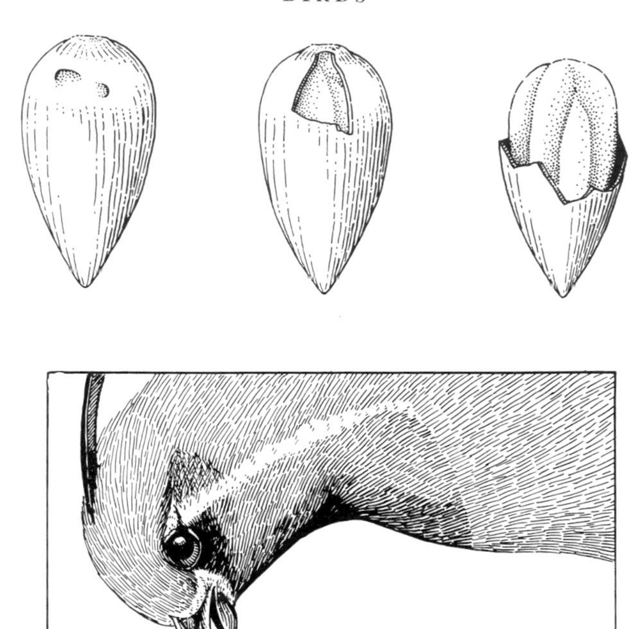

Fig. 25. Diagram of green scrub oak acorns showing method by which they are cut open at the base and shelled by the mountain quail.

Foraging is, however, not confined to the oak thickets. On April 8 a pair was watched feeding over what seemed to be bare gravel in a wash between groups of desert willows. This was at Indian Cove, at 3200 feet, the lowest point in the Monument that we have found this quail, except for Little Morongo Canyon, at 2500 feet, where the vegetative cover is higher zonally. A laying female taken on April 8 in the early morning had a well-filled crop, $\frac{4}{5}$ of which consisted of the fruiting capsules of a species of *Gilia*. The remaining $\frac{1}{5}$ were bone fragments, 14 in all, ranging up to 1 cm. in length. They derived chiefly or entirely from a jack rabbit, the quail obviously gleaning them from a weathered or scattered carcass. These may have been picked up for grit, but the selection of them in some number, coincident with laying, makes it seem possible they were taken for their calcium content. The mountain quail also takes some tuberous food dug from beneath the surface or uncovered in leaf litter. It also eats some insects. The diet is therefore not all highly desiccated.

The way in which the life of mountain quail centers about watering places is reflected further by the finding of their dusting spots near oases. Twice we have specifically noted dusting holes in the vicinity of water. Near the water below Lower Covington Flat many dusting places with quail feathers scattered in them were found on August 26. These round depressions were in damp gravel and sand, although at midday they might have dried out. Again at Stubby Spring on September 6 many dusting hollows were found in damp, soft earth by the water and also in dry places among the bushes nearby. Is damp sand and the coolness it affords actually sought for "dust-bathing" by this species? It is a possibility, yet the combination of conditions may only be fortuitous as a result of the birds' frequent trips to water.

Predation on this species of quail, like much else in their lives, probably centers at the springs and is reflected in the caution we have noted in the birds as they go to water. Near Black Rock Spring on September 4 we found feathers of a mountain quail that had been scattered there since the rain of two days earlier. Among the feathers on the soft ground was the very clear footprint of a coyote. All fleshy parts of the bird were taken. At this same spring on August 31, quail were flushed into an oak thicket 25 feet away. Almost immediately a rattlesnake rattled in alarm in the thicket. At once the quail gave their alarm notes, and a cottontail rabbit ran out of the thicket. The snake, although itself alarmed, was clearly a threat to the quail. On October 15 at Quail Spring a Cooper hawk stooped on a mountain quail in a dive estimated to be at a 45° angle. It seized the quail with both feet and arose abruptly with no evident pause, carrying off its prey.

The most spectacular incident of danger at watering places was observed at Stubby Spring in the evening of September 5 by Russell. Ten mountain quail walked single file to water and remained 5 minutes, drinking. They then retraced their route, again single file. A bobcat had evidently watched the quail also, and it ran down the slope, crossed the trail and hid on its lower side. The birds probably had a glimpse of the cat crossing the trail ahead of them, some 25 yards from the spring, for they stopped, milled around a little, but then seemed to forget the incident and proceeded upslope parallel to the trail and yet angling to get onto it again. The cat now scuttled forward a few feet and hid behind a rock three feet high. One of the quail hopped up on the rock and walked to the top. The cat and the quail sprang into the air, but the bird was free and escaped. Quickly the cat rushed for some of the other birds some six feet away. But again it missed, as the birds flew. The cat then walked slowly along the trail to the spring, the whole episode seemingly forgotten.

The beautiful loud, clear whistle of the mountain quail, chiefly if not entirely a call of the male, is usually heard in the spring in March and April. It is not exclusively the call of an unmated male, as is true of the *cow* call of quail of the genera *Lophortyx* and *Callipepla,* for we have heard it twice in October from birds in a flock, although it was then not full power. Moreover on April 3 in Little Morongo Canyon an isolated pair was watched running together between open oak clumps. One member of the pair was frightened by us and flew off, and the remaining bird, which had been giving the single whistle earlier, now began calling loudly and more regularly. Also on April 8 a male of a pair, the female of which was laying, was giving the loud, clear call. It ran toward us, obviously belligerent, as we imitated the assembly *cle, cle, cle, cle* whistle.

In April and May essentially all birds seen were in pairs. Evidence of several kinds indicates laying in late April. Thus a male on April 7 had a 10 mm. testis and a female on April 10 had a 3 mm. ovum and no brood patch, which would suggest that laying would have started soon, perhaps in about a week. Another female on April 8 had an egg in the oviduct with the shell formed, the second egg of a clutch as shown by examination of the ovary; there was no brood patch. Recently hatched flightless young (7.3 and 8.4 grams) were found on May 13, and only slightly larger young (10.9 grams) on May 20. Another bird on May 16 was flushed from some small young which were heard but not seen. And a brood just hatched was seen on May 21. But another brood seen on the 20th was estimated at one-third grown. This evidence would indicate in general that eggs are laid the last third of April.

By the time young are out, males in attendance on them may show partly reduced testes, as was true on May 20 when the male with the brood of one-third grown young was taken. This male also showed a bare brood patch of the kind typical of quail and which occurs regularly in males of this species as well as in females. On May 16 the female that had chicks had a similar brood patch.

The aforementioned young have all been seen within a half-mile of water and usually much closer. On May 13 the female taken at Indian Cove had several young, but only two of them could be found, as they hid in the dense tangle of low bushes and sedges along the trickle of water. Not far from here, lower in the canyon, two dead young were found, one newly hatched and another about a week old. There was no evidence of cause of death.

The mountain quail taken as fully grown birds from late August on, and chiefly in October, show a surprisingly low proportion of birds of the year judged by the standard marker feathers of this age group (see Leopold, 1939). Out of 26 birds, only 4 are immatures. Or, restricting consideration only to the October population of 1945, 4 out of 15 were immature. This reflects low productivity for quail in this year particularly concerned, but lack of immatures among the other 11 specimens suggests no different situation in subsequent years.

The weights of quail, exclusive of those preparing to lay or those heavily involved with molt, are as follows: 10 ♂♂, 233.6 (210–262); 13 ♀♀, 217.5 (189–240). A laying female weighed 284 grams on April 8. The four immatures referred to above are included, as they show no tendency to depart from the averages for the combined age groups.

The mountain quail taken in the last week of August all show molt. The body plumage consists of two-thirds to nine-tenths new feathers, the heads and necks showing more residual molt than other parts. All show molt of the outer primaries, the least advanced being an adult bird on August 23 in which the outer three primaries are still old. In the period from October 12 to 19 three adults show remnants of sheaths at the bases of the outermost primaries, but otherwise the molt seems to be completed by this period.

The mountain quail of the Joshua Tree Monument were found to be differentiated as a distinctive, pale-backed race, *Oreortyx pictus russelli* (A. H. Miller, 1946:75–76). The race shows the extreme in reduction of brown and olive colors dorsally for the species. This coloration is possibly the result of selection for concealing dull or subdued dorsal aspect in the comparatively open habitat of this desert mountain system. Possibly also it is a feature of a once more widespread desert mountain race, for the species occurred more widely in pre-Columbian times to the eastward in cave deposits in New Mexico (Howard and Miller, 1933), where it is now absent. The race *russelli* ranges west to Morongo Pass, beyond which in the San Bernardino Mountains the more richly colored *O. p. eremophilus* occurs.

RAILS AND SHOREBIRDS

Unlike migrant land birds, shorebirds that pass across the desert normally traverse the Monument without stopping. Few of the places that offer even satisfactory resting stations are large enough or stable enough as marshes to attract shorebirds regularly and in numbers. Killdeers and the common snipe are more likely to find significant foraging grounds in transit than are the other species, and consequently they may repeat their visits to oases once they have found them. Even such favorable areas as Barker Dam, the Twentynine Palms oasis, and Little Morongo Canyon with permanent water probably are not essential stops for foraging en route by killdeers and snipe but constitute merely attractive rest stops where water for drinking may be obtained. A flying shorebird in need of rest is probably barred instinctively from settling on dry hot desert sand or gravel, whether or not it actually requires moist surfaces for the comfort of its feet and for the proper conditioning of the thin bill and mouth membranes.

Shorebirds, as strong fliers, normally are on the wing for many hours at a time and do not require closely spaced settling areas along the desert crossing. Occasional exhausted individuals may find critical aid at the oases. But the visitants we have seen showed no sign of being in critical condition.

Either by day or night open water surfaces can be seen easily from several hundred feet in the air, and shorebirds doubtless are adept at recognizing them and the surrounding configuration of plant growth. The wet areas in the Monument may be located by shorebirds passing between the Salton Sea or the Colorado River Valley and the Great Valley of California or the Owens Valley or other waterfowl areas of the Great Basin. Possibly the main flight lines from the Salton Sea follow the Coachella Valley trough rather than cross the Monument, so that the suitable spots for resting would have a chance of attracting only strays or the lesser and more scattered flights to and from the Great Basin.

Records of the spring migration of shorebirds are somewhat more numerous than those for the fall. This is perhaps due in part to more regular observations at Twentynine Palms in spring than in fall, but it also may reflect the better condition of the stopping places at that season. Then again shorebird migration in general seems to be greater through the interior in spring than in the fall, when it tends to concentrate on the coast.

Beside the points of permanent water already cited, temporary bodies of water like Ivanpah Tank may draw migrants after rains. At this tank in 1950 shorebirds appeared within a few days after it was filled in September, although it had been dry for many months preceding. Such prompt usages of temporary water clearly indicate the way that migrants on the wing detect and seize an opportunity for stopping at wet places.

Virginia Rail. *Rallus limicola*

Description. A moderately small (usually 60–90 gram) rail with long (3 cm.), slender, red-brown bill. Back black with cinnamon feather borders; throat, breast, belly, and wing coverts cinnamon; chin white; sides of head gray; flanks and under tail coverts black, barred with white; wings and tail black, the latter barely exceeding the coverts in length.

Range. Breeds from southern parts of southern Canadian provinces south to northern Baja California, northern New Mexico, Missouri, and North Carolina; also Distrito Federal, México, and in South America in mountains of Ecuador and from central Chile and central Argentina to Straits of Magellan. North American race winters along Pacific coast and from Gulf states and southern Atlantic coast south through México to Guatemala.

Occurrence in Monument. Recorded once: Lost Palm Canyon, Sept. 14 *.

The rails in migration probably fly chiefly at night. The only member of this group we detected in the Monument was a Virginia rail that was flushed from a dry gravelly wash in Lost Palm Canyon 50 yards from a willow thicket and a tiny water hole hidden in the plant cover. Possibly the bird had run from this cover unnoticed before it flew. In any event the dense vegetation had probably attracted it in the course of its fall migration, and it was spending at least the midday period in the vicinity of this cover.

The bird was not fat and weighed only 56 grams, although its musculature was in good condition. The stomach was essentially empty, and probably this resting point offered little opportunity for foraging by a marsh bird of this type. The ovary showed remains of follicles that had been active in previous months and indicated that the bird was an adult.

The Virginia rails of North America are of the race *Rallus limicola limicola.*

KEY TO THE SHOREBIRDS OF THE MONUMENT

A. White of throat and breast marked by two black bands.................... *Killdeer*
A′. Two black bands on throat and breast absent.
  B. Rump at least partly white.
    C. Rump patch white; size large (140–180 grams).......... *Greater yellowlegs*
    C′. Rump black medially and white laterally; size small (18–20 grams).................................... *Least sandpiper*
  B′. Rump dark, concolor with back.
    C. Bill long (6–7 cm.); back with two longitudinal buff stripes.. *Common snipe*
    C′. Bill short (2–3 cm.); back gray without longitudinal buff stripes
      D. Wings with white patches showing when spread ...... *Spotted sandpiper*
      D′. Wings without white.............................. *Solitary sandpiper*

Killdeer. *Charadrius vociferus*

Description. A middle-sized (80–90 gram) shorebird of the plover group with short bill (less than 2 cm.) and moderately short legs (tarsus 3.5 cm.), marked by two dark rings across white of throat and breast and by rust red rump and base of tail. Forehead and ocular area white bordered below and above by black; white ring on back of neck; back dark brown; wings with white at base of inner primaries and secondaries; tail feathers with white terminal spots.

Range. Breeds from northwestern British Columbia, southern Mackenzie, northern Ontario, and southern Quebec south to Sinaloa, Guanajuato, Tamaulipas, and the West Indies; also Perú. Winters from southern British Columbia, Colorado, the Ohio Valley, and southern New York south through United States, México, and Central America to Venezuela, Colombia, and Chile.

Occurrence in Monument. Fairly frequent visitant, chiefly in periods of spring and fall migration. Recorded from: Little Morongo Canyon, Mar. 30–Apr. 3; Quail Spring, July 4; Barker's Dam, May 18; Ivanpah Tank, Sept. 12 *; Twentynine Palms, "March and April," Oct. 17; White Tanks, Sept. 12.

Killdeers are more likely to be seen in the Monument than are any other species of shorebird. Those noted have for the most part been clearly migrants or nonbreeding vagrants, but Miss Carter reported them in courtship behavior at the oasis at Twentynine Palms in March and April. Yet it is by no means certain that they nested. A summer vagrant noted in the Quail Spring area indicates the possibility of nonbreeding individuals occurring at times other than migrations. The sedge meadow in Little Morongo Canyon offers conditions that would seem to be favorable for nesting, but the birds seen there in late March and early April were not behaving as though nesting; indeed, six were seen in a group on March 31.

Miss Carter indicates that killdeers were common at Twentynine Palms, but in our experience one or two seen at a single oasis is the more usual situation. The birds are normally found only near water when settled, although this shorebird is less rigorously confined to moist ground for foraging than are other shorebirds. In the desert it seeks the few areas where surface moisture can be resorted to occasionally if not continually. At Barker's Dam and at Ivanpah and White tanks the birds were flushed from mud or damp gravel surfaces.

Killdeers may migrate at night, as they have been detected then on the wing by Miss Carter, and in March she saw birds circling over the open desert by day, occasionally alighting away from water.

The single specimen taken, on September 12, was a young of the year, undergoing postjuvenal body molt. Ivanpah Tank, where it was found, had been completely dry in the preceding summer until at least September 2, so the bird almost certainly arrived in that area subsequent to that date.

Killdeers of North America belong to the race *Charadrius vociferus vociferus.*

Common Snipe. *Capella gallinago*

Description. A middle-sized (90–110 gram) shorebird with long (6–7 cm.) bill and short legs (tarsus 3 cm.). Dorsal surface including head black, streaked and mottled with brown and light buff, the latter forming two longitudinal stripes on back and one medial stripe on head; no conspicuous white marks in wings or tail; under parts mottled brown and buff on throat and breast; sides barred with black; chin and belly white.

Range. Breeds from western and northern Alaska across Canada to central Labrador and south to southern California (mountains), northern Colorado, central Iowa, and northwestern Pennsylvania. In Eurasia from British Isles to Siberia and south to southern Europe and the Himalayas. Winters in the Americas from southern British Columbia, northern Colorado, Illinois, and North Carolina south through Central America and the West Indies to southern Brazil.

Occurrence in Monument. Occasional migrant, chiefly in March and April; one fall occurrence. Recorded from: Little Morongo Canyon, Mar. 31, Apr. 2; Twentynine Palms, Mar. 18, 21, 23, 25, 27, Apr. 2, 4, 7, 12, Oct. 30.

Snipe require not only damp ground but sedge or other low bordering vegetation for cover, even when stopping in migration. Few of the oases in the Monument provide this. In the sedge meadows in Little Morongo Canyon several acres of suitable marsh cover are present, and a single individual was seen there repeatedly. Miss Carter saw single snipe probing among weeds at the edge of the "swamp" at Twentynine Palms. Two were seen there on March 27.

Spotted Sandpiper. *Actitis macularia*

Description. A small (35–42 gram) sandpiper with fairly short (2 cm.) bill and legs (tarsus 2.5 cm). Dorsal surface olive gray, obscurely or moderately mottled with dusky; rump dark; tail with white bars laterally and terminally; wings black with white at bases of secondaries and primaries and on tips of secondaries; under parts white, either spotted prominently with black (breeding plumage) or unspotted but with gray wash on sides of neck and breast (winter plumage and immatures).

Range. Breeds from northwestern Alaska east to northern Quebec and Labrador and south in the mountains to southern California and southern Nevada; also to central Texas and northern Alabama. Winters from southern British Columbia, southern Arizona, and the Gulf coast south through Central America to Bolivia and central Argentina.

Occurrence in Monument. Occasional spring migrant. Recorded from: Twentynine Palms, Apr. 27; Barker's Dam, May 18.

Single individuals have been detected at the edges of open ponds as specified. The species should occur in the fall migration also. Normally this shorebird migrates solitarily rather than in flocks as do least sandpipers.

Solitary Sandpiper. *Tringa solitaria*

Description. A fairly small (45—55 gram), slender shorebird; bill straight, about 3 cm. long. Dorsal surface dark gray, but faintly mottled; sides of neck and anterior flanks streaked gray; breast sparsely streaked medially; under parts white otherwise; rump dark, concolor with back; wings black; tail barred black and white.
Range. Breeds from northern Alaska across Canada in the forest belt to Churchill, Manitoba, and to central Labrador south to southern Alberta and east to central Quebec, possibly into northern United States. Winters from northern Baja California and Gulf coast of United States south through México, the West Indies, Central America, and South America to central Argentina.
Occurrence in Monument. Occasional spring and fall migrant. Recorded from: Barker's Dam, Apr. 28; Ivanpah Tank, Sept. 12 *.

This migrant, as its name suggests, usually appears solitarily, although Williams (1938:258) saw two together at Barker's Dam. The single bird taken at Ivanpah Tank had found this temporary water area which had been available only since the rain of the week preceding. It was frequenting the muddy borders of the pond. The bird was a young of the year of the northwestern race *Tringa solitaria cinnamomea*.

Greater Yellowlegs. *Totanus melanoleucus*

Description. A fairly large (140—180 gram) shorebird with long (6 cm. tarsus) yellow legs, and slender, moderately long (5 cm.), straight bill. Back mottled gray and white or black and white; prominent white rump patch; under parts white, slightly streaked, or extensively streaked and barred, with dark gray according to plumage and season; wings black; tail barred with white and gray.
Range. Breeds from central southern Alaska east to Labrador and south to southern British Columbia and to Anticosti Island and Newfoundland. Winters from Oregon, southwestern Arizona, central Texas, and the Gulf coast south through México, Central America, the West Indies, and South America to Tierra del Fuego.
Occurrence in Monument. Two records of spring migrants: Twentynine Palms, Mar. 18, Apr. 10.

This species is reported by Miss Carter, who saw single individuals on two occasions at the "swamp" at Twentynine Palms which existed there in the 1930's. It is to be expected in the fall also and at other oases or tanks. Apparently very few interrupt their through migratory flight and find the suitable resting spots in the Monument.

Least Sandpiper. *Erolia minutilla*

Description. An extremely small (18—20 gram) sandpiper with short (18 mm.) bill and short, greenish legs (tarsus 18 mm.). Back streaked and mottled black and gray or (nuptial plumage) black and rust brown; under parts white with neck and breast mottled with dusky (darker in nuptial plumage); rump black, bordered with white; wings black, narrowly edged on posterior border with white.

Range. Breeds from northern Alaska east along Arctic coasts to central Keewatin, northeastern Manitoba, and central Labrador and south to southeastern Alaska, eastern Quebec, and Nova Scotia. Winters from coastal Oregon, southern Arizona, central Texas, the Gulf coast, and North Carolina south through México, Central America, and the West Indies to the Galápagos Islands, northern Perú, and central Brazil.
Occurrence in Monument. Recorded only in spring migration: Twentynine Palms, Mar. 7, Apr. 12.

Miss Carter reported these sandpipers in the "swamp" at Twentynine Palms in spring. There was a group of three on the first date indicated. It is characteristic of this species to appear in small flocks in migration, although single occurrences are not unusual. This species should occur in the fall also.

DOVES, ROADRUNNERS, AND ALLIES

White-winged Dove. *Zenaida asiatica*

Description. A dove larger and stockier than the mourning dove, with conspicuous white patches on the outer wing coverts and nearly square-ended tail banded terminally with white except on central pair of feathers. Back brownish; head, flanks, and belly plumbeous; throat and breast buff. Face blue, iris orange, and feet red.
Range. Breeds from southeastern California, southern Arizona, southern New Mexico, and the lower Rio Grande Valley in Texas south through México and Central America at least to Nicaragua; also in southern Bahama Islands and Greater Antilles, and southwestern Ecuador to northern Chile. Winters generally over the breeding range but chiefly south of California and Arizona.
Occurrence in Monument. Occasional vagrant and locally summer resident in small numbers. Recorded from: Quail Spring, May 21, July 2; Twentynine Palms, May 10, 12–17, 27; Cottonwood Spring, May 9 *, 13, 16, 17 *, June 22, July 11 [1].

White-winged doves that appear in the Monument are at the northwestern frontier of the species' range, and their occurrence must be regarded as that of border pioneers or vagrants. Some birds, perhaps not every year, do stay to breed, although no nests have been found. There is some evidence to suggest that the species is extending its range and numbers slightly at its northwest periphery in California, and yet there is a rather probable, although secondhand, record of an occurrence in the 1890's of this species at Twentynine Palms.

Miss Carter found white-winged doves arriving at Twentynine Palms on May 12 and 10 in successive years, and she estimated that six were present, calling in their characteristic fashion. At Cottonwood Spring they have appeared as early as May 9, and the song has often been heard there in May. Persons resident there have spoken of their presence through the summer in 1947 and 1950. A male taken at this spring on May 17, when at least three birds were seen, had a fully active testis, 14 mm. long, and it was calling in full voice from a rock spire above the spring. The specimen of July 11 also had enlarged testes. These circumstances strongly indicate breeding in this area, and breeding may also have occurred at Twentynine Palms. On the

Fig. 26. White-winged doves in the ocotillos.

other hand, birds seen at the more westerly station of Quail Spring, one on May 21 and two on July 2, were probably vagrants.

These doves, like mourning doves, seem to require water in this type of desert country, and they may be expected, if at all, in the vicinity of the better springs.

White-winged doves of the Monument belong to the race *Zenaida asiatica mearnsi.*

Mourning Dove. *Zenaidura macroura*

Description. A dove about 12 inches long, with pointed tail, the feathers white-tipped except the central pair. Body brown above, buff below, with vinaceous wash; neck and flanks blue-gray; secondaries and tertials spotted with black, and posterior auricular area with a black spot; sides of neck with iridescent purple. Immatures lack vinaceous and iridescent colors.

Range. Breeds from southeastern Alaska and southern British Columbia east to southern Quebec and New Brunswick and south to southern Baja California, the Mexican plateau, and the Greater Antilles; also Panamá. Winters over most of breeding range but largely migratory in northern part.

Occurrence in Monument. Locally common summer resident; a few remain in winter. Recorded from: Black Rock Spring, July 6, Aug. 29, 30, Sept. 2, 3; Lower Covington Flat, Apr. 9, 29[1], Aug. 22, 23, 24; Quail Spring, Apr. 22, 25, May 19—21, July 1, 3, 4, 6, Sept. 8, Oct. 14, 15; Stubby Spring, Sept. 5; Indian Cove, Apr. 7, 26, May 8, 13, 14, Sept. 5; Forty-nine Palms, Oct. 12; Twentynine Palms, Apr. 9, 11, May 1, 5, 6, 9, 17, 26[1], 31, July 12, 13; Pinyon Wells, Apr. 10; White Tanks, Apr. 26; Pinto Basin, Apr. 11; Cottonwood Spring, Feb. 7—9, Apr. 11, 25, 26, 29, May 12, 13, 14—18, July 7, 8, 14, Sept. 13, 14, 15; Lost Palm Canyon, Sept. 14.

Mourning doves occur in the Monument only in the general vicinity of springs and pools, as they require water daily in conjunction with their diet of small dry seeds, which they glean from the ground. Flights to water may be of several miles, probably two to six. This conspicuous species has been notably absent from large areas of otherwise suitable terrain where accessible surface water is lacking. Such for example was the situation at Pinyon Wells in October, 1945, when the only water present was deep in a mine shaft. But in the spring of 1960 doves were present and evidently breeding at Pinyon Wells, where they were coming to surface water then available in a trough.

The mourning doves become conspicuous in April with the arrival of many migrants, and they may be seen regularly through September and October. The one specific winter record we have is at Cottonwood Spring, where in the period from February 7 to 9 a bird was heard giving the cooing sequence or song at sunrise, with the temperature at 28° F. About springs, groups coming to water typically number 6 to 12 individuals; seldom are solitary birds seen at the water.

Cooing is commonly heard at oases in April and May when nesting is in progress, but usually the doves are silent from July through October. On April 9 near a water hole, a dove attempted to copulate with two other individuals in turn, each of which was unreceptive and flew off. We have not observed actual nests.

Fig. 27. Mourning doves coming in to an oasis.

An independent juvenile came to our camp above Black Rock Spring on September 2 and alighted on the black top of the field truck. It then walked out onto the horizontal surface of glass of the open tailgate and pecked twice at it. We presumed it reacted to this clear shining surface as though it was water.

Bartholomew and Dawson (1954) have studied especially the water requirements and ability to withstand heat in mourning doves. The species uses water heavily for evaporative cooling from the mouth and respiratory tract. At an air temperature of 102° F., doves without water experience body temperature higher than usual, and they begin panting when their body temperature reaches about 108° F. At these high temperatures a mourning dove drinks four times as much water as at an air temperature of 73° F. At such lower temperatures they can go 4 or 5 days without water or succulent food and suffer no permanent ill effects. At a single visit to water, a dove in 10 minutes can regain as much as 17 per cent of its body weight. These authors conclude in accord with field and laboratory observations that mourning doves have a capacity to endure elevated body temperature and extensive dehydration and that this, combined with their capacity to make up water deficits quickly and their ability to fly long distances, allows them to meet the demands of desert existence. We would emphasize, however, that good water supplies are very critical in this pattern of existence and that such are lacking in vast tracts of desert country. Thus the species is in fact limited in its desert adaptation.

The mourning doves of this area are of the race *Zenaidura macroura marginella*.

Yellow-billed Cuckoo. *Coccyzus americanus*

Description. A slender, long-tailed bird, dull brown above and gray to whitish below. Tail feathers except central pair, black with large white tips; wings dull rufous, the tips of the feathers blackish. Lower mandible yellow, upper mandible black.
Range. Breeds from southern British Columbia, North Dakota, Minnesota, southern Ontario, Quebec, and New Brunswick south to Cape region of Baja California, Sinaloa, Chihuahua, Nuevo León, Tamaulipas, Louisiana, the Bahamas, and the Greater Antilles. Winters from Colombia and Venezuela to Ecuador, central Argentina, and Uruguay.
Occurrence in Monument. Recorded once: Twentynine Palms, July 13.

One was seen by Miller at the oasis at Twentynine Palms as it flew past in clear view. It moved along the line of mesquites and cottonwoods. The date of occurrence, July 13, is in a period when the species should be on its breeding grounds, but the limited cover at the oasis is hardly adequate for the nesting of this cuckoo. The bird was probably a vagrant that had failed to reach its normal breeding habitat in the river valleys of the Pacific coast district.

Roadrunner. *Geococcyx californianus*

Description. A large, much elongated (20—24 inches), terrestrial bird, which seldom flies. Tail long and white-tipped, held up at an angle; head with erectile, slightly purplish crest; back gray-green, the feathers streaked with buff; under parts gray, tawny and streaked on breast, the belly and flank feathers loose and fluffy.
Range. Resident from the head of the Sacramento Valley in California, central Nevada, southern Utah, Colorado, southwestern Kansas, central Oklahoma, western Arkansas, and northwestern Louisiana south to Cape San Lucas, Baja California, Michoacán, Puebla, and Veracruz.
Occurrence in Monument. Resident in all sections; although conspicuous, not in fact numerous; normally only one or two seen in any single day afield. Recorded from: Quail Spring, Dec. 22 *, Feb. 11, Mar.; Stubby Spring, Dec. 4—6; Indian Cove, Apr. 7, May 8; Fortynine Palms, Apr. 7; 3 mi. S Twentynine Palms, May 7 [1]; Pinyon Wells, Oct. 12; near White Tanks, Mar. 18 [1]; Pinto Basin, Apr. 6, Sept. 13; Cottonwood Spring, May 1, June 3.

Roadrunners range both through the upper oak and piñon scrub and the open dunes and smoke tree-lined washes of the lower elevations. The species can often be detected by its distinctive tracks in the sand in which two toes (nos. 2, 3) are in front and two (1, 4) behind. The species seems to thrive in arid lands and is not bothered by heat or lack of water. Activity is reduced at midday. Also in late afternoon in cold weather roadrunners typically roost early.

Food consists largely of lizards and small snakes, and the former, especially,

Fig. 28. Roadrunner with a young collared lizard which it has just captured.

Fig. 29. Nest and two eggs of a roadrunner in a dense smoke tree in Pinto Basin, April 6, 1951.

need to be hunted in midmorning at the peak of their activity. Large insects and occasionally eggs and young of small birds are other items, all of which provide adequate moisture so that visits to springs are not necessary, even though the species may drink when water is available.

On May 8 a roadrunner was found dead in a pothole in the canyon at Indian Cove. The hole was 4 x 3½ by 3 feet deep, and the water was 1 foot deep at the time. The bird had been dead perhaps 24 to 48 hours. Larvae of *Bufo* were in the pool, and dead adults of this amphibian were also present. It seems likely that the roadrunner came to the pool to prey on the amphibians or possibly on mosquito larvae. What killed it is uncertain. Possibly it became waterlogged and could not jump up the steep rock walls of the pothole by springing through the water from the sandy bottom beneath.

We have heard the cooing or song call of males in the Monument in March and early April. On April 6 roadrunners were nesting in Pinto Basin. We first noted tracks of them on the sand dunes almost a half-mile from the nest. Tracks became more numerous as we approached the nest area in smoke trees at the east end of the dune. Here there were bands of tracks two feet wide radiating from the nest tree in at least five directions.

The nest (fig. 29) was situated in a large smoke tree, 5½ feet up, among the dense, spiny "foliage." The base of the structure was composed of smoke tree twigs and some woody sticks ¼ inch through. There was a lining of grass tufts, 1 to 2 inches tall, that had been pulled up or blown out so that the roots were intact. These tufts made a coarse, irregular lining, among which a few body feathers of the roadrunners were scattered. The nest cup was about 14 inches across and 3 inches deep, and it held two fresh eggs. Undoubtedly more would have been added later. The eggs were well shaded by the dense tree.

OWLS

Seven species of owls have been recorded from the Monument, but only two of them, the screech owl and the great horned owl, occur regularly and in numbers to constitute a significant part of the fauna. All species but the burrowing owl are nocturnal, and this type of existence seems especially well suited for life in the desert. By confining their activity as predators to the nighttime, owls escape extreme exposure to high temperatures and low humidity. By day they roost in shaded places or in cavities, and water loss is reduced by their inactivity then and in some instances by the relatively greater humidity of their roosting holes or crannies. Water is supplied by the vertebrate and insect food animals on which they subsist, each of these prey types having already solved for their predators, so to speak, the water problem. We know of no instance of an owl's seeking water at springs for drinking.

At night the abundant nocturnal rodent fauna of the desert is accessible to owls. Visibility by starlight and moonlight is especially good in the desert, and the night vision of these birds is clearly adequate for viewing prey, as Dice (1945) has shown experimentally. The auditory sense of owls, which is keen, is also important in locating rodents and large insects, and yet this may frequently be interfered with by desert winds.

The winter cold at night in the high deserts of the Monument actually may depress the activity of prey and also present dangerous extremes to the small species of owls themselves. The elf owl is migratory, leaving the area in winter, and the resident screech owls may at times of especial cold restrict their activity and food intake. Lower deserts are even more favorable for owls, therefore, than are those of the Monument.

Barn Owl. *Tyto alba*

Description. A large owl (stands 14 inches high; wingspread about 3 feet), whitish appearing, with conspicuous facial discs about the dark eyes; no ear tufts. Lightly mottled with dusky, especially dorsally, but this noticeable only at close range; upper surface light tawny; lower surface varies among individuals from nearly pure white to light buff. Tarsus feathered, the lower part only sparsely so.
Range. Virtually cosmopolitan except in arctic and antarctic areas.
Occurrence in Monument. Scarce resident. Recorded from: Indian Cove, Apr. 7; Smithwater Canyon, Apr. 10; Twentynine Palms, Mar. 21; Virginia Dale Mine, July 12.

On the evening of April 7 a barn owl was heard repeatedly calling overhead in the vicinity of cliffs at Indian Cove. On July 12 one was found roosting in a vertical mine shaft at Virginia Dale Mine in barren, rocky low hills.

At 10 A.M. it was sitting in full daylight 15 feet down in the 8 x 10-foot shaft, but it took alarm and dropped down about 30 feet into the darkness, where it was barely visible. In Smithwater Canyon one flushed from a steep, narrow and cliff-faced side canyon. It flew out, giving the characteristic screech, at 8:50 A.M. Probably there are a few other locations at mines or in canyon walls where this species is stationed, but the numbers must be small, as we have listened and watched for it carefully on many occasions without results.

The barn owls of North America are of the race *Tyto alba pratincola.*

Screech Owl. *Otus asio*

Description. A fairly small owl (stands about 8 inches tall), possessing "ear" tufts which, however, may not be visible at all times. Upper surface dark gray with inconspicuous blackish streaks and mottling; under surface light gray with dark streaks, the latter prominent on the breast. Feet feathered to base of toes. Eyes yellow.
Range. Resident from southeastern Alaska and southern parts of Canadian provinces south to the Isthmus of Tehuantepec in México.
Occurrence in Monument. Resident throughout in moderate numbers, although detected usually only by special search at night. Recorded from: Little Morongo Canyon, Mar. 30 *, Apr. 27 *; Black Rock Spring, Aug. 27, 30 *; Lower Covington Flat, Apr. 7, 8, Sept. 8 [1]; Quail Spring, Jan. 22 *, Feb. 11, Apr. 22, May 18, 21, July 1; Stubby Spring, Sept. 5, 9 *, Dec. 4 *; Key's View, July 16 *; Twentynine Palms, Jan. 26, Feb. 27; Pinyon Wells, Oct. 9, 16 *, 17 *; Live Oak Tank, Apr. 1 *, 6–8; Pinto Basin, Apr. 27; Cottonwood Spring, Feb. 7–10, May 6, 14 *, 15 *, July 11 [1], Oct. 21 *.

Screech owls have been discovered in essentially every camp location in the Monument where we have searched for them at night. Apparently they are absent only in the most open desert that is devoid of large shrubs. Catclaws, Joshua trees, scrub oaks, piñons, junipers, willows, cottonwoods, and rocky canyon walls all seem to afford adequate cover for screech owls—places in which they roost concealed by day and from which they hunt at night.

The whistled call, of increasing rhythm, is the note heard chiefly and the one, when imitated, which will most often elicit response from the bird. Although we have usually heard them, especially the lower-pitched voices of males, calling in response in April and May when they are likely to be especially aggressive because of breeding activity, it should be noted that this is also the time we have most often been in the field trying to find them. Birds also have been induced to call in August, September, October, January, and February. Calling, spontaneous or otherwise, is rare during windy weather and seems to be less frequent in bright moonlight than on dark nights. Since both these conditions are frequent on the desert, one gains the impression that these factors must depress or restrict the periods of activity of screech owls.

Screech owls actually seen by us at night while they were calling or hunting have been perched on dead flower spikes of Joshua trees, on the broken face of a rock cliff, in willow brush 3 feet from the ground, on top of a boulder

Fig. 30. Screech owl in concealment posture against the trunk of a juniper tree.

on a small ridge, in hanging limbs of a piñon 20 feet up, and on top of a 5-foot *Yucca mohavensis.*

In daytime we have twice discovered their roosts. On October 9, one was frightened from a well-worn hole in a growing limb of a scrub oak, 10 feet above ground, at Pinyon Wells. Also at this location on October 17 a bird was found in a rotted-out split trunk of a large piñon; this roost was open at the top, and the owl retreated down from the entry way to the bottom of the cavity a foot and a half below. Both these birds had been stimulated to call by giving screech owl whistles in broad daylight at 10:30 and 7:30 A.M., respectively. The owls would answer two or three times and then fall silent, thus revealing their approximate locations. In the second case the observer came close under the piñon, watching, and heard the scrape of claws as the owl, a female, scuttled down into the hole. Such cavities of course might well serve as nests, although we did not find active nests in the spring season.

The breeding season probably is early compared to that in more northern and coastal areas. A specimen of a female taken on January 22 had enlarged ovarian follicles, and a male taken in February had "quite large" testes. This suggests that laying may occur in late February or March.

Food is sometimes absent from the stomachs of owls taken soon after dusk. This was true at 8:30 P.M. on October 16, and a screech owl captured on March 30 in the early evening had only a few insect remains. On the other hand, a bird collected at 7:45 P.M. on August 30 had an entire, freshly taken juvenal cactus mouse (*Peromyscus eremicus*) in its stomach, and one taken at 1 A.M. on July 16 had a stomach filled with mouse hair, legs of orthopteran insects, and a scorpion claw.

The pattern of screech owl activity on the desert may be reconstructed from scattered visual and auditory evidence. The birds move from their roosts at dusk, before full darkness, and hunt close to the ground, primarily obtaining insect and rodent food from the ground surface. They may call before they are successful in feeding and also at any time during the night, but they are most inclined to be vocal soon after dark and in the morning when a small amount of light is beginning to appear in the east. Windy weather probably restricts and hampers foraging, which is done by sound as well as by sight. By day in the high-temperature periods, the screech owls are quiet, hidden in a cavity or in relatively dense, shaded vegetation.

This owl's general resemblance in color to the light gray and finely subdivided desert foliage, and its elongate concealment posture during fear, which entails narrowing of head feathers and elevation of ear tufts, doubtless are adaptive in partly protecting screech owls from predation by bobcats, coyotes, foxes, and larger owls. The species may be regarded as a highly successful small predator and nocturnal insectivore in the desert, enjoying the advantages in the desert generally received by owls through their nocturnal existence (see p. 83).

Fig. 31. Great horned owl in flight along the cliffs at Indian Cove, carrying a pocket gopher.

The specimens of screech owl taken in the Monument have been studied especially in the course of a revision of the desert populations of this group (Miller and Miller, 1951). At the east end of the Monument at Cottonwood Spring the birds are normal examples of the race occupying the lower Colorado and Imperial valleys, namely *yumanensis,* but west of this point the population is intermediate in color and size between *yumanensis* and *quercinus* of the coastal slopes. The birds fall along a cline of increasing wing length westward. In color they are variable, some being close to *yumanensis* and others as dark and coarsely striped as the grayer types of *quercinus.* None of the richly brown type of *quercinus* occurs.

Great Horned Owl. *Bubo virginianus*

Description. A large owl (stands 18 inches high; wingspread about 4 feet) with feathered "ear" tufts and yellow eyes. Upper surface mottled tawny and dusky; under surface with white throat patch bordered behind with black stripes on the breast; dark cross bars on flanks; belly white or light tawny. Feet and toes feathered.
Range. Breeds from tree limit in the Arctic in North America south to the Straits of Magellan in South America. Winters from southern Canada southward. Absent from the West Indies.
Occurrence in Monument. Resident throughout, but never commonly. Recorded from: Little Morongo Canyon, Apr. 1, 28; Black Rock Spring, Aug. 31; Upper Covington Flat, Feb. 11 [1]; Covington Flat area, Apr. 7, Aug. 22–29; Quail Spring, Jan. 18, 22, Apr. 22, May 18, 19–21; Stubby Spring, Sept. 5, 8, 11, Dec. 4–6; Indian Cove, Apr. 8, May 12, 14, Oct. 16; Twentynine Palms, Nov. 6, 7, 11; Pinyon Wells, Oct. 12; Split Rock Tank, Nov. 1; Live Oak Tank, Apr. 6–8; Pinto Basin, Oct. 13; Cottonwood Spring, Feb. 19, May 6, 14–19, July 8.

Horned owls may be heard at every camp, yet often in the distance, as the hooting carries far. Calling is noted at all seasons of the year, and frequently, as is typical of this species, the male and female call antiphonally, the female with the higher voice and normally a more complex rhythm. Bright moonlight does not seem to inhibit this species as it does the screech owl.

At Twentynine Palms Miss Carter saw birds at dusk on a fence post and in a dead tree top. One roosted there in a large willow. Through most of the Monument, roosting, as well as nest sites, may be expected in the cliffs and canyon walls. The calling at night seems usually to be in and near cliff faces, and yet the species has been found in the flats of Pinto Basin.

The horned owls of the desert section of California are all referable to the race *Bubo virginianus pallescens.*

Elf Owl. *Micrathene whitneyi*

Description. A minute owl (38–40 grams; stands about 4½ inches tall) without ear tufts and with yellow eyes. Upper surface gray with lighter tawny flecks and white and tawny spots on scapulars and wing coverts. Under surface mottled gray, tawny and white, indistinctly streaked on flanks and whiter on the belly. Tarsi and feet with obsolete, hair-like feathering.

Range. Southeastern California, central Arizona, southwestern New Mexico, and southern Texas south to central México (Puebla). Winters south of the United States.
Occurrence in Monument. Locally established at Cottonwood Spring, May 6 *.

On May 6, 1946, Loye Miller and A. J. van Rossem discovered this species at Cottonwood Spring, far west of its previously known range along the Colorado River (L. Miller, 1946). On this date, soon after sunset, a pair of birds was twice seen (pp. 284–285) as "the male appeared to bring food to the female. At these times the cricket-like trill which may be designated as the 'desire note' was given. . . . The male bird moved about more or less but the female appeared to remain in one closely restricted part of the [cottonwood] tree repeating the 'station call,' a single soft whistle. . . . The typical querulous note of the species was heard frequently, and on two occasions the male gave a note that was entirely new [to van Rossem and me]. . . . It was like the Pigmy Owl's metronomic whistle rendered at a much higher pitch and frequency but was more softly pronounced and in much shorter phrases." The specimens taken showed that the female was soon to lay.

In subsequent years reports (for example, 1958, Orians ms) have been received indicating that elf owls are still in evidence at Cottonwood Spring in the spring season, where they evidently nest in woodpecker holes in the cottonwoods.

The elf owls of this area belong to the race *Micrathene whitneyi whitneyi.*

Burrowing Owl. *Speotyto cunicularia*

Description. A fairly small (stands about 9 inches tall), long-legged owl without ear tufts; lives on or near the ground in the open. Above brown spotted with white; white superciliary stripe; under surface white with cross bars of warm brown on flanks and breast; dusky mottling on anterior throat separates pure white patches of chin and posterior throat. Iris yellow. Feet unfeathered.
Range. Unforested areas of North and South America from southern British Columbia and prairie provinces of Canada south to Tierra del Fuego. Extends to Revilla Gigedo Islands and West Indies.
Occurrence in Monument. Sparse resident. Recorded from: south side Pinto Basin at 2500 feet, May 21 [1].

A burrowing owl taken by a party from Long Beach State College near the road junction on the south side of Pinto Basin was in breeding condition on May 21.

The burrowing owls of western North America belong to the race *Speotyto cunicularia hypugaea.*

Long-eared Owl. *Asio otus*

Description. A middle-sized owl (stands about 12 inches tall), with long prominent "ear" tufts and conspicuously streaked and barred plumage ventrally. Upper surface dusky, com-

plexly mottled with whitish and tawny; facial disc tawny, black centrally and also at margins. Eyes yellow. Feet feathered, tawny.
Range. Europe and Asia north to about 60° N; south to northwestern Africa and northwestern India. In North America breeds from southern Alaska east to Nova Scotia and south to northwestern Baja California, southern Arizona, Chihuahua, Arkansas, and Virginia. Winters from southern Canada south to Sonora, Durango, Puebla, and the Gulf states.
Occurrence in Monument. Sparse resident and visitant. Recorded from: Lower Covington Flat, Apr. 8 *; Stubby Spring, Apr. 9; Twentynine Palms, Jan.; Eagle Mountain, 4500 ft., Feb. 19[1].

This species was unexpectedly rare. On April 9 in the morning one was flushed from junipers near Stubby Spring. It flew to an open Joshua tree and could not be approached within 50 yards. In January Charles G. Sibley was told of a group of seven roosting in thickets at Twentynine Palms. This is a type of concentration often seen in this species in winter and probably represents migrants to the area.

However, there evidently are some individuals that nest in the Monument. On the night of April 8 at Lower Covington Flat we became aware of the single, spaced hoots of this species far in the distance. Starting at 9 P.M. we traced the call to a point 2 miles away in scattered piñons on a hill consisting of a jumble of large boulders. The owl had called continuously in the moonlight until we arrived in its area at 11 P.M., the call carrying remarkably in the still air of the desert night. The calling owl was near the edge of a crown of a piñon and left as we shone its eyes. Soon it started calling again 100 feet away, where it was on a perch 15 feet up on a dead snag halfway to the top of a piñon. This bird, which had been calling for two hours, had an empty stomach and enlarged testis, 10 mm. long. The area back of the ears and the upper throat had swellings of lymphatic or glandular type, evidently associated with the heavy, nuptial calling. A second long-eared owl called in this area briefly.

The long-eared owls of North America have been subdivided racially (see American Ornithologists' Union Check-list, 1957) into eastern and western forms but we believe on poor grounds, as there is much individual variation and certain birds from New England are indistinguishable from those from California in the collection of the Museum of Vertebrate Zoology. Accordingly we use the racial name *Otus asio wilsonianus* as properly applicable to all western specimens as well as to those of eastern North America.

Saw-whet Owl. *Aegolius acadicus*

Description. A small owl (70–90 grams; stands about 6 inches tall) without "ear" tufts and with yellow eyes. Above, brown with white spots on nape, scapulars, and wing coverts; under parts white, streaked with brown. Feet feathered.
Range. Breeds from southern Alaska to Quebec and south to the mountains of southern California, southern Nevada, and southern Arizona, and in the highlands of México to

Oaxaca; also to Oklahoma, central Missouri, Ohio, West Virginia, and Maryland. Winters in breeding range, but moves south in part.
Occurrence in Monument. A rare winter visitant. Recorded once: Twentynine Palms, Jan. 23 *.

The record specimen here involved was found near cabins of the Twentynine Palms Inn (Carter, 1937). The carcass was deeply buried in leaves and completely desiccated. Another such winter occurrence from a nearby area is that recorded by L. Miller (1932*b*) near Desert Center, east of the Monument.

These saw-whet owls belong to the race *Aegolius acadicus acadicus.*

POOR-WILLS AND NIGHTHAWKS

Poor-will. *Phalaenoptilus nuttallii*

Description. About the size (8 inches long) of a lesser nighthawk, but wing shorter and rounded. Mottled dorsal coloring and ventral barring; no white in wing, but wing feathers prominently barred with warm buff; tail tipped with white laterally; throat white, bordered by blackish.
Range. Breeds from southern interior British Columbia, Montana, the Dakotas, and Iowa south to central México. Winters from central California, southern Arizona, and southern Texas south to Guanajuato.
Occurrence in Monument. Common resident in all areas. Recorded from: Little Morongo Canyon, Apr. 1 *, 3, Black Rock Spring, Aug. 29, 30, 31, Sept. 1 *, 2; Lower Covington Flat, Apr. 7, 9, July 9[1], 10[1], 17[1], Aug. 22 *, 23, 24, 25 *; Smithwater Canyon, June 27[1]; Stubby Spring, Apr. 11, Sept. 5, 8; Queen Valley, Oct. 12; Pinyon Wells, Apr. 10; Twentynine Palms, Apr., May, Oct.; $\frac{1}{4}$ mi. E junction roads to Split Rock and Live Oak tanks, Jan. 21; Pinto Basin, Apr. 10; Virginia Dale Mine, Apr. 23; Cottonwood Spring, Oct. 21–24.

Poor-wills are usually detected at dusk either by their clear two-parted call or by their presence on road surfaces, where they drop down almost flat to the ground in intervals between their low-sweeping foraging flights. They also alight on tops of rocks and occasionally as high as 20 feet off the ground in piñons, junipers, willows, and desert willows.

The birds start calling after sundown while it is still just light enough to read. This was true, for example, at Lower Covington Flat at 6:30 on August 25. Typically, first calls come from scattered stations on rocky slopes where the birds have been roosting through the day. After the first calls they begin insect-catching, flitting low over the vegetation and then frequently coming down to wash bottoms and roadways. At this time they often give a low *quup* note as they move about. Specimens taken 15 minutes after first calling may already have their stomachs full of insects caught in flight, thus indicating the great efficiency of food-gathering in these twilight flights. On April 1 at Little Morongo Canyon at dusk one was watched foraging over a warm gravel bank 50 feet above the canyon floor, flying low over the juniper tops. It moved back and forth within 10 feet of the ground, occasionally alighting.

Calling may continue off and on through the night, especially during moonlight, and there is a noticeable increase in calling again at dawn, with a second period of intense foraging then.

At most camps in hilly or mountainous areas two or three poor-wills were calling at once at dusk. On April 7 at Lower Covington Flat five were in voice simultaneously. Calling is perhaps most persistent in the spring months, but it is also common in August and September, and occasionally in October. The birds seem more definitely to respond to imitations of their calls in spring and summer and then will often come to a person, calling nearby or fluttering overhead to alight on the ground and give the *quup* note. Two were flushed at once by day on April 9 at a distance of about 15 feet; they alit again on the ground under shrubs 35 yards away.

In the near vicinity of Joshua Tree Monument in the Chuckwalla Mountains south of Desert Center, Jaeger (1948, 1949) made his original, stimulating observations which showed that poor-wills go into periods of torpor in the winter. The obvious adaptative advantage of this procedure is that the bird reduces its food utilization during times of cold weather when insects are difficult to obtain. The details of reduction in body temperature, metabolism, and food intake have been worked out experimentally and are

Fig. 32. Poor-will crouched at daytime roosting site.

reported in a series of papers by Brauner (1952), Marshall (1955), Bartholomew, Howell, and Cade (1957), and Howell and Bartholomew (1959). A poor-will can reduce its body temperature to as low as 40° F. and reduce its oxygen use at this time to as little as 3 per cent of its normal level. Torpor periods vary greatly in length from a few hours to many days, but Jaeger's bird, which was followed through parts of three winter seasons in the wild, in one winter apparently was torporous whenever checked by day from at least November 26 to February 14. However, there is no assurance that this individual was continuously torpid over this whole period. It, like experimental birds, lost weight during hibernation periods. For example, it declined from 45.6 grams to 44.5 grams between January 4 and February 14. Bartholomew, Howell, and Cade estimated that a nonhibernating poor-will requires about 8 calories a day and a torpid one 0.8 calories. One gram of fat would, they think, meet the energy need for 10 days during torpor at 50° F. Jaeger's bird may therefore have either used less fat because of lower average temperature or because it at times awoke and fed. This same bird a year later weighed 52.6 grams near the beginning of the hibernation season. Bartholomew, Howell, and Cade point out that 10 grams would sustain a torpid bird for 100 days. We note, moreover, that poor-wills are capable of depositing heavy masses of subcutaneous fat and that they may range in weight from 53 grams down to as little as 32 grams.

It is not at all certain yet what actually stimulates a poor-will to go into torpor, but it seems entirely clear that some individuals use this device to meet the problems (or to evade them) of food gathering and conserving calories used in keeping warm. Further field observations are much needed on torporous poor-wills in order to judge the degree to which they resort to hibernation. In the Joshua Tree Monument some poor-wills at least may be active in January on certain nights. We have no clear indication of migratory movements, although such is entirely possible involving birds from far northern areas or high mountain levels where even hibernation will not suffice for the winter situation.

Poor-wills taken as specimens in the Monument on April 1 had gonads (testis 3 mm.) well below breeding level, whereas in the last week of August and on September 1 we took adult males with testes up to 4 mm., which is submaximal but still not far into regression from the breeding state. At the same time a molting juvenile as well as adults were taken. The nesting period thus probably falls in late May and June, possibly even also extending into early July.

The poor-wills of the Joshua Tree Monument might be expected to show affinities to any one of the three races, *Phalaenoptilus nuttallii nuttallii, P. n. californicus,* and *P. n. hueyi,* which are typical of adjoining regions to the north, west, and east, respectively. In point of fact *P. n. nuttallii* and *californicus* are not sharply differentiated, and many individuals from the

Fig. 33. Lesser nighthawk passing overhead.

respective ranges of these races show overlap in characters. The usually blacker and broader barring ventrally in *californicus* is not seen in the specimens from the Monument, and the group as a whole matches rather well the race *P. n. nuttallii*. This is true despite the fact that our specimens come from western stations in the Monument. Two individuals, those from Lower Covington Flat, definitely suggest an approach to the race *hueyi* of the Colorado Desert. In them the back is much paler, with smaller black areas, than in the other specimens, although ventrally they show less reduction of black. Also in brown tones of the back they are not close to *hueyi*. Possibly birds from the southeast end of the Monument would be closer to or identical with *hueyi*, but specimens are lacking from these areas. The Covington Flat birds are not so close to *hueyi* as are certain intermediates earlier reported from the Providence Mountains of the eastern Mohave Desert to the north (see Johnson, Bryant, and Miller, 1948).

Lesser Nighthawk. *Chordeiles acutipennis*

Description. Long-winged, with short, weak bill and feet, typical of nighthawks (total length 8–9 inches). Plumage mottled and barred; white or buff on throat. Distinguishable from the common nighthawk of higher and more northern areas by location of white wing patch

nearer tip of wing rather than halfway between wrist and tip; also lacks sharp *peent* note of common nighthawk.
Range. Breeds from interior of central and southern California to southern Texas and south through México and parts of Central America to South America, extending to Perú. Winters from Cape region of Baja California and Sonora southward.
Occurrence in Monument. Summer resident in small numbers, especially in lower sections, from late April through July. Recorded from: Quail Spring, June 30; Twentynine Palms, Apr. 17, May 26, July 2, 12, 13; junction main road and Pinyon Wells road, June 30; Pinto Basin, Apr. 10; Virginia Dale Mine, July 12; Cottonwood Spring, Apr. 30, May 14–19, June 22, July 7, 8, 12 [1].

This nighthawk in late spring and summer is stationed for nesting in washes and flats in the relatively more open and lower parts of the Joshua Tree area. We have seen it principally at Twentynine Palms and at Cottonwood Spring at dusk. Doubtless insect life about these oases is especially attractive to it, and occasionally several individuals may be seen at once at these places. The occurrences at other points have involved single birds only.

At Twentynine Palms on July 13 and in Pinto Basin on April 10 nighthawks were giving their prolonged, purring trill as the species normally does from the ground in the vicinity of nesting areas. In the flats about the oasis they must surely have been raising young, although we found no eggs or juveniles.

The nighthawks are quiet in the heat of the day and become active at sunset, or even a little before, as at 5:45 P.M. on June 22. Much of their feeding must be accomplished in twilight hours, in the evening and at dawn, rather than during full dark.

In the evening we have seen them come frequently to open water, drinking on the wing by repeated passes over the surface. Thus at the close of a long, hot day they may replenish their water reserves as well as obtain water from their flying insect food. The great powers of flight of the species make it easy for it to move to water sources each evening, although we are by no means sure that all individuals are able to do so or require such visits.

The lesser nighthawks of this area belong to the race *Chordeiles acutipennis texensis.*

SWIFTS, HUMMINGBIRDS, AND ALLIES

Vaux Swift. *Chaetura vauxi*

Description. A small (16–19 gram) swift, entirely black or dark gray above and on belly; light gray on throat. Tail with feather shafts extended as spines.
Range. Breeds from southeastern Alaska, northern British Columbia, and western Montana south through Oregon and Washington to central California; also eastern and southern México south to Panamá. Winters chiefly in southern México and Guatemala; occasionally found much farther north in Louisiana and central California.
Occurrence in Monument. Spring and fall migrant. Recorded from: Black Rock Spring,

Sept. 4 *; Smithwater Canyon, Sept. 9 [1]; Barker's Dam, Apr. 28; Twentynine Palms, Apr. 27; Pinto Basin, Sept. 13; Cottonwood Spring, May 12, 13.

Swifts, like swallows (see p. 131), seem to face no particular difficulties in migrating across deserts. Airborne insects provide them moist food, and their great cruising power enables them to move readily to areas of local concentration of insects.

The dates for spring occurrences range from April 27 to May 13, and for fall migration, from September 4 to 13.

Vaux swifts normally are seen in groups, foraging for limited periods in the course of migratory movements. The groups seen in the Monument have consisted of 5 to 15 individuals. Yet on other occasions solitary birds have been noted, sometimes in association with white-throated swifts or with migrating swallows.

At Cottonwood Spring on May 12 about 8 were feeding low to the ground, from 5 feet up to a little more than the height of the cottonwood trees. At sunset a compact flock of about 15 was seen. Along the crest south of Black Rock Spring on September 4, about 5 were sweeping low over oak and manzanita brush and piñons; at the time they were not moving through but were circling and cruising back and forth, in a loosely organized group, and they evidently were foraging. After about 10 minutes they were joined by some barn swallows.

The specimen taken on September 4, of the race *Chaetura vauxi vauxi*, was a female in the middle of the molt of its primaries. It was fat and in good condition, and only a little body molt was in evidence. Judging from the wing molt and the wear of the old primaries, the bird was an adult. Apparently, as in the swallows, flight efficiency generally is so great that its impairment by molt is not sufficient to cause difficulty in migration.

White-throated Swift. *Aëronautes saxatalis*

Description. A moderate-sized (28–34 gram) swift with conspicuous white throat, center of breast, and posterior flanks, the remaining parts contrastingly black. Tail not forked and without spiny shafts.

Range. Breeds from southern British Columbia to Montana and northwestern South Dakota and south in the mountain areas through México to Guatamala and El Salvador. Winters from central California, central Arizona, and southwestern New Mexico southward.

Occurrence in Monument. Fairly common; seen throughout the year. Recorded from: Morongo Valley, Oct. 29; Black Rock Spring, Sept. 2, 4; Lower Covington Flat, Apr. 8, July 9 [1], Aug. 27, 28; Smithwater Canyon, Apr. 10, May 13 [1], 19; Quail Spring, May 19–21; Stubby Spring, Sept. 5, 7; Key's View, July 4; Barker's Dam, Feb. 18, Apr. 28; Indian Cove, Apr. 7, 8, May 13, 14, June 3; Fortynine Palms, Apr. 7; Pinyon Wells, Oct. 4, 13, 16; Live Oak Tank, Apr. 6–8; Pinto Basin, Apr. 12; Virginia Dale Mine, Apr. 24; Eagle Mountain, Oct. 19, 20; Cottonwood Spring, Apr. 25, June 3.

White-throated swifts, like swallows and other swifts, are dependent on airborne insects, which are fairly abundant at least in local concentrations throughout the arid desert lands. Wide cruising radius makes it possible for

Fig. 34. White-throated swifts seen from Key's View.

swifts to tap these resources. In winter local movements to lower warmer levels may be possible, and also this species is capable of going into torpor (see Bartholomew, Howell, and Cade, 1957), which may tide it over periods unfavorable for foraging.

This species of swift is frequently detected by reason of its loud, clattering call, especially when it is cruising along the faces of cliffs in which it roosts and nests in natural crevices. Although sometimes the birds are seen singly, typically they associate in groups of 4 to 50. In spring they may be seen in aerial copulation, the two individuals clinching and tumbling in the air, dropping rapidly. Such copulation was seen in the Monument on April 28 at Barker's Dam and on May 13 at Indian Cove. Both these locations provide good rock cliffs in which they could breed, and April and May are the months in which nesting might be expected.

On February 18 a "cloud" of over 50 of these swifts dropped down to the water at Barker's Dam to drink by skimming the surface. The flock then reformed, gained altitude gradually, and flew off. Such open expanses of water usable for drinking are rare in the area, but swifts can when necessary readily cruise to such places to obtain water.

The race of white-throated swift represented in this area is *Aëronautes saxatalis saxatalis.*

Costa Hummingbird. *Calypte coastae*

Description. A small (2.6–3.4 gram) hummingbird, light green above and pale gray to white beneath and entirely lacking rufous in the plumage. Male: crown, chin, and throat iridescent purple, the gorget extended laterally into long points; tail dull green, dusky laterally. Female: body plumage as in male, the head and throat areas dull green and whitish; tail dull green centrally, the lateral feathers gray basally, black distally, and tipped with white.
Range. Breeds from central California, central Nevada, southwestern Utah, Arizona, and southwestern New Mexico south to southern Baja California and at least to southern Sonora. Winters over most of breeding range from southern California and southwestern Arizona southward, reaching Sinaloa.
Occurrence in Monument. Common spring resident, breeding. Probably resident in small numbers the year around. Recorded from: Little Morongo Canyon, Mar. 31 *, Apr. 1, 28; Lower Covington Flat, Apr. 8; Smithwater Canyon, July 16[1]; Quail Spring, Apr. 10, May 14; Indian Cove, Apr. 7, May 13, 14; Fortynine Palms, Apr. 7; Twentynine Palms, May 13, 16; Pinyon Wells, Apr. 10; Live Oak Tank, Apr. 6–8; Cottonwood Spring, Feb. 7–10, Apr. 5 *, 25, 26, 29, 30, May 1, 13, 14, 15 *, 17 *, 19.

Costa hummingbirds are conspicuous in the Monument from early February through May. This is in general their nesting period. A pair was seen in copulation on April 1 in Little Morongo Canyon, and a laying female (egg in oviduct) was collected at Cottonwood Spring on April 5, when it was hovering about a restricted area in apparent defense of a nest site. Thus the species was very active in the early stages of nesting at the beginning of April. The aggressive nuptial performances of males have been observed from February 7 to May 19, and on February 7 at Cottonwood Spring females were thought to be prospecting for nest sites in the desert shrubs.

On April 7 a nest (fig. 35) with two fresh eggs was found at Indian Cove, situated 3½ feet up on a 2-inch horizontal limb of a desert willow at the edge of a tangle of this plant in the wash. The female settled on the nest, readily revealing its location. The nest was 1 x ¾ inches deep outside and ½ inch deep inside and was rather loosely constructed for that of a hummingbird. The female had a small but prominent iridescent gorget spot on the throat, and the lighter feathers of this area were yellow-stained from pollen. A territorial male took lookout posts within 35 to 100 yards of the nest on the tips of desert willows and of *Yucca mohavensis.*

From June to January we have few records of this hummer. In part the species may move away, as to higher elevations to the west, during the warmest segment of the summer. In winter in low areas in the desert mountains of this part of the state, occasional Costa hummers have been collected, and such occurrences may be expected in the Monument. We think that lack of conspicuous behavior in the nonbreeding season accounts chiefly for the failure to record the small number of year-around residents that prob-

Fig. 35. Female Costa hummingbird on her nest in a desert willow at Indian Cove on April 7, 1960.

ably are present. There is, however, rather clearly an influx of individuals in February and March.

The nuptial performance of males consists of a prolonged high-pitched whistle or hiss given as they swing down and up in a very open U-shaped arc. The birds do not go back and forth over this course, but circle about and retrace the course in the same direction; this aspect was carefully checked at Little Morongo Canyon on March 31. The note is also given during a short swinging performance of only 2 feet in length, in which the bird does go back and forth slowly. This was observed beneath a piñon tree on May 19 at Quail Spring. The note was the same as that during the long dive and could under these circumstances not have been produced by wing or tail feathers; we may conclude therefore that it is vocal. The more usual long or open arc performance may be observed throughout the day and even before sunup and after sundown, as noted on March 31. Between displays the males occupy lookouts on the bare tips of the highest vegetation available along washes.

The courting and copulating of the pair watched on April 1 took place in mesquite brush near a flowering bladder-pod plant (*Isomeris arborea*). The two birds were four feet from the ground in open twigs. The male approached the female from the front and made short, 2- to 3-inch darts at her from different angles within an arc of about 90°. Then he mounted while she perched and he dropped his tail and abdomen twice, apparently making quick if any actual contact at the cloaca. He then flew and again darted at her from the front, and the two buzzed off in close pursuit. This was in bright sunlight at 9 A.M.

In the nesting season in Joshua Tree Monument, Costa hummers are most commonly seen along canyons or washes. They often frequent encilias and mesquites, and at times when there are flower sources there may be up to half a dozen individuals within sight or sound at a time. The yellow-flowered bladder-pod (fig. 36) is especially favored as a nectar source by this species in March and April. This was conspicuously the case in Little Morongo Canyon, at Twentynine Palms as reported by Carter (1937:213), at Indian Cove, and at Cottonwood Spring. The plant, which has a very obnoxious odor to humans, seems to be most attractive to this species. The nectar obtained from flowers is of course not the only food source, for we watched a female in favorable light on April 1 sitting at the edge of a mesquite thicket and flying out repeatedly, 3 to 10 feet, to capture small flying insects.

Costa hummingbirds come to bathe at springs and doubtless also at times

Fig. 36. Bladder-pod bush in bloom; the yellow flowers are a favorite food source of Costa hummingbirds.

to drink (see Johnson, Bryant, and Miller, 1948:294). At Fortynine Palms on April 7 a male bathed in an inch-wide trickle of water as it flowed over and fell from a large boulder. The bird flew and fluttered head up into the stream, splashing on the move, and then backing off to hover. Occasionally it perched and shook and preened.

The period of the year when this desert species of hummer is most active and present in the greatest numbers in the Monument is that which is least severe with respect to high and low temperature extremes and one when flower and water sources are most likely to be at their best. The Costa hummer can doubtless tolerate high temperatures and low humidity better than most North American hummingbirds, but this tolerance may be achieved largely through behavioral adjustments and concentration locally and seasonally in places affording good food and some water within cruising radius. Certainly the main nesting season is scheduled earlier in spring than on the Pacific coast of California (compare Willett, 1933; Grinnell, 1914; Johnson, Bryant, and Miller, 1948) and avoids the intense summer heat that sets in on this part of the desert in June.

Anna Hummingbird. *Calypte anna*

Description. A hummingbird slightly larger (3.2–4.6 grams) than minimum for the group, entirely lacking rufous in the body plumage and tail. Male: forehead, chin, and throat iridescent purplish red; upper surface and tail green; under parts behind gorget greenish gray. Female: dorsal surface green; under surface dull gray with tinges of dull green; tail green centrally, laterally black with white tips, the bases of the feathers gray.

Range. Breeds in California west of the Sierra Nevada and southern coastal mountains from Humboldt, Shasta, and Tehama counties south to the Sierra San Pedro Mártir and San Quintín in northwestern Baja California; extends to Santa Cruz and Guadalupe islands. Winters over the breeding range; additionally, south in Baja California at least to Cataviña at latitude 29° 45′ N and east to southern Arizona and northern Sonora.

Occurrence in Monument. Fairly common winter visitant from September to April. Recorded from: Lower Covington Flat, Oct. 1 [1], Quail Spring, Jan. 23 *, 25, 26, Feb. 13, Apr. 22; Stubby Spring, Sept. 5, 6, 8 *; Barker's Dam, Feb. 18; Pinyon Wells, Oct. 12 *, 16; Eagle Mountain, Oct. 20 *, 21; Cottonwood Spring, Oct. 21, 23 *; Lost Palm Canyon, Oct. 22 *, 23.

There is no sure evidence that this hummingbird nests in the Monument, although the males when present in winter and early spring engage to some extent in aggressive territorial performances. On its breeding grounds west of the desert these displays of course mark the breeding stations of males, and nesting there occurs in the winter and early spring (see especially Williamson, 1956). But the displays seen in the desert must, on present evidence, be assumed to be unproductive of actual nesting and only a manifestation of the general aggressiveness shown by this species. No specimens are at hand to show development of the gonads in the winter months on the desert. Birds that we do have, taken in the fall in September, would not be expected to show this. Included are 5 females and one probable female. This latter bird, taken September 8, is in the late stages of primary molt and possesses no detectable juvenal secondaries (see Williamson); on the contrary, it has a larger than normal cluster of metallic throat feathers; it seems safe to conclude it is an adult. Two males taken on October 22 and 23 are, however, birds of the year in which the adult gorget is not fully developed and the juvenal lateral tail feathers persist. We may conclude therefore that the wintering population in the Monument includes both adult and yearling birds and that they arrive after the postjuvenal and adult annual molts are well past the early stages (in stages 4 or 5 of Williamson).

Male Anna hummingbirds have been heard giving the vocal aggressive note on September 6, October 12, 16, 21, January 26, and April 22. The full power-dive display was seen on January 26 and April 22. On this last date one darted at a white-crowned sparrow perched on top of a catclaw bush.

Feeding in the fall has been seen chiefly at red-flowering bushes of *Zauschneria latifolia* at several different localities. Two specimens taken had yellow pollen on the breast. Most occurrences of the species have been about oases where these flowers are present, or in the piñon woodland.

The principal occurrence of the Anna hummingbird in the Monument coincides with the less severely warm parts of the year (October to April) and indeed includes occurrences at stations in winter when there are temperatures below freezing each night. Insect and nectar sources are clearly adequate then for moisture and caloric needs; at least they are available and workable during the warmer parts of each day. The birds, not being in transit, find and set up quarters where sustaining food sources are available.

Rufous Hummingbird. *Selasphorus rufus*

Description. A small (2.8–3.5 gram) hummingbird with rufous flanks and base of tail. Adult male: dorsal surface rufous from nape to middle of tail; crown and forehead dull green; chin and throat iridescent scarlet; breast white; belly and flanks rufous; tips of tail feathers dusky. Females and young males: dorsal surface bronze-green; base of tail more or less rufous, green in midsection of feathers, and black subterminally, with white tips except on central pair (this plumage indistinguishable from that of Allen hummingbird except that outer tail feathers broader—2.8 mm. or more wide rather than 2.5 mm. or less); under parts dull white, rufous laterally, the throat spotted with green or sometimes with scarlet.

Range. Breeds from southeastern Alaska, southern Yukon, east-central British Columbia, and southwestern Alberta south through Washington and Oregon to Trinity Mountain region of northwestern California and to southern Idaho and western Montana. Winters from Sinaloa to Guerrero, state of México, and Querétaro.

Occurrence in Monument. Common migrant. Recorded from: Little Morongo Canyon, Mar. 31; Black Rock Spring, Sept. 1 *, 2 *, 4 *; Smithwater Canyon, July 10 [1]; Quail Spring, Sept. 10 *; Stubby Spring, Sept. 5, 9; Pinyon Wells, Oct. 16 *; Cottonwood Spring, Sept. 14 *, 15 *, 17 [1]; Lost Palm Canyon, Sept. 14 *.

This hummingbird is common in the fall migration for an extended period, principally from September 1 to October 16; probably later dates up to October 22 are involved, and there is one proven earlier occurrence on July 10. All birds seen in the fall are in female or immature male plumage, and

Plate I. Hummingbirds: Anna (above), rufous (below), and Costa (in flight).

GMC

identification as to species without specimens is therefore impossible, for the closely related Allen hummingbird also occurs. Records of spring migrants are surprisingly few. In the fall, migrants may linger for considerable periods about flowers in the canyons, but in the heat of early fall, especially, they are seen moving through the country in obvious migratory flights, stopping only briefly in the vicinity of springs.

This hummingbird seems to suffer losses while crossing the desert when water and flowers for nectar sources are scarce. On September 1 at the end of a severe heat period, a bird flew up the wash near Black Rock Spring. This was in the afternoon in strong sunlight and with humidity extremely low. The bird was so exhausted that it rested periodically on the ground. It finally came into the limited shade at camp. Here it was caught by hand, and it died in captivity in about an hour. The bird was a male of the year and was very thin; it weighed only 2.5 grams. It probably suffered from lack of energy reserves as well as water. Another came into this same camp in the piñons in exhausted condition the next day; it weighed only 2.3 grams. At Stubby Spring on September 9 one was found dead in the water tank; possibly it had arrived there in weakened condition and succumbed while attempting to drink in flight.

Dehydration must be rapid in a small species like this hummingbird. However, similarly small desert-dwelling hummingbirds, like the Costa, cope with the dehydration problem throughout the summer. Nectar and insect food, if regularly taken, should provide ample water for hummingbirds, but in migrating through strange regions, with flower sources unknown and sparse, many hummers must fail to survive. Of the 10 specimens of rufous hummingbird taken in the fall, 3 weighed only 2.3–2.5 grams and had no fat, conditions indicating failure, as in the instances at Black Rock Spring which constituted two of the three. Four distinctly fat birds taken in the period from September 14 to October 16 weighed up to 3.5 grams.

In the autumn the chief food source for rufous hummingbirds is the red tubular flower of *Zauschneria latifolia.* This plant normally blooms profusely in the fall in limited areas in canyons or near the edges of springs. Rufous hummingbirds have been seen stationed about these plants and feeding on them at Pinyon Wells, Cottonwood Spring, and Lost Palm Canyon. On September 14, 1950, at the last locality there were no less than 12 hummingbirds, apparently chiefly or entirely of this species, in 100 yards of canyon bottom bordered by this plant. They were chippering and threatening one another in constant competition over the food source. Similar concentrations seen here on October 22, five years earlier, doubtless involved rufous hummingbirds, although the one bird taken proved to be an Allen hummingbird.

Allen Hummingbird. *Selasphorus sasin*

Description. Like the rufous hummingbird (see p. 102), except adult males green on back, at least in middle section; lateral tail feathers narrower—2.5 mm. or less in females and young males.
Range. Breeds on Pacific coastal strip from southwestern Oregon to Santa Barbara County and the coastal islands of southern California. Winters from Sinaloa south to the Distrito Federal in México (the island race, *Selasphorus sasin sedentarius* of southern California is permanently resident).
Occurrence in Monument. Recorded once: Lost Palm Canyon, Oct. 22 *.

A single example of this species was taken from among a concentration of fall migrant hummingbirds at the beds of flowering *Zauschneria* in Lost Palm Canyon on October 22 (see p. 103). The bird could not be sexed, but judging from plumage it was an adult female with worn wing and tail feathers; no molt was in progress. It is a representative of the race *S. s. sasin.*

Belted Kingfisher. *Megaceryle alcyon*

Description. A large-headed, distinctly marked species (140—170 grams), with erectile crest and white collar. Dorsal surface and sides of head, back, and band across breast blue-gray; throat and belly white; flanks and anterior belly chestnut in female, white and gray in males; wings and tail black, edged with gray and finely spotted and barred with white, the primaries with a basal white patch.
Range. Breeds from northwestern Alaska, southern Yukon, southwestern Mackenzie, and across southern Canada to central Labrador; extends south to southern California, southern New Mexico, and the Gulf states. Winters from southeastern Alaska, western Montana, Wyoming, Nebraska, and Massachusetts south to the Cape region of Baja California and

through México and Central America to northern Colombia and Venezuela; also through the West Indies to the Guianas.
Occurrence in Monument. Occasional spring migrant. Recorded from: Twentynine Palms, Mar. 31, Apr. 9, 20, 28, May 9.

Like the herons and other water birds, kingfishers tend to stop at ponds whenever these are encountered in their migratory flights across the deserts. Probably little food for kingfishers is obtainable at such places, although amphibian larvae might prove attractive occasionally. The only place that fish are present in the area is in the stream in Little Morongo Canyon, but we do not happen to have seen kingfishers there. Lizards or even large insects might be taken in the course of desert crossings; still, a kingfisher would be expected to search particularly for food in the vicinity of water.

Frances Carter has reported single individuals about the oasis at Twentynine Palms on spring dates ranging from March 31 to May 9.

WOODPECKERS

Red-shafted Flicker. *Colaptes cafer*

Description. A large (130—160 gram) woodpecker, often ground-foraging. Upper surface gray-brown, lightly barred on back, the rump white; wings and tail black with red (sometimes salmon or yellow) shafts and under surfaces; wing coverts and inner flight feathers barred with black and light gray; throat gray, with red malar streaks in males only; breast with black crescent; posterior under parts light gray and white with black spotting.
Range. Breeds from southeastern Alaska, central British Columbia, west-central Alberta, southwestern Saskatchewan, and central North Dakota south to northern Baja California and through the Mexican mainland to Chiapas and Guatemala. Partly resident through most of breeding range; the partial migration results in spread to nonbreeding habitats in the desert.
Occurrence in Monument. Common winter visitant. Recorded from: Black Rock Spring, Oct. 16; Upper Covington Flat, Mar. 19 [1]; Smithwater Canyon, Jan. 28, Oct. 14 [1]; Quail Spring, Oct. 15, Jan. 21, 22, 23, Feb. 10, 11; Stubby Spring, Oct. 15, Dec. 4—6; Lost Horse Valley, Nov. 1; Fortynine Palms, Oct. 12; Twentynine Palms, Oct. 7, 16, Dec. 30, Mar. 27, Apr. 3; Pinyon Wells, Oct. 8, 10, 12, 13, 15 *; Split Rock Tank, Oct. 29—31; Eagle Mountain, Oct. 19, 20, 21; Cottonwood Spring, Oct. 13, 14; Pinto Wash Well, Oct. 16 [1].

Red-shafted flickers arrive as winter visitants regularly in early October and remain until early April; our earliest and latest dates of record are October 7 and April 3. They visit the piñon and yucca belts especially and also the oases lower down where there are trees. Usually they avoid the most open lower desert basins.

Trees and cliffs are used for roosting but much of the foraging is done on the ground, where the birds probe for ants especially, capturing them at nests or at runways with the aid of the sticky surface of their long extensible tongues. We have specifically noted ground-foraging on October 19 on Eagle Mountain, where a bird was thrusting its bill into the soil, and at Quail Spring on January 23.

Fig. 37. A male red-shafted flicker on the ground where it is foraging for ants.

At times as many as 10 individuals have been seen in the vicinity of oases, but more usually one or two are noted at a time. Water at the springs may be attractive to them, but we have not specifically noted drinking.

The insect food of the flickers and their presence only during the cool part of the year help the birds avoid the problems of difficult desert conditions.

The specimens taken in the Monument are of the widespread western race *Colaptes cafer collaris*, which breeds to the west in California and in the Great Basin and Rocky Mountain regions.

Gila Woodpecker. *Centurus uropygialis*

Description. A middle-sized (58–72 gram), stoutly built woodpecker. Head and breast gray, except for a dark red spot on the crown in males; posterior belly yellow; back, wings, and central and lateral areas of tail conspicuously and finely cross-barred, black and white.
Range. Resident from southeastern California, southeastern Nevada, southern Arizona, and southwestern New Mexico to southern Baja California, Jalisco, and Aguascalientes.
Occurrence in Monument. Vagrant. Recorded en route to Key's View, July 3.

This species, which occurs regularly in the nearby Imperial and Coachella valleys, may be expected occasionally as a vagrant to the Monument, although little or no tree growth suitable for its permanent residence occurs there. Several indirect reports of its presence in the Monument have come to us. Our parties noted only one specific occurrence, on July 3, without recording adequate detail; the area in which it was seen was that of Lost Horse Valley along the road to Key's View.

Acorn Woodpecker. *Balanosphyra formicivora*

Description. A conspicuously marked woodpecker (60–82 grams), black dorsally with white patches in wings and on rump. Forehead white; crown red in male, black with red posteriorly in female; chin black; throat white or pale yellow; chest black, streaked posteriorly and on flanks; belly white; wings and tail black except for the white patches on primaries. Iris white.
Range. Resident from western Oregon, California west of the Sierras, Arizona, New Mexico, and west-central Texas south to southern Baja California and through the highlands of México and Central America to western Panamá; also Colombia.
Occurrence in Monument. Occasional vagrant. Recorded from: Black Rock Spring, Aug. 30 *; Eagle Mountain, Oct. 19 *; Cottonwood Spring, May 13.

This species of woodpecker is resident in oak or oak and pine woods, where acorn supplies essential to it are available. Only rarely does it wander from such areas into the desert mountains. The first known occurrence in the desert was that recorded for Eagle Mountain (A. H. Miller, 1947). The bird was in an open piñon in an area at 4900 feet where there is a sparse dwarf woodland of piñons and scrub oaks. It had acorn mast in its stomach. The area on Eagle Mountain is inadequate to support this species in numbers or for any great period of time, but it is worthy of note that in crossing the desert this bird had found this small isolated area of scrub oaks and had utilized the acorns from it. Ordinarily the acorn woodpecker does not frequent dwarf oaks of this kind. The vagrant at Black Rock Spring lit in a Joshua tree, but oaks and piñons were nearby. The single bird reported from Cottonwood Spring was far from any growth of oak. The species of course takes insect food, both of flying and bark-dwelling types, in addition to its prevailing diet of acorns.

The vagrant on Eagle Mountain was being worried by a loggerhead shrike that was hopping about in the branches 6 feet over its head in the piñon.

The two specimens taken represent vagrants from different directions. The bird found on Eagle Mountain was of the small race *Balanosphyra formicivora formicivora*, resident in Arizona; the nearest population is resident in the Hualpai Mountains 90 miles distant across the Colorado River. The specimen, a male, appeared to be an immature and its plumage already showed wear; it weighed only 55.7 grams. The bird taken at the west end of the Monument at Black Rock Spring was of the race *B. f. bairdi* that is resident in the nearby San Bernardino Mountains; it was a male in fresh plumage; it weighed 69.0 grams and possibly was an adult.

Lewis Woodpecker. *Asyndesmus lewis*

Description. A large (85—110 gram) woodpecker, generally black-appearing at a distance. Dorsal surface black, with greenish and purplish iridescence, broken by a gray collar; throat and breast intermixed gray and rosy red; wings and tail black.
Range. Breeds from southern British Columbia and western Alberta south to southwestern California (Kern County), central Arizona, and southern New Mexico, extending east to northwestern Nebraska and eastern Colorado. Winters south and east to northern Baja California, northern Sonora, and western Texas.
Occurrence in Monument. Winter visitant in small numbers. Recorded from: Smithwater Canyon, May 19; Quail Spring, Mar. 19, Apr. 21, 22, Oct. 21 *; Lost Horse Valley, Dec. 4—6; Pinyon Wells, Oct. 13.

Lewis woodpeckers because of their habit of perching on exposed tree tops and their crowlike flight with slow wingbeats are likely to be noted even when only a few are present. The species winters in small numbers in the Monument, our earliest and latest dates being October 13 and May 19. On the earliest date at Pinyon Wells three were moving together eastwardly, possibly in migration. Typically single birds or two or three at a time are noted and not flocks.

Quail Spring has been the point of most frequent occurrence. Near here in Smithwater Canyon a bird was seen on May 19 on the top of a large cottonwood near the small stream. It may be that these woodpeckers are attracted to the water sources. For food they depend much on flying insects and also on insects and acorns taken from the trees.

In some areas this species varies noticeably in winter occurrence from year to year. The occurrence in the Monument has not, however, been restricted to occasional favorable years. Our records involve 1928, 1942, 1945, 1946, and 1947, the latter three years being those when most of the winter field work was undertaken.

Yellow-bellied Sapsucker. *Sphyrapicus varius*

Description. A middle-sized (42–52 gram), fairly slender woodpecker. In race *nuchalis,* crown, nape line, and throat red, the sexes alike; black line divides nape and crown areas; sides of head black with two white lines, one above and behind the eye, the other extending from base of bill below eye and ear. In race *daggetti,* head, throat, and breast red, the black and white pattern limited to area in front of eye. Back black with two somewhat broken white lateral stripes; wing coverts with conspicuous white patch; rump white; tail barred centrally and white-bordered laterally, otherwise black. Flanks dusky, imperfectly barred; belly dull yellow or yellowish white.
Range. Breeds in the boreal forests from southeastern Alaska southeast to southern Labrador and south to the mountains of southern California, central Arizona, and northern New Mexico; in the Mississippi Valley to Missouri, and in the Alleghenies to Tennessee and North Carolina. Winters south to northern Baja California, western Panamá, the Gulf coast, and the Greater Antilles.
Occurrence in Monument. Winter visitant in moderate numbers. Recorded from: Upper Covington Flat, Nov. 4 [1], 5 [1]; Lower Covington Flat, Mar. 19 [1], Nov. 6 [1]; Smithwater Canyon, Feb. 17 [1]; Stubby Spring, Dec. 4–6; Pinyon Wells, Oct. 13; Cottonwood Spring, Sept. 13, 15 *.

This sapsucker may be expected as a winter visitant from mid-September until March in areas where there are trees in which it can drill holes for tapping the flow of sap beneath the surface. In the Monument this requirement seems to restrict the bird largely to the piñon belt, as Joshua trees are not structurally suitable for such a feeding procedure. In the cottonwoods, willows, and mesquites of the oases, this woodpecker also finds adequate food trees.

On September 15 a female was taken at Cottonwood Spring in the east side canyon. The condition of its ovary indicated that it had bred in the preceding spring, and it had probably just arrived from its breeding grounds. The bird had started a fresh series of drillings in a tamarisk tree that had been planted here near the willows and cottonwoods. The places where it had drilled and peeled off bark were entirely dry and had thus not as yet produced sap for the bird. Extensive sap workings have been seen in the piñons.

Since this woodpecker is largely a sap-feeder in winter, it encounters no moisture problem in the desert so long as adequate trees for drilling are available. The species is not present during the severe warm periods, and the cold desert nights present no problem to it, for it is adjusted to a boreal breeding habitat.

Three of the specimens taken are of the race *Sphyrapicus varius nuchalis,* a well-marked form known as the red-naped sapsucker, which breeds in the Rocky Mountains and Great Basin conifer and aspen belts. The nearest breeding stations are in the White Mountains of east-central California. Three other specimens are of the race *S. v. daggetti* that is summer resident in the Sierra Nevada and high mountains of southern California.

Williamson Sapsucker. *Sphyrapicus thyroideus*

Description. A middle-sized (48–58 gram) woodpecker in which the sexes are very differently colored. Male: black above and on breast with conspicuous white patch on wing coverts; narrow white lines above and below ear coverts; chin red, belly yellow; flanks barred black and white. Female: barred gray and black on back, breast, and flanks; head and throat gray; belly yellow.
Range. Breeds in boreal forests from southern British Columbia south in mountains to southern California, central Arizona, and northern New Mexico. In winter south to northern Baja California, Jalisco, and western Texas.
Occurrence in Monument. Winter visitant in small numbers. Recorded from: Upper Covington Flat, Oct. 8 [1], Dec. 10 [1].

Parties from Long Beach State College obtained specimens of this species on Upper Covington Flat and reported (Rainey MS) seeing the species in this area on two other occasions. The specimens belong to the race *Sphyrapicus thyroideus nataliae,* which breeds in the eastern Great Basin and Rocky Mountain regions.

Ladder-backed Woodpecker. *Dendrocopos scalaris*

Description. A small (32–40 gram) woodpecker, conspicuously cross-barred with black and white on back, wings, and sides of tail. Crown flecked with red and nape red in males, the hind neck black; crown, nape, and neck black in females; young of both sexes with sparse red flecks on crown only; sides of head black with white stripes above and below eye; under parts gray, streaked with black on sides of breast and spotted and barred on flanks; center of tail black; lores dusky.
Range. Resident from southeastern California, southern Nevada, southwestern Utah, and southeastern Colorado south through México including Baja California, the Tres Marías Islands, and Cozumel Island, to Chiapas and British Honduras.
Occurrence in Monument. Permanent resident; widespread but sparse population. Recorded from: Little Morongo Canyon, Apr. 1 *, 2; Morongo Pass, Apr. 28; Black Rock Spring, Apr. 3, 4, Aug. 30, 31, Sept. 1, 3, 4 *; 3 mi. S Warren's Well, Aug. 31; Upper Covington Flat, Apr. 9, Aug. 23, 28 [1]; Lower Covington Flat, Apr. 7, 24 [1], June 26 [1], Aug. 23, 26, 27, 28, Nov. 26 [1]; Smithwater Canyon, Apr. 8; Quail Spring, Jan. 22 *, 23 *, 25, Feb. 10, 11, 13 *, Apr. 10, 11, May 19–21, July 4, Sept. 8 *, 9 *, 10 *; Hidden Valley, Feb. 18; Stubby Spring, Apr. 9, 10, Sept. 6 *, 7, 10; Lost Horse Valley, Feb. 10, Nov. 1 *; Key's View, July 4, 15; Indian Cove, May 13–14; Twentynine Palms, Jan. 19; Pinyon Wells, Oct. 12, 17 *; Split Rock Tank, Oct. 29–31; Live Oak Tank, Apr. 6–8; Eagle Mountain, Oct. 19; Cottonwood Spring, Feb. 7–14, 8 *, Apr. 1, 5 *, 25, 26, 29, 30, May 13, 14, 15 *, June 3, 22, July 7, Sept. 13, 15 *, 17 [1], 23 *; Lost Palm Canyon, Oct. 22.

This small species is a truly desert woodpecker, more so than any other member of its family in North America, tolerating and seeking open desert vegetation on dry exposed slopes and flats. Although it will readily live in cottonwoods and willows, and especially in more southern parts of its range in México may occupy less severely arid areas, it always seems to do well in dry scrub country. In the Joshua Tree Monument it occurs wherever there is woody vegetation of more than low bush height. Only open tracts of creosote bush and sand dunes are avoided. Small lines of catclaw, ocotillos, small yuccas, and palo verdes are visited for foraging, for the species will readily

feed on slender trunks and stems and down to ground level. However, ladderback woodpeckers are seen more regularly in the Joshua tree forests, in the piñon and juniper belt, and in the trees at oases.

Except when individual birds have become accustomed to the presence of people at a continually occupied camp site, they are unaccountably wary, often leaving a feeding location in a Joshua tree 100 yards ahead of an advancing person whom it obviously sees and fears. The single *check* note, most often used by the species, carries far over the open desert and serves to keep members of a pair in touch throughout the year. This note is extended at times into a staccato, raspy chatter or sputter, the counterpart of the piping call sequence of the downy woodpecker. Often as a bird takes wing, in its bounding flight, the single call is given.

The birds feed, so far as we have been able to see, on woody surfaces, both hard and soft. The birds do not regularly feed among foliage, as the related Nuttall woodpecker does, but doubtless they glean food from small twigs and yucca spines, as well as taking it beneath bark surfaces by drilling. Specifically in the Monument we have seen them pecking in catclaw bushes, on willow and cottonwood trunks, on cholla cactus "wood" lying on the ground, on erect ocotillo stalks, on slender flowering stalks of yuccas as well as on dead limbs, in low oak shrubs (*Quercus dumosa*), and on a fallen piñon snag. We suspect, but have no proof, that near the ground, and in fallen limbs, they take termites for food.

Although a pair of these woodpeckers is likely to be present about each oasis, we have not seen them come to water and believe that they do not regularly seek it or require it. Moist subsurface insect food should provide fluids for them, and probably also moist flowers, stems, and seed pods of yuccas and fruits of cactus can be utilized, as this has been noted to a limited degree in other parts of the species' range.

In April nesting is underway. A female which had not yet laid was drumming on April 5, an action to be expected even more of the male in the pre-laying period. In a later year, at this same station, Cottonwood Spring, a ladder-backed woodpecker entered a nest hole 50 feet up in a large cottonwood on April 29. On still another occasion here on May 15 a male had a brood patch and the testes were below maximum, at 6 mm. length, thus indicating that regression from full breeding state had started. In Little Morongo Canyon on April 1 a pair was seen frequently in the willows near the small stream. The birds had a nest hole one foot from the broken top of the 8-inch trunk, 30 feet above ground. Occasional drumming was heard here. The female was taken on this date and had a yellow ovum 3 mm. in diameter, no empty follicles, and an enlarged oviduct but no brood patch. The male likewise had not yet developed a brood patch and had a 9 mm. testis. Thus this pair would have had its first egg laid in this nest within the next four or five days. On April 8 in Smithwater Canyon a pair had a nest

Fig. 38. A pair of ladder-backed woodpeckers at a nest hole in a Joshua tree; male outside the entrance.

in a partly dead cottonwood near the almost dry water seep. The hole was 25 feet up in a dead snag 15 inches in diameter. The female first flew quietly to this hole, thrusting her head in tentatively, then working up to the top of the tree, where she drummed ten times and also gave three series of piping notes. The male gave only the single *chup* note, and it was lower-pitched than that of the female. Although the female later entered the hole, the incubation of a set may not yet have started. Fresh chips of wood from excavation could be found on the ground below.

This species is never seen in concentrations, but only in pairs or family groups, and often only solitary individuals are noted, for, if paired, they do not readily reveal this fact, as the two members may be in vocal communication only at considerable distance. A given pair may be followed a mile or more through the Joshua trees, and they evidently have large territories. Thus the population is sparse. About any one camp, one or rarely two pairs may be found. In Little Morongo Canyon, for example, members of the pair near camp ranged at least 600 yards upstream, and the next pair was a mile distant up canyon from camp.

Ladder-backed woodpeckers in the first ten days of September are in the late stages of the annual molt and are then replacing the outermost two or three primaries.

The ladder-backed woodpeckers of the Mohave and Colorado deserts are regarded as belonging to the race *Dendrocopos scalaris cactophilus*, for the proposed subdivision of this population (van Rossem, 1942) into a Colorado Valley race, *yumanensis*, and a western Mohave Desert race, *mojavensis*, both distinct from *cactophilus* typical of south-central Arizona, has not been supported by other investigators (Grinnell and Miller, 1944; A.O.U. Check-list, 1957). Van Rossem claimed that the birds of the Colorado River valley are smaller and paler than those of the Mohave Desert and paler than typical *cactophilus*. We have sought to obtain as many specimens as possible from the Monument in order better to examine this question. Again we find the purported color difference to be unreal or to consist of only slight trends which would not permit a separation of more than half the individuals from the respective areas. This is particularly true of ventral coloration when fresh plumages are considered. There is a tendency for more individuals with extremely broad white dorsal bars to occur along the Colorado River valley than elsewhere, but such types are not confined to this population and many normally barred individuals occur there too. In size, as best shown in wing length, there is increase to the west of the Colorado River valley. Measurements we have taken in the series at the Museum of Vertebrate Zoology generally conform to van Rossem's tabulation, but it should be noted that his figures overlap too much to permit racial separation at the conventional 75 per cent separation level for subspecies. Nevertheless, the trend of differentiation is real and is interesting to note. Ten males from the Monument average

107.1 for wing length and range from 104 to 109 mm. Comparison with van Rossem's table shows that this conforms to his figures for populations of the western desert border. There is no tendency whatsoever to approach the values for the small-sized population of the Colorado River valley.

FLYCATCHERS

Key to the Flycatchers of the Monument

A. Size large, 15 to 25 cm. long, the size of a house finch or larger.
  B. Tail relatively short, not extending far (less than 3 cm.) beyond wing tips; under surface with light longitudinal area bordered by dark flanks .... *Olive-sided flycatcher*
  B′. Tail long, extending far beyond wing tips; under parts without contrasting longitudinal areas.
    C. Head, breast, and flanks black.......... *Black phoebe*
    C′. Head largely gray.
      D. Tail rusty brown .... *Ash-throated flycatcher*
      D′. Tail black or black and white.
        E. Under parts brown posteriorly.......... *Say phoebe*
        E′. Under parts gray and yellow.
          F. Tail black .... *Cassin kingbird*
          F′. Tail black with white lateral border.......... *Western kingbird*

A′. Size small, less than 15 cm., smaller or less bulky than house finch.
  B Red or pink, at least on flanks and belly.......... *Vermilion flycatcher*
  B′. No red in plumage.
    C. Head and flanks blackish, the white of belly and lower belly a V-shaped patch bordered by dark flanks; no eye ring.... *Western wood pewee*
    C′. Head and flanks gray or greenish gray; light eye ring present.
      D. Under parts uniformly yellow-green.......... *Western flycatcher*
      † D′. Under parts not yellow-green, or if greenish not uniformly so. . *Traill flycatcher, Hammond flycatcher, dusky flycatcher, gray flycatcher.*

† These species of flycatchers of the *Empidonax* group can seldom be identified by sight in the field with any assurance of accuracy. Records must depend on specimens collected, and these in turn must be thoroughly analyzed for a number of characters. For the principal features used in such analysis, see species accounts and full treatment of the subject by Johnson (1963).

SUMMER RESIDENT FLYCATCHERS

Western Kingbird. *Tyrannus verticalis*

Description. A large (35–45 gram) flycatcher with gray head and back and black wings and tail, the latter distinctly edged on the sides with white. Belly yellow; crown of adults with vermilion patch which is usually concealed by gray feather tips.

Range. Breeds from southern British Columbia east to southern Ontario and south to northern Baja California, Sonora, northern Chihuahua, central Texas, and Missouri. Winters chiefly from southwestern México (Guerrero) to northern Nicaragua.

Occurrence in Monument. Transient and also summer resident from late March to early September. Scarce and local. Recorded from: Black Rock Spring, Sept. 4 *; Upper Covington Flat, Apr. 7 [1], 30 [1], May 29 [1], June 16 [1], July 23 [1]; Lower Covington Flat, May 14 [1], 28 [1], June 27 [1], Aug. 28 *; Quail Spring, Apr. 10 *, July 1; Queen Valley, June 12 [1]; Squaw Tank, June 16 [1]; Twentynine Palms, Mar. 23, 30, Apr. 3, 4, 5, 18, 20, May 6, June 2, July 13; Virginia Dale Mine, July 14; Cottonwood Spring, May 14.

Earliest and latest records are March 23 and September 4. The number of pairs that settle to nest in the Monument is variable from year to year. The Joshua tree forests afford suitable places for nesting as do also the vicinities of oases. Miss Carter's records and ours at Twentynine Palms seem to suggest summer residence there, and a pair with a nest was found at Quail Spring on July 1. Rainey (MS) saw an active nest in June in the Covington Flat area.

Kingbirds are seen most frequently during the migratory periods in late March and early April and in late August and early September.

Two specimens taken on August 28 and September 4 were juveniles, probably in migration, the bird of the latter date being conspicuously fat. Neither of these juveniles had proceeded significantly with the postjuvenal molt, as only occasional body feathers were showing replacement. Evidently the species accomplishes most of its molting after departure from the breeding range.

The nesting pair at Quail Springs on July 1 apparently had young, as the adults were frequently visiting the nest. As each parent approached the nest, the other would flutter up and give a flight song or twitter. The nest was 15 feet up on top of a clump of spines in a Joshua tree, where it could not be inspected.

Large insects flying in the open are the food resource for this kingbird and also for the Cassin kingbird. Water is not necessary, although its presence, as at oases, probably results in a better food supply for these species than does open and unvaried desert terrain.

Cassin Kingbird. *Tyrannus vociferans*

Description. A large (40–52 gram) flycatcher, in appearance very similar to the western kingbird (see p. 113), but tail without definite white edging and head, back, and breast darker gray, the throat contrastingly white. Notes lower pitched, nasal, and more powerful, not twittering.

Range. Breeds from central California, southern Utah, southeastern Montana, and western South Dakota south to Guerrero. Winters from central California south to Guatemala.

Occurrence in Monument. Apparently a scarce spring migrant and also summer resident locally. Reported from: Upper Covington Flat, May 19[1], June 2[1], 16[1]; Lower Covington Flat, May 29[1]; Twentynine Palms, Mar. 23, Apr. 2, 6, 11, 12, May 5, May 21–June 2.

C. A. Harwell listed early spring observations in the course of visits to the general area of Twentynine Palms, and Miss Carter records the species from that locality, mentioning nesting observed from May 21 to June 2. Twentynine Palms does not afford the habitat normal for the nesting of this species, and yet such oases may attract it. Parties from Long Beach State College found Cassin kingbirds in the Covington Flat area in 1960 and 1962. One obtained on June 16 had an ovum 3 mm. in diameter and evidently would soon have laid; males taken at the same time had testes 14 and 15 mm. long.

Fig. 39. Western kingbird (left) and Cassin kingbird (right).

Cassin kingbirds of the United States and northern México are of the nominate race *Tyrannus vociferans vociferans.*

Ash-throated Flycatcher. *Myiarchus cinerascens*

Description. A fairly large (27–34 gram) flycatcher with long tail marked with cinnamon on the inner webs of all but the central pair of feathers. Back grayish brown; chin, throat, and breast light gray, the belly pale yellow; crown feathers somewhat elongated and at times elevated into an irregular, short crest (fig. 40).

Range. Breeds from southwestern Oregon and eastern Washington east to northern Texas and south to Michoacán. Winters from southeastern California and central Arizona south to El Salvador.

Occurrence in Monument. Widespread and common summer resident. Recorded from: Morongo Pass, Apr. 28; Little Morongo Canyon, Apr. 28; Black Rock Spring, July 6, Aug. 29; Upper Covington Flat, Apr. 9; Lower Covington Flat, Apr. 9[1], 15[1], 29[1], 30[1], May 13[1], 15[1], July 16[1], Aug. 15 *, 22, 25; Smithwater Canyon, Apr. 8; Quail Spring, Apr. 22, May 19–21, July 1, 2; 6 mi. E Quail Spring, Apr. 23; Key's View, July 15; Indian Cove, Apr. 7, May 13, 14; Twentynine Palms, May 10, 17, June 2, July 13; Live Oak Tank, Apr. 6–8; Squaw Tank, June 16[1]; Pinto Basin, Apr. 11, 27; Virginia Dale Mine, Apr. 25; Cottonwood Spring, Apr. 26, May 1, 13, 14, June 3, July 7, 8, 11[1].

Ash-throated flycatchers are resident in the Monument from early April through August. They are especially noticeable in April, when spring migrants apparently augment the population that is just settling for breeding. While in summer residence, ash-throated flycatchers favor the piñon belt and the Joshua tree "forests" but will settle also along sparse lines of smoke trees and ironwoods on open alluvial fans and about all oases and canyons. The birds do not flock, and pairs are isolated and scattered so that they are not often in sight or sound of one another.

The species requires cavities for nesting, and this is probably a reason for its limitation to vegetation of sufficient size to provide holes in their woody trunks. Also, this flycatcher favors working through the open foliage of trees and bushes to capture flying insects at fairly low level.

The one nest on which we have detail is that reported at Twentynine Palms by Miss Carter, who states that two miles south of the oasis on June 2 a nest was built within a gatepost pipe, 14 inches below the top. The pipe measured 6½ feet in height and 3½ inches inside diameter. The nest on that date contained one well-grown young and three unhatched eggs. She implies that these eggs and those of a second clutch reported to her were killed by the summer heat. This does seem possible in such an unnatural and hot location as that afforded by the interior of an exposed metal pipe.

Aside from such a hazardous situation as that just described, ash-throated flycatchers seem quite able to withstand high summer temperature. They are rather inactive at midday. We have no evidence that they supplement the moisture of their insect food by coming to water, which anyway would not be available to most nesting pairs in the Monument.

On August 15 a full-grown, independent juvenile was collected that had started the postjuvenal body molt, but only about a third of the new plumage was in evidence; there was no wing or tail molt in progress. Such an individual must certainly migrate at least to lower levels, where the species winters, before completing the postjuvenal molt, as our records for the species in the Monument stop with August 29. Doubtless some stragglers could be found after this date, but their rarity indicates that the main exodus has taken place before September.

The ash-throated flycatchers of the Monument area belong to the race *Myiarchus cinerascens cinerascens.*

Black Phoebe. *Sayornis nigricans*

Description. A moderate-sized (16—21 gram) flycatcher, dusky and black except for white belly and under tail coverts; edgings of wing coverts and secondaries gray.
Range. Resident from northern California east to central Texas and south to northern Argentina.
Occurrence in Monument. Vagrant; also occasionally nests in a few favorable localities. Recorded from: Little Morongo Canyon, Mar. 30, 31 *, Apr. 1, 2; Upper Covington Flat, Mar. 19[1]; Quail Spring, July 3, 4; Twentynine Palms, Feb. 23—28, Apr. 4, 12, July 13.

Fig. 40. Ash-throated flycatcher (above), olive-sided flycatcher (right), and Say phoebe (below).

This phoebe, which is dependent on mud for nest building and favors moist areas over which to search for flying insects, finds few suitable localities for even temporary residence in the Monument. Along the small stream in Little Morongo Canyon, three were seen in the course of a mile and a female was taken on March 31 in which the ova were definitely beginning to enlarge. There is every reason to suppose that this bird was starting to nest in this suitable location. Occurrences at Quail Spring and at Twentynine Palms in July may have represented vagrants, although in the latter area nesting might be possible. Miss Carter observed one or two black phoebes here in late winter and early spring in two successive years, but in each year they disappeared in early April.

The specimens are of the northern race of the species, *Sayornis nigricans semiatra.*

Say Phoebe. *Sayornis saya*

Description. A medium-sized (18–24 gram) flycatcher, light brown above with contrasting black tail. Throat and breast ashy, the belly, flanks, and under tail coverts tawny.

Range. Breeds from central Alaska and Yukon southeast to southwestern Manitoba and south to Michoacán and Hidalgo. Winters from northern California, northern Arizona, and central New Mexico south to Puebla and Veracruz.

Occurrence in Monument. Resident. Population sparse but widely distributed. Recorded from: Little Morongo Canyon, Mar. 30–Apr. 3; Upper Covington Flat, Oct. 8 [1], 10 [1]; Lower Covington Flat, June 26 [1], July 9; Quail Spring, Jan. 21 *, Apr. 10, July 2; 6 mi. E Quail Spring, Apr. 23 *; Indian Cove, Apr. 7, 8 *, May 13, 14; Twentynine Palms, May 14; 3 mi. S Twentynine Palms, May 15 [1]; Squaw Tank, June 16 [1]; Pinyon Wells, Apr. 10; Live Oak Tank, Apr. 6–8; Pinto Basin, Feb. 18, Apr. 5, 6; Virginia Dale Mine, Apr. 24; 1 mi. N Sunrise Well, Sept. 28 [1]; Cottonwood Spring, Feb. 7–11, May 13; Lost Palm Canyon, Oct. 22.

Say phoebes are resident at all times, but there may be local movements that take them away from their roosting and nesting headquarters at cliffs or buildings; and some migrants from the north doubtless reach or pass through the Monument. On April 6 we saw a phoebe far out in the open part of Pinto Basin where no nesting sites exist; such a bird may well have been a spring migrant, as local resident individual should have been near their nests at this time.

Say phoebes forage in very open terrain, "floating" lightly on the wing from one exposed perch to another. Insects seem available for them both over open creosote flats and in denser desert woodland and along cliff faces.

Nests are placed on solid supporting ledges of cliffs or banks and more recently on buildings under eaves. The nest site is usually thus protected from wind and from sun during part or most of the day. On April 8 at Indian Cove near the canyon mouth a male was singing (testis 7 mm.), and a second individual was flushed from a shaded cavern in the boulders. The bird stayed close by and evidently had a nest concealed somewhere on a ledge or in a crevice in the rocks. Miss Carter reports nests under eaves of buildings at Twentynine Palms. On May 14 an accipitrine hawk took three newly hatched Say phoebes from a nest under the eaves of a porch at this locality.

Say phoebes seem well adapted to desert winds and open places by their lightness on the wing. Flight appears effortless, and the birds drift with the wind rather than beating directly against it. Although they can stand bright sun and heat, they use cliffs for retreats in midday, and, as already stated, these serve essentially, too, for nest protection and roosting. The nest-building instincts of this phoebe do not require mud, and the species is thus not tied to the vicinity of water.

For some years the racial status of the Say phoebes breeding in the Colorado Desert area of California has been in doubt (Grinnell and Miller, 1944: 256), since wintering examples of the race *Sayornis saya quiescens,* typical of Baja California, have been taken there. Shortage of birds in relatively unworn plumage representing the breeding population has hindered analysis. We do not yet have enough material from the Monument to enable us to reach a firm conclusion. The bird of April 23 is rather clearly not *quiescens,* but it was not known to have been nesting. The breeding specimen of April 8 makes some approach to early spring examples of *quiescens,* but it is less ashy than most and fits better with breeding *Sayornis saya saya* as represented in the Mohave Desert area to the north.

MIGRANT FLYCATCHERS

The flycatchers that are nonresident and nonbreeding consist of eight species, seven of which are normal migrants moving between breeding grounds to the north and west and wintering grounds south and east. Three of these migrants perform exceptionally long migrations, traveling into South America or at least in the main beyond México into Central America. These are the Traill flycatcher, western wood pewee, and olive-sided flycatcher. The movements of the eighth species, the vermilion flycatcher, are much more local, and the occurrences of it in the Monument represent northern pioneers or birds moving short distances east and west in southern California; we probably should class this species as a local vagrant.

The migrant flycatchers divide into two groups, the long-winged, vigorous fliers, the western wood pewee and the olive-sided flycatcher, and the somewhat weaker, shorter-winged fliers, the species of the genus *Empidonax.* The latter are more inclined to forage en route through the desert than is the long-winged group, and in them definite migratory flights of individuals are less often seen.

Only among the *Empidonax* flycatchers—and notably the Traill flycatcher faced with a long migration—were individuals encountered that showed signs of succumbing to the ordeal of desert crossing in the period of severe autumnal heat and drought. As in the migrant warblers, food and essential water must be derived from insects. There is only uncertain recourse to free water at springs. Severe heat and sun exposure seem quickly to dry the soft bill, mouth borders, and feet of the small flycatchers, and considerable foraging, as well as recourse to shade during maximum daily heat periods, may be required to offset this The insect supply to be drawn upon by flycatching methods appears generally to be good in the desert at times of migration, although it is not so good as in more humid regions, and perhaps in extremely dry years when vegetative growth has been poor in spring it could be severely low. Unlike swallows, the small flycatchers cannot quickly move about to points of local insect concentration.

The desert crossing by flycatchers is of a length and direction equivalent to that of most species of warbler (see p. 202). Shade-seeking or riparian species like the western, Hammond, and Traill flycatchers must, as do some warblers, make considerable adjustments and preparation for the desert trip. The deposition of a fat reserve, with its provisions for energy and moisture, is an important undertaking. Usually migrant flycatchers are heavily fat and this reduces the need for foraging during migration, although in some species it doubtless does not come anywhere near eliminating that need.

In the responses to desert conditions, the several migrant flycatchers do reflect some elements of their instinctive use of cover which they display on the breeding grounds. Thus, Traill flycatchers stay low down and noticeably seek canyon or wash bottoms even though these are dry and but vaguely suggest the riparian conditions preferred by the species. Contrarily, western wood pewees and olive-sided flycatchers move in the open, farther above ground as is their general custom.

The schedule of migratory passage shows some distinct contrasts among species (see table 5). Some of the spans of occurrence are undoubtedly incomplete (see details in species accounts), but it is noteworthy that the three species engaging in long migrations, the Traill and olive-sided flycatchers and the western wood pewee, do not extend their season in the Monument into October. They quickly move on and in the spring reappear late. On the other hand, the Hammond flycatcher in our experience did not appear in

TABLE 5

SPAN OF MIGRATORY SEASONS OF SEVEN SPECIES OF FLYCATCHERS IN JOSHUA TREE NATIONAL MONUMENT

Species of Flycatcher	*Spring*	*Fall*
Traill	——	Aug. 23–Sept. 14
Hammond	Mar. 31–May 15	Oct. 13–Oct. 23
Dusky	(Apr. 29)	Sept. 6–Sept. 10
Gray	Apr. 23–May 13	——
Western	Apr. 2–June 3	Aug. 25–Sept. 15
Western wood pewee	May 5–May 31	Aug. 23–Sept. 15
Olive-sided	May 17–May 24	Aug. 29–Sept. 14

fall until October, and it has a prolonged occurrence in the spring; the western flycatcher shows a prolonged spring migration also.

Correlated with the migratory schedule and with the total length of the migratory movement is the physiologic adjustment made in the molts that normally follow the breeding season. Certain species that either travel especially far or begin the fall movement early forego an immediate postbreeding molt and carry this out on the wintering grounds, whereas some other species undergo postjuvenal molt or postnuptial molt before fall migration. The details appear to be different in different species of *Empidonax* flycatchers (see species accounts), but the important point is that physiologic adaptations rooted in endocrine and nutritional mechanisms have developed in fairly good conformity with the migratory schedule in the several species and that, unlike birds such as warblers, the *Empidonax* group shows an important differentiation among very similar appearing species—species that on morphologic grounds merit the designation of sibling species.

Flycatchers are perhaps second only to the warblers as important elements in the migratory land bird fauna that traverses the Monument. In general they are nearly as numerous as the warblers, although, owing to the inconspicuousness of the *Empidonax* group, they are not as readily noticed, and there are of course fewer species. At no time, as is true occasionally of warblers, are there waves or concentrations of more than a half dozen individuals to be counted. Ordinarily only one to ten of a given kind can be detected in a day's hunting, even at times of peak abundance.

Traill Flycatcher. *Empidonax traillii*

Description. The closely similar species of *Empidonax* flycatchers of the Pacific coast of North America show the following characters in common: small size (9–12 grams) and relatively long tail, which is periodically flicked; two light-colored wing bars on wing coverts; light eye ring; feathers of crown somewhat long, occasionally elevated to form blunt tuft; color generally gray to pale greenish. The species *traillii* is safely distinguishable by color and structure only when in hand: bill broad (6 mm. at nostril); throat white, fairly well contrasted with gray breast band and pale yellowish posterior under parts; back more or less brownish gray or olive; tail tip slightly rounded (fig. 41).

Range. Breeds from central Alaska across Canada to southeastern Quebec and south to southern California, southwestern Arizona, northwestern Arkansas, and West Virginia. Winters in Central and South America chiefly from Guatemala to northern Perú, Bolivia, and northern Argentina.

Occurrence in Monument. Common fall migrant in late August and early September. Not detected by us in spring, although probably occurs at least occasionally then. Recorded from: Black Rock Spring, Aug. 30 *, 31 *, Sept. 3 *, 4; Lower Covington Flat, Aug. 23 *, 24 *, 25, 26, 27 *, 28 [1], 29 *; Quail Spring, Sept. 8, 9 *; Stubby Spring, Sept. 6 *, 7 *, 9, 11 *; Cottonwood Spring, Sept. 13; 4½ mi. W Shaver Summit, 1500 feet, Sept. 14 *.

In the fall migration Traill flycatchers are seen commonly; they usually are engaged in foraging and are not traveling conspicuously. Indeed, at times they spend many hours, perhaps several days, about vegetation bordering washes or springs. They favor low situations about large chrysothamnus bushes and the bare twigs in the lower or middle levels of piñons, junipers, and the occasional willows. Quite in contrast was the action of one bird taken as it flew high over the tops of piñons in a direct flight southeast across Lower Covington Flat in the early morning of August 23. This obviously was a migratory movement.

The conspicuous tarrying for foraging suggests that daily food intake is required, or that many individuals are still building up fat reserves for the long travels yet to be accomplished to the distant wintering ground. The trip from California to Central America probably cannot be made without some refueling and, when suitable way stations are encountered, it is to be expected that they will be used, even those of marginal suitability in the desert.

The canyon bottoms in the higher levels of the Monument afford the nearest parallel in this area to the stream course or lake border habitat of the breeding grounds of Traill flycatchers, even though they are greatly different in moisture conditions. But in the desert washes, flying insects near the ground and partly within the vegetation are readily seen by man, and the capturing of them through the instinctive actions of this *Empidonax* can be carried out from the low perches of the border vegetation.

Traill flycatchers may seek water when it is available in order to supplement the moisture in their food, although the only clearly observed instance of going to water was concerned primarily, if not solely, with bathing. On September 9 at Stubby Spring, several *Empidonax* flycatchers were working about the green chrysothamnus bushes and the lone willow tree. Most of those taken or seen here were Traill flycatchers, although western flycatchers also were present. What were apparently Traill flycatchers were seen bathing from the wing at the surface of the deep water of the cement tank. The surface was circular with a diameter of about 10 feet. In each instance the bird flew from a perch on bushes overhanging one side of the tank to the center of the open water, landed with a splash and took off immediately. On returning to the perch, it ruffled and shook its plumage in fashion typical of bathing, and industriously preened itself. Two individuals did this, and each returned for a second wetting. There were several places about the tank and at seeps in the vicinity where the birds could bathe in shallow water. The remains of one unidentifiable *Empidonax* was found floating in the tank on this day, but the cause of death could not be ascertained.

On August 23 in Smithwater Canyon below Lower Covington Flat, at a small, shallow pool, a Traill flycatcher was noticed fluttering. Closer approach showed it was held in the mouth of a large, speckled rattlesnake (*Crotalus mitchelli*) that had captured it and was lying 1½ feet from the water. The bird fluttered repeatedly, and, when the snake was shot, the bird remained in the snake's mouth, but soon freed itself and fluttered off, but was in turn collected. Its breast was clotted with dried whitish material, and the wings were smeared with saliva. We judge that the bird had been in the rattlesnake's mouth for some time but that it must not have received much if any venom in the body. The flycatcher was an adult male without fat and had testes 3 mm. long. Its weight was only 9.5 grams, somewhat below normal, and it may have been particularly eager to get to the water, where the snake was lying in waiting.

In the favorable foraging places as many as six individuals might be seen within 50 yards of a spring, and along the wash in Lower Covington Flat we found one on the average of every 50 yards in a half-mile stretch. Such numbers were noted from August 23 to September 4.

Our failure certainly to record the species in spring must reflect scarcity, at least, at that season. Possibly the spring migration in this section of the state largely avoids the relatively high desert areas comprising the Monument. Also most of the movement may occur in late May and early June, when we have done little or no collecting.

Some Traill flycatchers seemed to be in poor condition. Whereas the normal weight of the species ranges from 10 to 12 grams, or even 13 grams when heavily fat, one taken on August 23 weighed only 7.9 grams, one on August 31, 8.4 grams, one on August 24, 9.0 grams, and one on August 27, 9.1 grams; all were adult females and showed no fat. The first two, at least, were in a precarious state. Thus, of 11 adults and 6 immatures taken, 2 and perhaps 4 were likely to fail in migration.

In this species the adults migrate in worn breeding plumage, only a few showing some feather replacement in progress on the throat. The wing and tail molt may take place as late as February on the wintering grounds in Colombia (specimens in Museum of Vertebrate Zoology), although whether the birds involved are adults or young of the year is not known. The birds of the year migrate in juvenal plumage, as demonstrated by the six samples taken. Thus the two age groups are decidedly different in appearance when in fall migration, the fresh-plumaged juveniles being much more richly yellowish beneath than the adults.

The deferment of postbreeding molt in this species has several significant aspects. It permits migration in the late summer and fall immediately after completion of nesting, which itself is usually late in July or in August in this species. Otherwise the birds would be forced either to delay departure or to combine molting and migrational activities, a circumstance which might prove a serious drain on energy resources. The possible disadvantage of the schedule is that fall migration of adults must be carried on when the plumage is in poor condition. This presents no difficulty as far as the reduced body feathering is concerned, as temperatures are high and the bird soon moves into the tropics. But the wing feathers, if badly worn, could lower flight efficiency. The fact that the flight feathers are no more than eight months old owing to the winter molt means that they are usually less deteriorated by August than those in most passerine species which have a postnuptial molt in late summer. The adults we took in any event showed no instances of extreme wear of the remiges.

All Traill flycatchers collected in the Monument are referable to the western race *Empidonax traillii brewsteri*. We do not regard the proposed Great Basin race, *E. t. adastus*, as recognizable (A. H. Miller, 1941), and the specimens from the Monument do not show any division into color types suggesting the distinctions claimed for Great Basin and coastal races.

Hammond Flycatcher. *Empidonax hammondii*

Description. See general description of flycatchers of this genus (p. 121). The Hammond flycatcher is distinguishable from its close relatives with certainty only through museum comparison of specimens: bill short (8–9 mm.), narrow (4.5 mm. at nostril), and dark; outer primaries long, the wing relatively sharp-pointed; tail short and slightly forked; back greenish or olive-gray, not brownish; under parts without distinctly outlined white throat area, yellowish generally, especially in fall; breast and throat more or less gray.
Range. Breeds from central Alaska and southern Yukon south through the coniferous forests of the coast and mountains to northwestern California and the southern Sierra Nevada, northern Utah, southwestern Colorado, and northern New Mexico. Winters from Sinaloa and Nuevo León in northern México, rarely from southern Arizona, south through México and Central America to El Salvador and northern Nicaragua.
Occurrence in Monument. Common spring and fall migrant. Recorded from: Morongo Pass, near county line, Apr. 28 *; Little Morongo Canyon, 2500 ft., Mar. 31 *; Lower Covington Flat, Apr. 8 *, 9 *; Upper Covington Flat, May 13 *; Smithwater Canyon, Apr. 23 *, 30 *, May 13 *; Pinyon Wells, Oct. 13 *, 14 *, 15 *; Cottonwood Spring, May 15 *, Oct. 23 *.

In the fall this species was taken in October. Probably migration also occurs earlier, though the species may largely avoid or quickly cross the desert areas in late summer and early fall movements, for it was not taken in our considerable collecting of *Empidonax* flycatchers in August and early September. In October they were seen chiefly in the scrub oaks and piñon cover along canyon bottoms, as at Pinyon Wells. At this time of year temperatures were moderate and humidity not so low as in early fall—circumstances that are probably generally tolerable for this flycatcher, which in the breeding season adheres to dense, shaded, and fairly humid tracts of conifers.

In the spring migration, the dates of occurrence were scattered from March 31 to May 15. Birds were detected chiefly about springs or along moist canyon bottoms at middle elevations, as in Little Morongo Canyon and at Cottonwood Spring.

When trees are available, this *Empidonax* may forage from perches at middle heights in them, showing more inclination to move up into the trees than does the Traill flycatcher.

Unlike the Traill flycatcher, the Hammond flycatcher molts immediately following the breeding season. Autumnal migrants show evidence among the adults of having undergone a complete postnuptial molt, and among the immatures of having carried out a postjuvenal molt involving at least the body plumage. The relatively less extensive migration and the lateness of movement in the fall make such a molt schedule feasible; or conversely, the molt schedule permits a late and leisurely migration. Accordingly migrants show much more wear of the flight feathers in spring than do Traill flycatchers.

No migrants were in poor condition. Most showed some fat deposit, although the amounts usually were not large. Males taken on March 31 and May 15 had testes only 2 mm. in length, and thus both birds, despite the spread in dates, were far from breeding condition.

Dusky Flycatcher. *Empidonax oberholseri*

Former name. Wright flycatcher.
Description. See general description of flycatchers of this genus (p. 121). The dusky flycatcher is distinguishable from its close relatives with certainty only through museum comparison of specimens: bill moderately short (10—11 mm.) and narrow (5 mm.); outer primaries relatively short, the wing thus rounded; tail relatively long and double-rounded; back olive-gray; under parts gray with pale yellow on abdomen.
Range. Breeds from northwestern British Columbia and central southern Yukon south through interior mountain areas to southern California (Santa Rosa Mountains), southern Nevada, central Arizona, and central northern New Mexico. Winters from southern Arizona, Coahuila, and Nuevo León south through México.
Occurrence in Monument. Sparse fall and spring migrant. Recorded from: Lower Covington Flat, Apr. 29[1]; Stubby Spring, Sept. 6 *, 10 *.

This species is an abundant breeding bird in the San Bernardino and San Jacinto mountains bordering the desert. However, the long experience of ornithologists with this flycatcher in southern California has indicated that it disappears from its breeding grounds in late summer without moving through the lowlands. The two record specimens from the Joshua Tree Monument are therefore important indications of the largely unknown fall migration. Their occurrence may suggest in part the direction and time of movement, although the rarity of the species in the Monument indicates that we have not intercepted the main flight there.

One of the fall birds was taken in the vegetation at Stubby Spring, the other in the oak and piñon belt, $3\frac{1}{2}$ miles to the east, both at elevations greater than 4500 feet. Both were young of the year judging from skull and plumage; the one taken on September 10 showed early postjuvenal body molt in progress, whereas the other had not yet started such molt.

The single example of a spring migrant is a specimen in the collection of Long Beach State College which has been identified by Ned K. Johnson.

Gray Flycatcher. *Empidonax wrightii*

Former name. *Empidonax griseus.*
Description. See general description of flycatchers of this genus (p. 121). The gray flycatcher is distinguishable from its close relatives with certainty only through museum comparison of specimens: bill relatively long (11—13 mm.) and narrow (5 mm.), light-colored base of

mandible contrasting with dark tip; wing relatively rounded; tail either double-rounded or nearly square-tipped; back gray and belly whitish, not yellow; lateral rectrix with distinct white outer border.
Range. Breeds from central Oregon and southwestern Wyoming south through Great Basin and Rocky Mountain areas to southern Nevada, central Arizona, and central western New Mexico. Winters from southern California and central Arizona to Michoacán and Puebla.
Occurrence in Monument. Spring migrant. Recorded from: Lower Covington Flat, Apr. 29 [1], 30 [1]; Smithwater Canyon, Apr. 23 [1], May 13 [1].

Among specimens of *Empidonax* flycatchers obtained by parties from Long Beach State College were six examples of this species taken in late April and early May. These have been identified by Ned K. Johnson. Evidently they represent a period of spring migration in which the species is moderately common as a transient. Our parties have not detected it, but we have collected few small birds in this particular part of the spring period.

Western Flycatcher. *Empidonax difficilis*

Description. See general description of flycatchers of this genus (p. 121). The western flycatcher is distinguishable from its close relatives in western North America as follows: general coloration greenish and yellowish; under parts greenish yellow from throat to belly; upper parts dull green; wing bars yellowish; bill broad (5.5 mm. at nostril).
Range. Breeds from southeastern Alaska south through coastal districts to mountains of Baja California, through Rocky Mountains from Montana southward, and through mountains of México to Guatemala and Honduras. Winters from Baja California, northern Sonora, and Coahuila south through México and western Central America.
Occurrence in Monument. Common fall and spring migrant for protracted periods. Recorded from: Black Rock Spring, Aug. 31 *, Sept. 1 *, 2, 4; Lower Covington Flat, May 19 [1], Aug. 25 *, 26 *; Quail Spring, May 14, June 3, Sept. 8 *; Stubby Spring, Sept. 6 *, 9 *; Twentynine Palms, Apr. 2, 3, 4, 9, 22, 27, 28, May 1, 6, 30; Pinto Basin, Apr. 11 *, 12 *; Virginia Dale Mine, Apr. 25 *; Cottonwood Spring, Apr. 26, May 14, 15 *, June 3, Sept. 13 *, 15 *.

Western flycatchers show a prolonged spring migration; probably several different breeding populations are involved, which partly accounts for this. Fall migration extends from late August to mid-September and probably later, though evidence from specimens is not available to prove suspected occurrence in mid-October. In spring more often than fall, the characteristic location or identification note of the species, the call with a distinct upward inflection, is given which aids in identification of birds not collected.

More frequently than is true of the Traill flycatcher, this species is seen in piñons and scrub oaks, foraging under and about the foliage canopy. Yet western flycatchers also appear regularly along washes and in willows and cottonwoods about springs.

A migrant taken on May 15 at Cottonwood Spring and another in Pinto Basin on April 12 were birds that still showed immature skull condition, indicating that they had hatched in the previous year. Although the first was not fat, it was not in poor condition; the testis was 4 mm. long, indicating close approach to breeding stage. On the other hand, the bird of April 12 had a testis only 2 mm. long and was fat.

Fig. 41. Traill flycatcher (left), western flycatcher (below), and western wood pewee (right).

In the fall migration, adults (3) were taken only in August. Six immatures were taken from September 1 to 15. No individuals were in distinctly poor condition, although some lacked fat.

Both the material from the Monument and that from Pacific coastal areas shows that the young of the year undergo a considerable postjuvenal body molt before fall migration, whereas the adults do not molt at all on the breeding grounds nor in the course of migration. The fall migrant adults taken in the Monument were in badly worn plumage. Thus the physiology of molt is different in this species than in either the Hammond or Traill flycatchers. The young molt as in the Hammond whereas the adults molt as in the Traill, although apparently not delaying it so late as the winter period. The extent of the migration of the Hammond and western flycatchers is similar, whereas that of the Traill is much more extensive.

The western flycatchers taken in the Monument all belong to the Pacific Coast race *Empidonax difficilis difficilis.*

Western Wood Pewee. *Contopus richardsonii*

Other name. *Contopus sordidulus.*
Description. Moderately small (11–13 gram) flycatcher, with head and flanks blackish. Throat light gray; upper breast dark gray; lower breast and belly whitish, forming V-shaped

patch; back sooty; wing coverts faintly marked by light buff bars. Wing tip long, pointed, the flight feathers black.

Range. Breeds from central eastern Alaska, southern Yukon, southern Mackenzie, and central Saskatchewan south to mountains of Baja California and through México and Central America to western Panamá. Winters in Central America and South America south to Bolivia and east to central southern Venezuela.

Occurrence in Monument. Common migrant, chiefly in May and in late August and September. Recorded from: Black Rock Spring, Aug. 30, Sept. 3 *, 4 *; Upper Covington Flat, May 15 [1]; Lower Covington Flat, Apr. 30 [1], May 19 [1], 26, Aug. 23 *, 26, 27 *, 28; Quail Spring, Sept. 9 *; Stubby Spring, Sept. 5; Indian Cove, May 13; Twentynine Palms, Apr. 15, May 5, 7, 17, 21–31; Cottonwood Spring, May 13, 14, Sept. 13, 15 *; Eagle Mountain, May 14.

Wood pewees are common and fairly conspicuous as migrants through the desert. The spring dates extend from April 15 to May 31, but late migrants are to be expected even into June. The greatest numbers are seen in the first half of May. In the fall, migration is in full force by the last week in August and continues until September 15 at least, but apparently it does not extend into October. During peak movements 3 to 12 birds may be seen at one station in a day, and the numbers may surpass the total of all species of *Empidonax* flycatchers noted at the same time.

In line with the instinctive foraging behavior displayed on the breeding grounds, the migrating wood pewees take exposed lookout posts, usually high above ground. Tops of piñons, particularly dead snags, and dead juniper and yucca spikes are favored. Such locations make the species easily detected. Foraging from these lookouts may be noted, but the perches seem to serve at times only as resting points from which the bird will then take off in a long flight, not always clearly related to migratory direction, but certainly indicating a tendency to move along rather than stay about a given station. The wood pewees normally are silent while passing through in the fall, but in the spring the buzzing identification note is often uttered. Concentrations about springs or along canyon bottoms is not conspicuous in this flycatcher.

In line with the early dates of fall migration and with the long migratory movement that is undertaken, adult wood pewees move south in worn breeding plumage as do Traill flycatchers. The birds of the year are essentially in full juvenal plumage at this time, although there may be a rather inconsequential partial body molt, chiefly on the throat, before departure. The two adults taken in transit showed some beginning of body molt, but there was none involving the critical wing or tail feathers, which are changed on the wintering grounds. The five immatures showed no molt in progress. Thus there is no heavy physiologic strain or chance of reduction of flight efficiency as a consequence of molt during migration, as is true of the violet-green swallow (p. 134).

No western wood pewees were found in distinctly poor condition, although one immature lacked fat and weighed only 10.0 grams, whereas fat birds, adults and immatures, weighed 12.5 to 13.4 grams.

The western wood pewees of this area all belong to the northern race *Contopus richardsonii richardsonii.*

Olive-sided Flycatcher. *Nuttallornis borealis*

Description. A large (27—33 gram) flycatcher, with tail relatively short, extending only 2—3 cm. beyond wing tips when bird is perched. Crown and back black and dark gray; throat and center of breast and belly white, bordered by conspicuous dark gray or olive-gray sides; tail and wings black with two inconspicuous buff wing bars on wing coverts (fig. 40).
Range. Breeds from central Alaska eastward through transcontinental Canadian forests to Newfoundland and south to northern Baja California, central Arizona, central northern New Mexico, northern Wisconsin and Michigan, and in Appalachian Mountains to eastern Tennessee and western North Carolina. Winters in northwestern South America, south to Perú.
Occurrence in Monument. Sparse migrant in spring and fall. Recorded from: Black Rock Spring, Aug. 29, 30 *; Lower Covington Flat, May 13 [1], Aug. 25, 27, 28 *; Stubby Spring, Sept. 6 *; Twentynine Palms, May 24; Cottonwood Spring, Apr. 26, May 1, 17, Sept. 14.

Seldom is more than one olive-sided flycatcher noted in a day, although on August 28 two were seen moving across a ridge top in a southeasterly direction. On May 17 several migrants were moving in a westerly direction over rolling terrain and along washes at Cottonwood Spring.

This species stays in the open, free of heavy foliage and branchwork, and usually it is not seen foraging for more than a few minutes in any one area.

The three autumnal specimens taken were all birds of the year engaged in their long autumnal migration, and, like young migrating wood pewees, they were in essentially full juvenal plumage.

Vermilion Flycatcher. *Pyrocephalus rubinus*

Description. A small flycatcher marked with brilliant red or (females and immatures) pink and yellow. Adult males: under parts, crown, and forehead vermilion; back, ocular and auricular areas, and wings dark gray or black; tail black with narrow white borders laterally and terminally. Females: under parts white, streaked with dusky on breast, the posterior flanks and belly pink and yellow; dorsal surface gray, the wings and tail dull black, marked with white as in male. Immature males variously intermediate. Juveniles as in females but belly yellow, usually without pink.
Range. Breeds from southeastern California, southwestern Utah, southern Arizona, southern New Mexico, and central Texas south through México and Central America, and in South America south to Chile and Argentina; also Galápagos Islands. Largely resident, but in north, at least, disperses in winter westward and eastward to limited degree; probably also some southward movement.
Occurrence in Monument. Vagrant or casual visitant, chiefly in early spring. Recorded from: Twentynine Palms, Feb. 6 *, 22—27, Mar. 22—24, 27, 28, Apr. 3—5, 12.

Only a single individual has been seen at a time in the vicinity of Twentynine Palms, although in the spring of 1935 Miss Carter was able to distinguish three different individuals seen on different days. On April 3 a male appeared and sang, fluttering in the air, about mesquite thickets. This bird stayed only until April 5 and, although thus showing evidence of breeding activity, it must have moved off or have succumbed, for Miss Carter who was present in the area through May would certainly have recorded it later if it had become established for nesting. In March and April this flycatcher is commonly occupied with nesting in the Coachella and Colorado River valleys. The vagrants in the Monument at this time might, then, be regarded as individuals making as yet unsuccessful efforts to pioneer for nesting in the area. Or they could be individuals, late in their return to normal breeding areas, that use the oasis at Twentynine Palms as a temporary station in the course of ill-defined spring migratory movements.

As a desert or arid-land species with a large range in tropical America, this flycatcher would seem to find no problems of high temperature in the Joshua Tree Monument that would prohibit its residence there. The species does seem to favor bottom lands and old lake basins, usually, perhaps always, those with water sources. In these respects the high desert terrain of the Monument would be unsuitable. Even at Twentynine Palms, where water and mesquite thickets are present, the elevation with correlated cold periods in the early spring may be a deterrent to breeding. Some pioneer nestings in similar areas in the Mohave Desert area to the north have taken place, however (Jaeger, 1947:213; Fish, 1950:137).

The vermilion flycatchers of southern California belong to the race *Pyrocephalus rubinus flammeus.*

LARKS

Horned Lark. *Eremophila alpestris*

Description. A slender, long-winged, ground-dwelling passerine bird. Head marked in male by black forehead and lateral crown streaks, the latter somewhat elongated; anterior face, suborbital patch, and broad anterior breast area black; frontal, superciliary, and throat areas whitish to yellow. Similar head markings of female variously developed, but black of crown usually lacking. Back and sides pinkish brown to clay color, more distinctly streaked with dusky in female; posterior breast and belly white; tail black except central pair of feathers, which are brown; lateral rectrix white-bordered. Bill and feet black, the latter with elongate, fairly straight spur on hind toe.

Range. Northern Hemisphere, from arctic areas south to northern Africa and middle Asia, and in North America to southern México, southern Texas, and the northern parts of Gulf states; disjunctly in Colombian highlands. Partly migratory.

Occurrence in Monument. Scarce resident locally. Winter visitant probably in somewhat larger numbers. Recorded from: Upper Covington Flat, Nov. 4; Pleasant Valley, Apr. 10 *; Pinto Basin Apr. 12 *.

Horned larks require open, sparse, more or less flat ground over which they forage on foot, flying readily, however, from one location to another. In the deserts of North America they are limited in the main to old drainage sinks where the sparse annual vegetation develops sufficiently to provide food, chiefly insects, in the spring and summer season.

There are few such suitable areas in the Monument. Pleasant Valley (fig.

15) northeast of Pinyon Wells was found to be occupied by a few nesting larks in 1960. In this year rain had stimulated extensive growth of filaree and sand verbenas about the playa in this valley. Here in an area of 2-inch-tall filaree a pair was found stationed where there was a scattering of small 8-inch rocks, which along with a 3-foot shrub they used as lookout posts. Both birds fed quietly at midday, although when first encountered the male was engaged in a song flight high in the air. The female had a brood patch that was partly developed but not yet edematous. Dissection showed that the first egg of a 3- or 4-egg clutch would have been laid the next day. At least one other male was seen, but apparently there were very few larks nesting in this basin.

In Pinto Basin on April 12, 1960, near the dune in an area of sparse *Larrea* on coarse sand a few larks were present. The only forbs here were widely spaced *Achyronychia cooperi.* A lark collected at this point had two nymphal grasshoppers in its bill. A half mile away a pair was encountered in similar terrain. The female was taken and proved to have a brood patch and ova 7 and 12 mm. in diameter. Evidently it was laying. The male circled high overhead, almost out of sight and moving over a course 100 yards long, eventually to return to the same ground station. This basin this year had many low forbs and, in places, tall lush verbena and primroses, more than we have before encountered here. In previous visits in spring to this spot no larks have been seen. Possibly they are nomadic to some degree, settling to nest in spots like this that are favorable only in a particular year.

Winter flocks of horned larks have been reported without detail by others visiting the Monument. These may be migrants wintering in the area or wandering groups of the local race. The breeding birds collected belong to the race *Eremophila alpestris ammophila,* the resident race of the Mohave Desert. The nesting in the Monument represents a southern extension of the known breeding range of this race. The paler race *E. a. leucansiptila* occurs nearby to the south in the basin of the Salton Sea. One winter-taken specimen in the collection of Long Beach State College is also of the race *ammophila.*

SWALLOWS

Four species of swallows have been observed as migrants in the Monument. One of them, the cliff swallow, also nests in small numbers. Probably three other species of swallow migrate through the area, although we have not detected them. Of these latter, the most likely to occur is the rough-winged swallow.

Swallows apparently find no difficulty in carrying on a necessary minimum of foraging on airborne insects while crossing the desert. Insects are common in the air in local concentrations, even though they are not as plentiful as over damp ground, water surfaces, and moist woods, and they presumably are of types appropriate as food for swallows. Swallows of the two more common species, the violet-green and the cliff, have been seen making direct migratory flights without deviations for foraging. Their flight equipment permits long, efficient traverses, in the course of which the birds may intercept or easily seek out occasional insect concentrations because of generally superior flight capability. That a considerable safety reserve in energy and flight efficiency is available to swallows for migration through less favorable foraging areas is indicated by the heavy deposits of subcutaneous fat which most show and by the fact that at least one species is able to undergo wing molt in the course of its travels. The magnitude of the problem of desert crossing for swallows seems, then, not to be great and indeed may be little different from that of flight over other regions. Wide cruising radius and great flight power minimize the problem of long distances across barren lands, and water sources for drinking, although utilized when encountered in the course of travel, probably are not critical on a daily basis because the insect food taken is high in moisture content.

A bird like a swallow that is almost continually on the wing by day has a constant cooling factor from convection. This with respiratory evaporation and freedom from the heating effect of close proximity to ground radiation should mean that desert heat constitutes no great problem even though the swallow is continually exposed to solar radiation while airborne. A general solution of the cooling problem while engaged in the exercise of flight has already been evolved by swallows. Probably desert air does not impose much additional burden on the mechanism. The hottest air masses and ground reflection can be avoided or minimized by changing elevation and by cruising about.

Migrating swallows in the Monument may be seen most often moving along high ridges or over the high plateaus in autumn and at exposed water surfaces in the open where they may concentrate temporarily. All movements of a direct kind which we have noted have followed roughly the northwest-southeast axis of the principal mountain system of the area.

Violet-green Swallow. *Tachycineta thalassina*

Description. A strikingly marked swallow with clear white under parts, the white extending upward onto rump, although not joining in midline. Cheeks white or gray; upper parts green in adults, the rump and inner wing purple in race here involved; upper parts of juveniles and immatures gray or dull green; tail with slight fork only.

Range. Breeds in forested and wooded areas from central Alaska and southwestern Yukon south through Pacific coastal districts to southern Baja California and through Rocky Mountains (extending also to Black Hills) and mountains of México to Oaxaca; occupies desert areas in southern Baja California and southern coastal Sonora. Winters chiefly in México and northern Central America, sparingly north to central coastal California and southern Arizona.

Occurrence in Monument. A common fall migrant, chiefly in October, along the higher ridges. Seen less frequently in spring migration. Recorded from: Black Rock Spring, Sept.

Fig. 42. Species of swallows. Violet-green (upper left), tree (upper right), barn (lower left), and cliff (lower right).

3 *; Upper Covington Flat, Oct. 8 [1]; Lower Covington Flat, at 5600 feet, Aug. 28; Quail Spring, Oct. 15; Pinyon Wells, Oct. 14 *, 16, 17; Pinto Basin, Oct. 14; Virginia Dale Mine, Apr. 24; Eagle Mountain, Oct. 19, 20; Cottonwood Spring, May 13.

Migratory movements were conspicuous in autumn in the piñon belt, where, at high points, long direct flights could be watched. The direction was generally eastward, sometimes even a little north of east on Eagle Mountain, perhaps in order to adhere to the crests in that particular area. On October 14 a flight was passing eastwardly along a ridge at Pinyon Wells. It was watched for two hours, and it may have lasted longer. The birds traveled singly, although often only 50 to 100 yards behind or to the side of another, and they seldom turned or seemed to catch insects. About 15 birds passed by per hour, and this may have gone on all day. Indeed, on the night of October 17, while hunting owls in the area between 8 and 10 P.M. by nearly full moonlight, violet-green swallows were occasionally heard calling on the wing along the same flight line.

At other times in this same period, these swallows ceased their direct flight and circled about for insects. This was noted especially on October 16, when six were flying about a small peak near Pinyon Wells where two white-throated swifts also were flying. Once one of the swallows dove at a swift, driving it away momentarily.

The spring migration has been noted at Cottonwood Spring, where on May 13 a group of about 30 was noted at midday; these moved on quickly.

Both our specimens taken in the fall were immature birds that had been hatched in the preceding spring season. The one taken on October 14, a female, was conspicuously fat and weighed 15.4 grams. The one taken on September 3, a male, weighed only 11.2 grams and was evidently not fat. Both were undergoing molt of the flight feathers, the September bird still retaining the outer four primaries from the juvenal plumage, the October bird the outer three. The tail of the September bird also was molting, but the body molt had not progressed far. The fact that the taxing process of molt takes place while in migration and involves the major flight feathers without impairing flight efficiency to a dangerous degree indicates that these swallows have ample safety factors with respect to flight mechanics and nutrition during the migratory movement. The insect supply in the air along mountain crests probably is enhanced by updrafts of air at these points. The strictly insect diet, largely of dipterans, supplies moisture so that drinking is not mandatory.

The violet-green swallows of this area belong to the race *Tachycineta thalassina lepida,* the only form occurring north of the Mexican boundary.

Tree Swallow. *Iridoprocne bicolor*

Description. A black and white swallow with the white of the under parts not extending onto sides of rump nor to ear coverts and eye region. Back blackish, varying from glossy blue-black or green-black to dull sooty (immatures); tail but slightly forked.

Range. Breeds from northwestern Alaska, northeastern Manitoba, and Newfoundland south to southern California, western Colorado, northeastern Louisiana, and Virginia. Winters in suitable areas from central California, southern Arizona, Virginia, and the Gulf coast of the United States south to southern Baja California and through mainland México to Honduras and Nicaragua.
Occurrence in Monument. A single report: identified at Barker's Dam, May 21.

This swallow probably passes through the Monument not infrequently. Miss Carter detected it only once as one flew in company with cliff swallows over the surface of the water at Barker's Dam.

Barn Swallow. *Hirundo rustica*

Description. A long, fork-tailed swallow with rust-red or rich buff throat and dark, steel blue-black back and rump. Posterior under parts dull white or rust-red; forehead dark rust-red, buff, or whitish; sides of breast black, the dark areas sometimes meeting to form necklace; tail with subterminal white spots on inner vanes of all but central tail feathers.
Range. Breeds in holarctic region from northern tree belt south to central México, western Florida, northwestern Africa, Egypt, northwestern India, and northern China. Winters in South America and in Old World in South Africa, East Indies, and Micronesia.
Occurrence in Monument. Occasional in spring and fall migration. Recorded from: Black Rock Spring, Sept. 4 *, Oct. 16; Pinto Basin, Apr. 6.

On April 6 a solitary barn swallow flew overhead in the middle of Pinto Basin, not pursuing a direct line of flight. It circled around over the smoke trees once.

On September 4 after a period of rain, a group of Vaux swifts was circling about the mountain crests near Black Rock Spring where a wind was sweeping upslope. After about 10 minutes they were joined in foraging by a small group of swallows that flew with them low over the oak and manzanita brush and piñons. Most or all these were barn swallows, one of which was taken. The bird was in juvenal plumage, with no molt in progress; it is of the subspecies *Hirundo rustica erythrogaster,* the North American form.

Cliff Swallow. *Petrochelidon pyrrhonota*

Description. A square-tailed swallow with conspicuous light buff rump patch and (in races here involved) conspicuous white forehead. Crown and back black, glossy blue-black in adults; neck with gray collar, the gray extending into back area as light, even whitish, longitudinal streaks; throat and face chestnut with black area in center of lower throat and upper breast; chestnut of face reduced or lacking in juveniles; breast otherwise dark gray or brownish gray; posterior under parts whitish.
Range. Breeds from central Alaska, northern Mackenzie, Ontario, and southern Quebec south through United States, except southeastern section, to Veracruz and Oaxaca in southern México. Winters chiefly in southern Brazil, Paraguay, and central Argentina.
Occurrence in Monument. Present chiefly as a spring and fall migrant, in which status it is fairly common. Nests, at least in some years, in small numbers at one known locality. Recorded from: Upper Covington Flat, Aug. 24; Barker's Dam, May 21; Indian Cove, May 13, June 3; Twentynine Palms, May 8; Pinto Basin, Apr. 5, 26.

This swallow performs an exceptionally long migration, traveling far into South America each winter season. Migration across the desert probably is a problem of no magnitude considering the distances and diverse terrain it encounters in its total migratory flight. Moreover, the species is capable of nesting in desert areas, at least in the vicinity of water sources, where high summer temperature and low humidity prevail.

Migratory flight was seen exceptionally well in Pinto Basin on April 5. Here on the floor of this great open area the flight of a bird could be watched continually for a distance of half a mile. Three swallows were seen in the course of 1½ hours, pursuing a straight course west-northwest over the basin. They flew without any side-to-side irregularity and sang constantly, an impressive and attractive performance. Two were in sight at once, and they could have kept in touch by the song. One flew within 30 yards of the ground but the others were 100 yards in the air. Do they sing all the way in these direct migratory flights in spring? The song, though perhaps serving to keep scattered birds in contact, was the full-toned, song performance, the creaking and gurgling notes given about nest sites, and was much more than would be necessary for keeping in touch with other migrant individuals. One gained the impression that there was an excess of energy and full readiness to engage in the activities of a nesting colony once the bird arrived at a breeding site.

In the fall, on August 24, similar straight flights in a southeasterly direction were seen on Upper Covington Flat over the tops of the tree yuccas. No song was given then. Some flights were noted in which two birds traveled near one another.

As with other swallows, occasionally migrants stop to circle about to forage. On May 21 this species was seen at Barker's Dam flying about over the surface of the water. One found dead at Twentynine Palms on May 8 succumbed to unknown causes. It is not entirely certain that these May records relate to birds in transit, although they probably do. Barker's Dam might provide adequate conditions for a nesting colony.

The one known breeding station in the Monument is at Indian Cove. Here there is permanent water. About the rock potholes and tanks in the canyon there is mud essential for the nest construction of this species. However, conditions seem to be marginal, for but few swallows nest here, whereas cliff swallows normally nest in large colonies. On May 13, 1945, a few cliff swallows were seen at Indian Cove, but no nests were discovered. On June 3, 1950, several were present, the date rather surely indicating summer residence. On April 8 in the following year an old nest of construction unmistakably that of this species was found near the lower water holes in the Rattlesnake Canyon drainage at Indian Cove; it had probably been built the previous spring. No birds were present on that date, but the entrance of the nest had droppings on it suggesting that young had been in the nest the summer before. The nest was 12 feet up on the side of a great boulder protected by an overhang from rain and most of the day from sun. It was the usual flask-shaped nest made of mud pellets and was still largely intact, although it was at least 10 months old. There had been no other nests on this same rock face, nor could other signs of nests be found on the surrounding cliffs. The nesting group here must certainly be small, although not necessarily confined to one pair.

In other areas cliff swallows occasionaly abandon nest sites and colonize new locations. In the Monument we might expect that, in years with favorable water and mud supplies located near cliffs, other nesting colonies would develop, only to be abandoned in drier years. Probably in some years no cliff swallows breed in the Monument. The inadequacy of nest-building material and water surfaces for drinking rather than any lack of food supply probably is the factor governing residence here in summer.

No specimens of cliff swallows have been preserved from the Monument. The birds that nest are probably of the race *Petrochelidon pyrrhonota pyrrhonota,* which breeds in adjacent regions of California. This race and the Great Basin race *P. p. hypopolia* may both be involved in the migratory movements.

JAYS, TITMICE, AND ALLIES

Scrub Jay. *Aphelocoma coerulescens*

Description. A long-tailed jay, the wings and tail bright blue. Upper surface of head and neck blue; back brown; sides of head marked with narrow white superciliary line, the ear coverts dark, dull blue; throat white, obscurely streaked and outlined laterally and behind with blue; breast and belly light gray.
Range. Resident from southwestern Washington, southeastern Oregon, southern Idaho, southern Wyoming, and western and southern Colorado south to Guerrero, Oaxaca, and Veracruz; also central Florida.
Occurrence in Monument. Common resident. Recorded from: Morongo Pass, Apr. 28; Little Morongo Canyon, Mar. 30—Apr. 2; Black Rock Spring, Apr. 3, July 6, Aug. 28, 29, 30 *, 31 *, Sept. 1 *, 2, 3, 4, Oct. 16; Upper Covington Flat, Aug. 24 *, 28; Lower Covington Flat, Apr. 2 [1], 8, 9, 15 [1], 29 [1], May 14 [1], 15 [1], 19 *, 29 [1], July 16 [1], Aug. 19, 22 *, 23 *, 24 *, 25, 26 *, 27 *, 28 *, 29, 30, 31, Oct. 23 [1], Nov. 26 [1]; Upper Covington Flat, Feb. 17 [1], Mar. 18 [1], Apr. 15 [1], May 29 [1], June 11 [1], 16 [1], 17 [1], 25 [1], 26 [1], July 24 [1], Aug. 28 [1], Nov. 4 [1], 18 [1], 19 [1]; Smithwater Canyon, Jan. 8 [1], Feb. 12 [1], May 19, Aug. 23, 24, 29, Nov. 19 [1]; Quail Spring, Jan. 18, 19, 20, 21, 22 *, 23, 25, 26, Feb. 10, 11, 12 *, 13, Apr. 10, 22 *, May 14, 19 *, 20, June 3, July 1 *, 2, 3, Oct. 15, Dec. 4 *, 19 *; Stubby Spring, Sept. 5, 6 *, 7 *, 9 *, 10 *, 11 *, Oct. 15, Dec. 4 *, 5 *; Lost Horse Valley, Feb. 10, Nov. 1—2; Key's View, July 15, Oct. 31; Pinyon Wells, Oct. 9, 12, 13 *, 14 *, 15, 16 *, 17; Split Rock Tank, Oct. 29—31; Live Oak Tank, Apr. 6–8, May 9 *, Dec. 6 *; Eagle Mountain, May 16 *, Oct. 19 *, 20 *.

Scrub jays occur throughout the piñon and scrub oak habitat of the higher levels of the western section of the Monument and also in the same habitat on isolated Eagle Mountain at the southeastern extremity of the area. Although they may range into Joshua trees where, as commonly, these are intermingled with scrub oak and junipers, we have not found the species out in pure stands of Joshua trees, nor even in the vegetation of oases in low rocky canyons where oaks are absent. Thus they do stay, perhaps significantly, above the areas of most intense summer heat.

Scrub jays in the Monument seldom are seen in flocks; when they are, the flocks appear to be loose, transitory aggregations occasioned by visits to water, to favorable food trees, or to a source of excitement. Sometimes immature birds associate in small groups. Usually the species is in pairs, at all periods of the year, although the second member of a pair may not be seen unless one searches carefully, for much of the time these jays are unaccountably shy in the open habitats of the area. The largest groups we saw ranged from 12 to 15—groups that assembled by individuals coming from different directions to Black Rock Spring in the August heat periods. These assemblages were in no sense cohesive; in fact, evidence of antagonism and enforced spacing within them was noted. At such times there were as many as 50 individuals in several groups or scattered as individuals within a 100-yard radius.

Scrub jays often become excited by the presence of a predator. This we witnessed in relation to the great horned owl when jays came cautiously to imitated calls of this species. Sometimes this was the only means by which we were able to collect samples of these wary jays, especially on Eagle Mountain. We also found jays scolding Beechey ground squirrels on August 26, and an excited group was found on July 3 when apparently a snake had drawn their attention. Once a jay came in quietly close by the commotion caused by a wounded mountain quail.

Like all members of the jay group, scrub jays are omnivorous, and yet they depend heavily on acorns for food and they also take piñon nuts. Acorn gathering and carrying is a conspicuous autumnal activity (see especially Grinnell, 1936). The jays have a soft *chuck* note which they give commonly as they hop about in the oak foliage searching for acorns. We recorded acorn gathering especially on October 13 and 14 at Pinyon Wells in the scrub oak bushes, and occasionally a jay was seen in flight carrying an acorn, doubtless to a point where it would bury it for possible later retrieving. On October 14, we saw one pry into piñon cones in search of nuts. A bird taken on Eagle Mountain on October 20 was carrying an acorn and had acorn fragments in the stomach.

In spring and early summer when acorns and nuts are not available, unless from stores or ground gleaning, we suspect that animal food, chiefly large insects, is the staple diet. This extensive resort to animal food would minimize the water balance problem for the species much of the time.

In the heat of summer, drinking water is often sought, although it apparently is not available to all populations, for we do not have evidence that these jays travel regularly to springs from distances of over a mile. At a water seep in Smithwater Canyon in the last week of August scrub jays visited frequently. On the 29th at 7:30 A.M. a jay came to the spring here and found that the damp sand borders had caved in on it, leaving no exposed water. The water level was $\frac{1}{4}$ to $\frac{3}{4}$ inches below the surface. The jay did not work

Fig. 43. Scrub jay on lookout post.

with bill or feet to dig down to the water but thrust or drilled its bill into the damp sand repeatedly; between times it stood up alert in the lookout posture. On August 30 and 31 at Black Rock Spring jays seemed very eager to come to the open water, although only two or three at a time hopped along the ground to the water to drink, the others hanging back and waiting turn; at the spring proper they threatened one another, at distances of 3 feet. On September 3, after an intervening rain and a break in the hot weather, we noted that there were no longer any concentrations of jays at this water source; nor were there any here on April 3. But again they were at the water here on July 6. Visits to water tanks may, however, be made in winter, as at Quail Spring on January 19 and 22. In Smithwater Canyon on August 24, the jays bathed as well as drank.

Nesting must begin in early April, as we found a family of fully grown young at Quail Spring on May 20. On Eagle Mountain, at 4400 feet, on May 16, a brood of three young with tails only slightly short of full length was perched together in a bush in a canyon bottom. The female parent taken here had a dry or inactive brood patch and was an adult bird, judged by the shape of her tail feathers (see Pitelka, 1945). The other parent was scolding 200 yards away. Two females taken on May 19 at Lower Covington Flat and Quail Spring in the same year were first-year birds and had no brood patches. In view of the date, they evidently undertook no breeding that spring.

A large sample of jay specimens was obtained in the fall because of the need of working out the geographic variation in this area. This group taken in August, September, and October permits estimation of the age composition of the population at that time. A surprising preponderance of adults is revealed: 40 adults and 3 first-year birds. All but 8 of these were taken in 1950, and no immatures were among those of that year. It seems very unlikely that we were selectively collecting adults, as they probably are in general more wary than first-year birds. The evidence suggests that there was a poor production of young in that year. On the contrary, in 1960, parties from Long Beach State College took ten juvenal jays in June and July in the Covington Flat area.

Pitelka (1951: 235–239) has reported in detail on the geographic variation of the scrub jays of the Monument area. The population of the isolated piñon and oak belt of Eagle Mountain is recognized as a lighter-colored, duller, less purplish blue race, *Aphelocoma coerulescens cana,* with less heavy bill basally than *A. c. obscura,* which occupies the Monument from the Pinyon Wells section westward. There is a gap of unsuitable habitat of about 20 miles that intervenes, and one of 70 miles between *cana* and *A. c. nevadae* to the north.

The large series of *A. c. obscura* from the middle and western sections of the Monument "is of interest in that general coloration is dark and tends to be more similar to [typical] *obscura* [of Baja California] in color of under parts than is true to the west and southwest" in southern California. In these latter areas intergradation with the light under parts of *californica* from the north is evident. "In hue of blue, most of the specimens [from the Little San Bernardino Mountains] differ from both *obscura* and *californica* in averaging less purplish." This tendency "seems to be the result, in part at least, of genetic interchange with" [*cana*] to the east. But because of continuity in the habitat to the west, the influence from that direction would be expected to be greater and apparently is so in fact. "The possibility remains that the [partial] resemblances between [Little San Bernardino birds and *cana*] . . . may be the result of selective action of more or less similar environments as much as they may be the result of genetic interchange between the two populations" (Pitelka, *loc. cit.*).

Common Raven. *Corvus corax*

Description. All black, with tail wedge-shaped rather than fan-shaped. Bill deep and massive. Total length about 22 inches.
Range. Holarctic region in general, extending south through western United States, México, and Central America as far as Nicaragua.
Occurrence in Monument. Sparse permanent resident. Recorded from: 2 mi. W Joshua Tree, Oct. 16; Twentynine Palms, May 5; Split Rock Tank, Oct. 30; Pinto Basin, Apr. 12; Cottonwood Spring, May 18.

Ravens have proved to be surprisingly scarce in the Monument in view of their widespread occurrence in desert regions and about cliffs in the barren mountains of the southwest. However, they often are seen scavenging along the highways in the nearby Coachella Valley.

On May 18 a pair came into the cottonwoods at Cottonwood Spring but proved to be shy. Miss Carter saw the species only once in the course of two winter seasons at Twentynine Palms, when a pair flew over the desert to alight on a fence near the oasis. Other occurrences noted have been of single individuals. The abundant lizards of the Monument as well as large insects should provide good foraging for this species.

Ravens of this section of North America are usually regarded as belonging to the race *Corvus corax sinuatus*.

Piñon Jay. *Gymnorhinus cyanocephalus*

Description. A nearly uniformly dull blue jay, the upper surface of head and neck darker; throat with poorly defined streaks of whitish. Bill long and straight. Tail relatively short for a jay.
Range. Breeds east of Cascade Range and Sierra Nevada from central Oregon to northern Baja California and east to eastern side of Rocky Mountains from central Montana to western Oklahoma and New Mexico. Winters casually to coast of California and to Nebraska, Kansas, and Chihuahua.
Occurrence in Monument. Permanent resident; locally common. Recorded from: Black Rock Spring, Aug. 29, Sept. 1 *; Upper Covington Flat, Apr. 9, Aug. 24 *, 26, 28; Lower Covington Flat, Apr. 7, 9 *, May 19, Aug. 23, 24 *, 25, 26, 27, 28; Smithwater Canyon, May

19; Quail Spring, Sept. 8, 9, Oct. 15, 21 *; Stubby Spring, Apr. 10, 11, Sept. 6, 7 *, 8, 11 *, 26; Lost Horse Valley, Apr. 22, Dec. 4–6; Key's View, Mar. 21, 23, 28, 30; Pinyon Wells, Oct. 10; Split Rock Tank, Oct. 30.

Piñon jays have been seen in all periods of the year and may be presumed to be resident. However, the species is nomadic, locally, and at times over greater distances, and thus part of the birds seen in the Monument may represent dispersals from known nesting areas along the east face of the San Bernardino and San Jacinto mountains. There always seem to be birds present in the area bounded by Upper Covington Flat, Lost Horse Valley, and Key's View, and nesting occurs, probably each year, in the piñons and junipers of this mountain complex.

In April of 1960 we found this species nesting in the western part of Lower Covington Flat in juniper and Joshua tree woodland, at 4800 feet. Here on the 9th a jay flew quietly to a yucca top and later circled about and disappeared in a large juniper (fig. 45). On approaching we heard young call, a wavering whistle or rasp, not a hiss. The young were in a nest 7 feet up in a dense crown tuft of juniper foliage which was heavily laden with berries. The four young lifted their heads as the branches were parted. They were probably about a week old, with eyes still closed and feather tracts beginning to show, but no feathers appreciably broken from the sheaths. The body was black, without down, and the mouth lining red.

The nest was a deep cup with strong stick base and rim, and it had a felted but rough lining of shredded bark. The dimensions were 7 by 10 inches deep outside, the cup 5 by 3½ inches deep inside. The nest was well concealed and shaded in the tight strong clump of foliage.

The parent that had brought food was taken and was a female with an inactive brood patch. Two days before in this place a group of about 50 noisy piñon jays had been moving about, chiefly associated in two's and three's. One was flushed from a juniper near a partly built nest. Courtship feeding was seen in one pair. On the later date a noisy flock was heard in the distance, but at the hour when we found the nest only the pair belonging to it was in the immediate vicinity. We suspect that other nests were scattered about within a few hundred yards.

Piñon jays are at all times gregarious, even about nest sites, as just indicated. The members of flocks maintain contact by constant calling, a quavering, nasal *caw*, that is softened and musical in the distance. Typical flock action in foraging was noted on August 24 on Upper Covington Flat in a large level expanse of Joshua trees and junipers. Here at 7 A.M. a group of about ten was feeding on the ground, spaced 30 to 50 feet apart. They moved along in a rolling fashion, the members at the rear flying over and past those ahead of them on the ground. They called chiefly when in flight. The group was followed in a circular course for more than a mile in the period of an hour. On September 20 when feeding in piñons, piñon jays were scattered

Fig. 44. Piñon jays in Lost Horse Valley

Fig. 45. Nest tree of a piñon jay. The nest location has been exposed by bending down a berry-laden limb of the juniper, revealing some of the stick structure; Lower Covington Flat, April 9, 1960.

two or three to a tree over a half-mile stretch, but apparently all were in touch by voice with the whole group as well as with those close at hand. While moving about in the forage area, they become very wary when pursued and will not pass over a person who is exposed to view. The birds fly with a fairly slow rate of wing beat like that of a crow or Clark nutcracker, but with seemingly more flight speed and power.

The largest flocks we have seen were those estimated to consist of 125 to 175 individuals on October 15 at Quail Spring, but flock estimates of 50 to 100 have been made at other times in the late summer and fall.

A flock maneuver frequently seen consists of a very compact group flying in direct line, presumably to or from a roost. The birds are then bunched almost as closely as are large shorebirds, and a striking chorus of voices is heard. When such groups alight, if not spread out for feeding, as many as 50 may congregate in a single bush or tree. A group of this size was seen in a catclaw bush beside the water tank at Quail Spring on October 15. At our camp at Lower Covington Flat a flock of about 50 flew down canyon between 7 and 10 A.M. on several successive mornings in late August; they returned between 3 and 4 P.M.

On August 27 at 7:45 P.M. after all daylight was gone, but during full moonlight, a flock was heard calling in flight on its usual course over Lower

Covington Flat. Again on September 8, at Stubby Spring, at 7:30 P.M. in full dark, a chorus of jays, presumably moving, was heard. The meaning of these nocturnal movements is not clear; possibly roosting flocks have been disturbed, or they may voluntarily shift about. In any event, even on moonless nights, there is enough light on the desert for such a bird to see its way to new locations.

When feeding on the ground, piñon jays walk rather than hop and in this way move about searching for insects. One taken on August 24 from a ground-foraging group had grasshoppers and beetles in its stomach. On September 7 and 11 near Stubby Spring birds were feeding in the piñons, and specimens taken had nothing but piñon nut meat or some form of mast in the stomach. One had an intact but shelled piñon nut in the stomach, and others had pine pitch on their feet.

Piñon jays go to water to drink, and perhaps they do so regularly when it is available and especially if the fairly dry pine nuts constitute their diet at the time. On October 15 at Quail Spring these jays came to the water tank at 8:20 A.M. This was the period of the morning when flocks often moved in this direction from Lower Covington Flat, and such visits may have been a daily occurrence in summer and fall. A species like this, which has inclination and easy capacity for long flights, could readily go to the few and distant water sources in these mountains.

In our experience, piñon jays are confined in the Joshua Tree Monument to the piñon–juniper belt of the upper levels of the western section. We have not seen them in lower desert vegetation or southeast of the vicinity of Pinyon Wells.

The specimens we took were all judged to be adult rather than first-year individuals by reason of skull examination or by the shapes of the tips of their primaries and rectrices. The characters of the feathers are those employed by Pitelka (1945) for the genus *Aphelocoma;* they seem to apply satisfactorily to *Gymnorhinus.* We find from study of molting young from the San Bernardino Mountains that juvenal flight feathers are not replaced in the postjuvenal molt; thus juvenal feathers in the wing would be retained as markers through the first year of life.

The adult molt is under way in July and is well advanced by late August and early September. Three birds taken on August 24 and one each on September 1 and 11 all had progressed to the molt of the outer three primaries, and all but one, on August 24, had dropped the old tenth primary. Body molt was still much in evidence, with patches of old plumage showing in one case (August 24) on the belly and in four of the five about the neck and throat. Three others taken on September 7 and 11 had completed the wing and tail molt but had scattered pinfeathers in the head and neck area.

Brodkorb (1936) advocated racial subdivisions of this species which in 1944 we (Grinnell and Miller) were unable to substantiate, nor did the division receive endorsement by the fifth edition of the American Ornithologists' Union Check-list (1957). Since 1944 we have obtained a good series of eight breeding adults from Powder River County, Montana, not far from the type locality of *Gymnorhinus cyanocephalus*. These, when compared with near topotypes of Brodkorb's *Gymnorhinus cyanocephalus rostratus* from the San Bernardino Mountains and with birds from the Monument, show the bill differences reported by Brodkorb. We cannot closely duplicate his measurements, but this apparently is because of differences in measurement techniques, and there is a little more overlap in bill length between racial groups than he indicated, but a reasonably consistent difference exists, the Californian birds having bills that are larger and usually straighter and more attenuated. We cannot offer evidence to support Brodkorb's third race in the southern Rocky Mountain area and thus are not in a position to map the racial distribution of the species throughout. But the extremes in eastern Montana and California appear separable. Accordingly, the name *Gymnorhinus cyanocephalus rostratus* is used for the piñon jays occurring in the Monument.

Clark Nutcracker. *Nucifraga columbiana*

Description. A crowlike bird, somewhat larger than a jay, with body plumage all gray. Wings black, with white patch at posterior border of secondaries; tail black centrally, the lateral three pairs of feathers entirely white.
Range. Resident from central interior British Columbia and southwestern Alberta south through the higher mountains to northern Baja California, central Nevada, central Arizona, and central New Mexico. In nonbreeding season in some years descends to lowlands, spreading west to Pacific coast, east to central Texas, and south into northern México to Chihuahua and Nuevo León.
Occurrence in Monument. Occasional visitant. Recorded from: Lower Covington Flat, Aug. 25 *, 1950; Quail Spring, Feb. 11, 1947.

Davis and Williams (1957) have shown that Clark nutcrackers leave the high conifer forests, where they are normally in residence, at times of low food supply of pine nuts. These irruptions correlate with severe and widespread failure of cone crops in their normal habitat following two or more years of large crops, during which the breeding population of nutcrackers apparently increases. In these irruption periods food in the conifers of lower belts may be especially sought, and the piñon belt of the Joshua Tree Monument is potentially such a source. Irruptions begin in late summer and early fall, and the occurrence recorded in the Monument on August 25, 1950, fits this situation, 1950 having been a general invasion year.

On this date a bird flew across the piñons above Lower Covington Flat. Soon after we came upon a nutcracker, possibly the same one, settled two-thirds of the way up in a piñon. The forest here at the time had a good supply of half-grown green cones which we suspect had attracted the nutcracker. The bird was an adult male in new plumage.

The occurrence at Quail Spring on February 11, 1947, was not in a known year of irruption; apparently more than one bird was present on this occasion. However, it should be realized that there are resident populations of this species nearby in the San Bernardino Mountains and that some vagrancy as far east as the piñon belt of the western part of the Monument might be expected in any year.

Mountain Chickadee. *Parus gambeli*

Description. A moderate-sized (11—12 gram) member of the titmouse or chickadee group. Top of head and throat and upper breast black contrasted with white sides of head and white superciliary line; remainder of plumage gray.
Range. Resident in interior mountains from northwestern British Columbia and southwestern Alberta south to northern Baja California, southeastern Arizona, southeastern New Mexico, and southwestern Texas.
Occurrence in Monument. Winter visitant in small numbers. Recorded from: Black Rock Spring, Apr. 3 *, 4 *, Aug. 28, 29, 30, Sept. 2; Upper Covington Flat, Nov. 19 [1]; Lower Covington Flat, Jan. 7 [1], Aug. 24 *, 25 *, Sept. 9 [1]; Stubby Spring, Sept. 5, 8, 11 *, Oct. 15, Dec. 4—6 *; Split Rock Tank, Oct. 30.

Mountain chickadees occur in the piñon belt in winter, arriving as early as August 24 and remaining until April 4 at least. We had thought that the species might be resident, as it occasionally will breed in good piñon woodland below its normal habitat in the Transition Zone and Boreal Zone pine forests. However, specimens taken in April were from winter flocks, and, although song was heard occasionally from the flock, the testes of three males taken then were only 1½ mm. in length and thus showed no significant recrudescence.

In the late summer, the occurrences were of single individuals detected moving through or near piñons at 4500 to 5600 feet. At this time, as later, also, they would respond to imitated chickadee call notes and come toward a person accordingly. The two specimens taken on August 24 and 25, an immature and an adult, had some pinfeathers on the head, but no wing molt showed in them and certainly in the adult this had taken place and had been completed.

From October on, our records show birds in groups of 2 to 8; they were only occasionally found, but always in such associations. On April 3 and 4 two, apparently different, groups were found in piñons, each time in a flock with which a black-throated gray warbler and ruby-crowned kinglets were loosely associated.

The chickadees in the Monument were at one time suspected of representing a race other than that which is abundant in the adjoining San Bernardino Mountains, namely, *Parus gambeli baileyae*. However rechecking of fairly adequate samples with recently taken fresh plumages of *baileyae* leads to the conclusion that they are that form, and Behle (1956) in his review of the races of the mountain chickadee has come to a similar conclusion. Apparently

Fig. 46. Pigmy nuthatch (left), red-breasted nuthatch (right), and plain titmouse (below).

a short-range dispersal from the center of heavy population in the San Bernardino Mountains results in a small number moving down eastward, surprisingly soon after the breeding season. These mountain chickadees have not been detected farther east in the Monument than Split Rock Tank. We did not find them southeast of here in the high country at Pinyon Wells, where, however, suitable winter-range habitat occurs.

Plain Titmouse. *Parus inornatus*

Description. A fairly large (13–16 gram) titmouse, gray-brown dorsally and uniform light gray below. Head with distinct, pointed crest.

Range. South-central Oregon, southern Idaho, southwestern Wyoming, and south-central Colorado south to the Cape district of Baja California, southern Nevada, southeastern Arizona, northeastern Sonora, and western Texas.

Occurrence in Monument. Common resident of upper levels of western section. Recorded from Black Rock Spring, Apr. 3, 4 *, July 6, Aug. 28, 29, 30 *, 31 *, Sept. 1 *, 2 *, 3 *, 4 *, Oct. 16; Upper Covington Flat, Jan. 28 [1], Aug. 28; Lower Covington Flat, Jan. 8 [1], Apr. 8, 29 [1], May 10 [1], Aug. 22, 23 *, 24, 25, 26, 27 *, 29; Smithwater Canyon, Mar. 24 [1], May 13 [1], 19; Quail Spring, Jan. 22, 23 *, 26, Feb. 13 *, Mar. 20, Apr. 11, 22, May 19 *, 20, July 1–5, 4, Oct. 15; Stubby Spring, Feb. 10, Apr. 9, 10, Sept. 5, 6 *, 7 *, 8 *, Nov. 1, Dec. 4 *, 5 *; Lost Horse Valley, Feb. 10, Dec. 5 *; Pinyon Wells, Sept. 2, Oct. 9 *, 12 *, 13 *, 15 *, 16, 17 *, 18 *; Live Oak Tank, May 9.

Plain titmice are confined to the higher-zone vegetation of the Monument. We have found them only in piñon, juniper, and scrub oak cover, and they have not spread across the gap in such cover to isolated Eagle Mountain in the southeastern section of the Monument. Their occurrence is therefore essentially from the 4000-foot level upward, since only locally and to small extent do they range lower, as at Quail Spring at 3700 feet.

These titmice are always seen singly or in pairs. We have found no tendency to flock, even in fall and winter. In suitable bush and tree cover at Pinyon Wells an estimated 50 individuals were seen in the course of a morning's cruise on October 13, but always the birds were spaced out and pair associations were definitely noted on this date.

This species becomes disturbed by imitations of screech owl calls and can be brought close in by this means. At such times the movement of two together, presumably male and female, can be noted readily.

Foraging occurs in the smaller branch work of scrub oaks, piñons, and junipers, but not in our experience on the smooth stems of the associated manzanitas. Tapping or pounding is often heard, as the birds break up food or explore the bark of small limbs for insects. On October 15 one was watched pounding vigorously in a juniper bush 1 foot above the ground. Thus they often feed low down and commonly drop to the open ground to hop about in foraging over the surface.

Singing may occur sporadically in the fall, as we noted on October 13, but usually in fall and winter the varied forms of the wheezy location note are heard. Singing is normal by early April, as noted on April 3 and 10, when nesting is under way. On April 4 at Black Rock Spring a laying female was taken. The bird would have deposited her first egg the next day to start the set. The brood patch had not yet become vascular. The mate of this female had an 8 mm. testis.

We have not examined nests of this species in the Monument, but tree cavities are required for nests; for this reason, as well as others, adherence to the woodland habitat is necessary.

Titmice by staying at upper levels in the mountains and by having available the shelter of dense vegetation can avoid the greatest extremes of summer heat. A largely insectivorous diet affords water, and we have not seen titmice go to the springs that are available in the piñon belt.

The population of titmice in the fall period is composed of a surprisingly large proportion of adults as judged by skull condition. Since signs of immaturity seem always to persist until October in this species, we think this separation of age groups in samples taken between August 23 and October 20 is reliable. By these means we find the ratio is 19 adults to 3 immatures.

All titmice taken in the last week of August and the first week of September were molting and consistently were at a stage where all old primaries had been dropped or only the outermost, number 10, remained. The body plumage was predominantly new superficially, but many feathers were still in sheaths. In the period from October 12 to 18 birds were finished with the primary molt or at most had sheath remnants at the bases of the feathers, but some few body feathers were still growing. Two immatures on September 2 and October 14 showed evidence that at least some of the primaries had been replaced in the postjuvenal molt.

The plain titmice of the Monument have been recognized as a distinct endemic race, *Parus inornatus mohavensis,* which differs from *P. i. transpositus* of adjoining southern California in less olivaceous and brownish back and crown. Also, the under parts are paler. The bill averages longer than in coastal races (see A. H. Miller, 1946). "Although it is thought that *mohavensis* has developed as a gray-backed type in an arid area through modification of a coastal brown-backed titmouse, the loss of brown and the increase in average bill size suggest intergradation with the gray titmice of the interior. . . . There is, however, a gap of about fifty miles of unsuitable desert terrain between the ranges of *mohavensis* and *ridgwayi* [of the Providence Mountains], a formidable barrier for a strictly resident species. On the other hand, there must be continuity with *transpositus* in the juniper belt in the vicinity of Morongo Valley along the east flank of the San Bernardino Mountains. *Mohavensis* in its total of color characters is distinctly closer to the coastal complex of races than to *ridgwayi*." The most probable origin of its distinctive characters is then by selective modification toward *ridgwayi* of the coastal stock rather than through occasional interbreeding with vagrants of typical *ridgwayi.*

Verdin. *Auriparus flaviceps*

Description. A small (6–7 gram) member of the titmouse family, with yellow head and gray body, the belly paler than the back and almost white; bend of wing with small chestnut patch.
Range. Southeastern California, southern Nevada, southwestern Utah, western and southern Arizona, southern New Mexico, and south-central Texas south to Cape district of Baja California, Jalisco, and Hidalgo.
Occurrence in Monument. Resident; widespread, but sparse population. Recorded from: Little Morongo Canyon, Mar. 30, Apr. 3; Smithwater Canyon, May 19; Quail Spring, Feb. 11, May 19–21, July 1–5, 19–21, Sept. 8 *, 10 *; Stubby Spring, Sept. 10; Lost Horse Valley, Nov. 1; Indian Cove, Apr. 7, 8 *, May 12, 13, 14; Fortynine Palms, Apr. 7; Twentynine Palms, Jan. 5, Mar. 21, 25, July 12, 13, Oct. 11; Pinyon Wells, Oct. 10 *; Live Oak Tank, Apr. 6–8; Pinto Basin, Mar. 18[1], Apr. 5, July 14, 15, Sept. 13 *, 15 *; Virginia Dale Mine, July 14; Cottonwood Spring, Feb. 7, Apr. 5, 25, 26, May 1, 6 *, 13, 14, 17 *, June 3, 22, July 7, 8, 14, Sept. 13, 14 *, 15 *, Oct. 21, 22, 23 *; Lost Palm Canyon, Oct. 22 *, 23.

Verdins are more often detected by their distinctive thorn-covered, globular nests than by sight of the birds, although experienced ornithologists find them by their high-pitched, far-carrying notes. They have a distinctive, bleating call note, a *thup* alarm note, and a two- or three-parted, thin, whistlelike song.

This species is typical of open, relatively low desert terrain and in the

Monument is found below the piñon woods and oak brush, which are the habitat of its higher-zone relative, the plain bushtit. Verdins do, however, follow the catclaw, smoke tree, and desert willow growth of the canyons and valley bottoms up into the edge of the piñon belt and thus through much of the Joshua tree-filled basins. Thus in small areas of suitable habitat they may be found as high as 4200 feet, as in Lost Horse Valley. Lower down they are found in mesquites, as well, and they range in foraging out into the creosote bushes, although this plant does not afford adequate locations for nests.

Verdins, because of their small size and great exposure to hot sun and wind in their desert scrub habitat, would seem to be particularly vulnerable to desiccation (see Bartholomew and Cade, 1956). Yet the species is obviously highly successful in its desert existence. Some reduction in activity is noticed in midday, but verdins can be found moving about even in the hottest part of the day. The nests, which are persistent and are used and built for roosts as well as nesting, may aid at times as a retreat during the day, but they are in hot exposed places and must warm up internally. They would offer protection from drying wind and direct sun, however. Doubtless the principal factor in the verdin's maintenance of proper water balance is its insectivorous diet. We have never seen it come to drink at springs. The possibility remains that this species has special water-conserving mechanisms and heat tolerance of an unusually high order. Such have not been searched for as yet through physiologic experimentation.

Verdins are nonflocking, and each pair may range over large areas, often making long, fairly powerful flights in the open, especially when alarmed and pursued. In the Monument, the most satisfactory estimate of numbers was made in the vegetation of the wash below Indian Cove. In this strip of suitable cover, we estimated that on April 8 there was a pair present about every $\frac{1}{4}$ mile over a linear distance of $1\frac{1}{2}$ miles, thus about 6 pairs in all. We doubt that they are ever more concentrated than this in areas in the Monument. Roosting nests and residual unused nests may give a false impression of numbers. Thus in Pinto Basin, in a few scattered smoke trees (*Parosela spinosa*) near the road, 8 nests were found, probably the efforts of a single pair of birds over a period of a year or two.

The nests typically are placed in thorny shrubs and are constructed on the outside of thorny sticks of the same or other available shrubs. The side entrance is a short tube or is sheltered by a canopy so that the bird must fly up and scramble in at an angle; the narrow, thorn-bordered entrance makes it difficult for a person to probe within the nest. The inner part of the finished nest, and especially the breeding nest, is a well-felted globe protected by the thorny shell, and breeding nests have a soft, smooth lining of feathers. On April 5 at Cottonwood Spring a nest was lined with Gambel quail feathers. In Pinto Basin nests were lined with duck feathers, but we believe these came from a trash dump nearby and not from ducks occurring naturally in this formidable desert flat. Nests we have seen in the Monument range from 4 to 10 feet above ground. The thorny smoke trees and catclaws seem especially favored, and yet one was situated 10 feet up in a desert willow, which is not thorny. We saw one nest 5 feet up in a cholla cactus which was 6 inches in diameter. The shell of the outside was made of catclaw twigs, and these and cholla spines guarded the entry way.

Miss Carter noticed verdins building a nest in a mesquite on March 21 at Twentynine Palms, but, although it may well have been a breeding nest, this cannot be certainly assumed. Egg-laying may be expected in late March and in April. The nest in the cholla had 4 eggs, a completed clutch, on April 7 at Indian Cove. On April 25 at Cottonwood Spring two, and possibly three, flying young were noted and an adult was seen to feed one of them. Such youngsters would have hatched from eggs laid in the last third of March.

The nests provide an excellent shelter for the nestlings both from predation, by reason of difficulty of tearing them open, and location among fine, often peripheral branches of a bush, and because of shielding from sun and wind afforded the delicate naked young. The nests also may, as we have already surmised, serve the adults in similar fashion, but it should be noted that night roosting in the nests also is a protection from heavy winds and from the severe cold of winter nights. The instincts that guide the standard nest architecture thus are probably an adaptation of critical significance to the existence of this small desert bird. Occasionally, however, the thorns of the bird's own nest may entangle it, so that the verdin must learn to move carefully through the entryway it has constructed (A. H. Miller, 1936).

The annual molt extends well into September. Adults taken from September 13 to 15 show either heavy involvement of the outer three or four primaries or near completion of these feathers coupled with little residual body molt. An immature taken on September 8 was in heavy postjuvenal molt and was replacing the rectrices and remiges, the outer two primaries (numbers 9 and 10) still being old. Thus this species appears to undergo a complete postjuvenal molt.

The verdins of the Monument and adjacent northwestern deserts are of the race *Auriparus flaviceps acaciarum.*

Plain Bushtit. *Psaltriparus minimus*

Other names. Bushtit; common bushtit.
Description. A small (5 gram) member of the titmouse group, brownish gray above, the crown darker than the back (in local race); light gray below. Eye of female white or ivory, that of male black. Tail long.
Range. Resident on Pacific coast and in Great Basin from southwestern British Columbia, southwestern Idaho, southwestern Wyoming, and western and southern Colorado south to Cape district of Baja California, northern Sonora, southern New Mexico, and western and central Texas.

Plate II. Plain bushtit (above) and verdin at its nest.

GMC

Occurrence in Monument. Common resident of upper vegetation belts. Recorded from: Morongo Pass, Apr. 28; Little Morongo Canyon, Mar. 31 *, Apr. 1, 2 *; Black Rock Spring, Apr. 3, 4, Aug. 29, 30 *, 31, Sept. 2 *, 3, 4 *, Oct. 16; Upper Covington Flat, Nov. 18[1], Dec. 16[1]; Lower Covington Flat, Apr. 8, 9, Aug. 23 *, 25, 26, Dec. 10[1]; Smithwater Canyon, Feb. 17[1], Apr. 23[1], May 19; Quail Spring, Feb. 13 *, Mar. 20, Apr. 11, 23, May 19, 20 *, July 1–5, Sept. 8, 9 *; Stubby Spring, Apr. 9, Sept. 6, 7 *, 10, Oct. 15, 29, Nov. 1 *, Dec. 4 *, 5 *; Lost Horse Valley, Feb. 10, Oct. 31 *; Key's View, July 15; Barker's Dam, Feb. 18 *; Indian Cove, Apr. 7, May 13 *, 14; Pinyon Wells, Oct. 9, 12 *, 15 *, 17 *, 18 *; Live Oak Tank, Apr. 6–8, May 9 *; Eagle Mountain, May 16 *, 19 *, 20 *, 21 *.

Bushtits, like plain titmice and scrub jays, adhere chiefly to the piñon-juniper woodland and associated oak scrub, but, unlike the plain titmice, bushtits are more inclined to range lower into other types of brush, and in some localities they overlap the ecologic range of the related desert-adapted verdin. Thus in Little Morongo Canyon, bushtits ranged down to the mesquite and catclaw that is interspersed with willows in the canyon bottom and were nesting there. Verdins were singing in the same area. At Quail Spring on September 8 a flock of 10 bushtits was in the catclaw thickets of the valley bottom, and a verdin was associated with them for a short time near a nest of the verdins. At Indian Cove, as low as 3200 feet, a family group of bushtits was found in May, but this was at the lower edge of oak and piñon in the canyon.

Bushtits are highly gregarious at all times except when isolated as pairs or trios engaged in nesting. From August to March invariably they have been seen in flocks. Our specific estimates of the size of such flocks has been as follows: 10, 10, 10, 12, 12, 12, 15, 15, 15, 15, 15, 20, 20, 20, 25. Even in the breeding season small remnants of flocks may be expected. While flocking the birds may be relatively stationary at a favorable food tree or, conversely, they may move along in a straggling, vocally cohesive formation that may in a minute or two cover 200 yards.

Foraging is almost always in the foliage layer of the trees dominant in their range—piñons, junipers, scrub oaks, and mountain mahogany (fig. 47). But as noted, they will range into catclaw and other low, open brush, and also into willow and desert willow when these are available. On one occasion, unexpectedly, we noted them dropping to the ground to forage beneath or beside scrub oaks.

Bushtits are insectivorous and, despite their small size, they do not require other water sources. We have not seen them come to the springs to drink.

Nesting takes place in March, April, and May insofar as our records indicate. On April 1, in Little Morongo Canyon, a pair already had a nest nearly or quite complete. It was the typically long, saclike structure of the species and was situated 15 feet above ground in a willow. The outside was decorated with patches of white material. A torn-open nest was found in

Fig. 47. Bushtit habitat on Eagle Mountain consisting of mountain mahogany, scrub oak, and piñon.

this canyon, at a different location, the next day. It also was in a willow tree, 10 feet up, in sparsely leafed twigs; paper scraps had been worked into the outside.

At Indian Cove on April 7 a pair of bushtits had a nest hanging in fairly open foliage of a cottonwood in the canyon bottom. The nest was 11 feet above ground and held 5 fresh eggs. The nest lining was a deep feather fluff in which mountain quail feathers were prominent. The nest measured 10 inches long by 3 wide outside. The female (white-eyed) returned after our inspection, working vigorously inside to repair the slit in the pouch we had made to examine the eggs. A nest below Lower Covington Flat on April 8 was 7 feet up in the dense needles of a piñon. Its exterior was largely covered with oak catkins and the dried flowers and seedheads of a curly blue-flowering plant. This nest held 6 eggs, more than half through incubation.

Males taken on March 31, April 2, May 16, and 20 had testes 4½ to 5 mm. long, which is functional breeding size for this species. On the other hand, a male with a family at Indian Cove on May 13 had a 3-mm. testis, indicating it was past its breeding period. On May 20, also, a family of fully grown young was seen.

Bushtits, like other members of the titmouse group, often will respond to the calls of small owls, showing disturbance but not the extreme alarm that they do in the presence of a hawk; they come to the source of the owl call, and they usually are less demonstrative and less consistently responsive than plain titmice. On October 12 at Pinyon Wells a flock clearly came to imitated screech owl calls.

At Black Rock Spring in a flock encountered on September 2, three out of five bushtits taken had a white neck patch which in some instances was visible as the birds moved about. On August 30 a similarly marked bird was taken here at exactly this point, from what most probably was the same flock. All birds thus marked were adults. One adult and one immature taken with them did not have this albinistic feature. The marking involves, at the minimum, white at the bases of the crown feathers and in maximum development forms a white neck collar dorsally and a white, rather than light gray, lower throat. The four "mutants" were equally divided as to sex. The significance of this situation lies not so much in the occurrence as such of an abnormality but in the fact that it showed up in high proportion in a single flock with probable history of familial relationships. In short, it suggests the arising or the first expression locally of a mutation in one or several interrelated broods.

Bushtits in the last week of August are only halfway through the molt. Immatures on August 23 and September 2 still had old primaries 9 and 10, although they were progressing with complete molts of remiges and rectrices, and both adults and immatures in the first week of September had ragged body plumage, with much new feather growth in evidence. Traces of molt on the head were noted in skins taken as late as October 21. Thus the molt probably extends over a long period, from early July, presumably, to mid-October.

As in some other resident species of the Upper Sonoran areas of the Monument, immatures are scarce in the population among the late summer and fall samples taken. In 1950 the ratio was 6 adults to 3 immatures, and in October of 1946 it was 19 adults to 3 immatures, a combined ratio of 25 to 6. Although 1950 marked the end of a succession of especially dry years which might have been thought to have led to low reproduction, the comparison with 1946, which was not as extremely dry, does not bear out such an hypothesis.

The bushtits of the Monument are an endemic race known as *Psaltriparus minimus sociabilis*. The characters of this race and their probable origin were fully analyzed earlier (A. H. Miller, 1946), and the newer specimen material, chiefly fresh plumages from western localities of the Monument, correspond with the earlier samples from Pinyon Wells.

Bushtits, unlike plain titmice, occur on Eagle Mountain and may be presumed to move occasionally across unfavorable low desert terrain more readily than do the plain titmice.

The coastal bushtits and those of the Great Basin constitute two very differently colored groups of races. In the coastal races, the pileum is dark and contrasting with the back, the dorsum is dark gray, and the wing is short, whereas in Great Basin races there is a concolored pileum, a light gray dorsum, and longer wings. Three of fifteen bushtits taken on Eagle Mountain have the pileum ashy gray as in the race *providentialis* of the southern Great Basin. In other respects these birds and all the other specimens from the Monument resemble more closely the coastal races, although differing from them in lighter pileum and dorsum and in progressively longer average wing eastward through the Monument. It seems "fully possible that out of an initially diverse parentage [= *providentialis* and *minimus*] the *sociabilis* population might have retained and established its particular combination of features in which there are high gene frequencies for dark, contrasting pileum (from *minimus*), for dark gray dorsum (from *minimus*), for grayish as against more brownish hue (some factors from *providentialis*), and for fairly pallid sides (some factors from *providentialis*); length of wing and tail seem to reflect a multiple factor situation with genes from both parental types persisting and yielding intermediate averages."

"Some of the color characters might have been attained merely through modification of *P. m. minimus*" by selection. However, "in light of clear evidence of some dual ancestry," it is best to suppose that a "wide variety of genes was assembled by interbreeding of diverse stocks, from which state certain genes subsequently have gained predominance to produce a new combination of prevalent characters. The amplitude of variation, except in dorsal pattern on Eagle Mountain, is no greater than in other races of bush-tits" (A. H. Miller, 1946).

White-breasted Nuthatch. *Sitta carolinensis*

Description. A large (16—18 gram) nuthatch, black or dark gray on top of head and neck, with conspicuous white face and under parts. Back blue-gray; flanks gray and brown; tail gray centrally, black with subterminal white patches laterally.
Range. Resident from southern British Columbia, southeastern Alberta, southern Manitoba, southern Ontario, southern Quebec, and central Nova Scotia south to southern Baja California, Guerrero, Oaxaca, central Veracruz, and central Florida. Generally absent in the Great Plains area of the United States.
Occurrence in Monument. Occasional winter visitant. Recorded from: Black Rock Spring, Sept. 4; Lower Covington Flat, Nov. 18 [1]; Quail Spring, Oct. 15.

This species has been detected only in or near the piñons in the western part of the Monument. It is not regularly migratory in this part of California, but some autumnal dispersal takes place and probably the occurrences in the Monument represent such scattering or vagrancy in winter. We have seen the birds only briefly.

Two specimens in the collection of Long Beach State College are of the race *Sitta carolinensis aculeata* as represented in the adjoining coastal mountains of southern California.

Red-breasted Nuthatch. *Sitta canadensis*

Description. A small (9–10 gram), strikingly marked nuthatch, with black crown, white superciliary line, and black line running through the eye to the neck. Back blue-gray; breast cinnamon to rich rusty; tail gray centrally, black laterally, with white subterminal patches and gray tips (fig. 46, p. 148).
Range. Breeds principally in transcontinental coniferous forests from southern Alaska to Newfoundland and south in the mountains to central coastal and southern California, southeastern Arizona, eastern Tennessee, and western North Carolina; also Guadalupe Island.
Occurrence in Monument. Occasional autumnal migrant. Recorded from: Black Rock Spring, Aug. 29, 30, 31 *, Sept. 4; Lower Covington Flat, Aug. 28, Oct. 23 [1]; Stubby Spring, Sept. 11, Nov. 1.

This species in its southward migration extends to the lower Colorado River valley, and it may yet be found in midwinter in the Monument. It occurs at this season in the Providence Mountains to the north in the Mohave Desert. We have found it in the Monument only in the fall. Dates range from August 28 to November 1. Single individuals only have been noted, and about six in all have been seen.

Apparently red-breasted nuthatches are attracted by camps, either from curiosity or because of the possible shelter from bright sunlight which they afford. We have had them come to our tables under tarpaulins stretched for shade, and they have even entered our parked cars. They are attracted to shining objects, conceivably mistaking them for water.

One that was seen on August 31, at a time of intense heat, flew up a wash, stopping to rest under small oaks in the shade. The bird flew weakly, although when it was taken it was not thin; it was an immature of the year.

The birds that came into camp were very tame, but their familiarity and adherence to camp may have been a reflection of their special desire for shade and water. Doubtless the fall heat and drought is rather severe for this boreal, shade-seeking species, even though it is insectivorous and can gain considerable moisture from the food it takes. Three individuals have been seen near or at springs, although none has been watched actually taking water.

Pigmy Nuthatch. *Sitta pygmaea*

Description. A small (9–11 gram) nuthatch that is usually gregarious. Crown brownish gray and nape whitish; back blue-gray; stripe through eye and ear area black or dark brown; under parts whitish or light buff; tail gray and white medially, black with subterminal white spots and gray tips laterally (fig. 46, p. 148).
Range. Resident in mountainous areas from southern British Columbia, northern Idaho, western Montana, central Wyoming, and southwestern South Dakota south to northern Baja California and through the Mexican highlands to Jalisco, Michoacán, Puebla, and central Veracruz.
Occurrence in Monument. Occasional autumnal vagrant. Recorded from: Black Rock Spring, Aug. 30, Sept. 1 *, 2 *, 3 *; Lower Covington Flat, Aug. 23 *, 25 *, 27, 28 *.

Pigmy nuthatches normally are permanent residents of coniferous forests of the Transition Zone. It was surprising therefore to find them occasionally in the piñons of the Monument, where they evidently represented an autumnal scattering from adjacent higher mountains to the west. In the Monument we have never seen them expect in piñons, and then usually in the larger tracts of this pine. The piñons provide foraging facilities in the branches and needle tufts similar to those in the conifers of higher zones. Also abnormal was their occurrence solitarily on six occasions. Once two were present, and on another occasion at least four were seen. One juvenile came into camp on August 23, but otherwise this species did not seem to be attracted to camps and oases as the red-breasted nuthatches are.

All nine of the specimens taken were juveniles or immatures—birds hatched in the preceding spring. Possibly young birds alone are involved in this autumnal dispersal. Four of the specimens are in juvenal plumage (August 23–September 2), and four are in fresh postjuvenal body plumage (August 25–September 3). Only one (August 25) was actively engaged in postjuvenal molt. One juvenile taken on September 2 weighed only 7.0 grams and may have been in weakened condition.

The specimens are referred for convenience to the race *Sitta pygmaea melanotis*. In the pale and gray crown color of the immatures, they suggest *S. p. leuconucha*, but in dimensions they accord best with *melanotis* of the San Bernardino Mountains. The two races, which are actually segments of a cline, are more or less arbitrarily divided at San Gorgonio Pass, those south of the Pass being a little closer to *leuconucha* than are those north of it. Since the populations from the San Bernardino Mountains and the San Jacinto Mountains overlap in characteristics, it is not possible to indicate from which mountain area the birds taken in the Monument were derived. Either or both of these high mountain masses could have been the point of origin.

WRENS

House Wren. *Troglodytes aedon*

Description. A small (8–9 gram) wren, with short tail. Back gray-brown, faintly barred; under parts gray, obscurely mottled; tail, wings, and flanks barred; no white line over the eye.

Range. Breeds from central British Columbia and Alberta, southern Saskatchewan and Manitoba, central Ontario, and southern Quebec south to northern Baja California and through Mexican highlands to Oaxaca; southeastwardly to northern Texas, northwestern Arkansas, northeastern Tennessee, and northeastern Georgia. Winters from southern edge of breeding range in the United States south to southern Baja California, lowlands of south-central México, the Gulf coast, and southern Florida.

Occurrence in Monument. A regular but inconspicuous spring and fall migrant. Recorded from: Little Morongo Canyon, Mar. 31; Lower Covington Flat, Aug. 25 *, 29 *, Sept. 9[1]; Smithwater Canyon, Apr. 23[1]; Quail Spring, Sept. 9 *; Stubby Spring, Sept. 5, 6 *; Indian Cove, Apr. 7; Fortynine Palms, Apr. 7 *; Twentynine Palms, Apr. 26; Pinyon Wells, Apr. 10, Oct. 12–18; Cottonwood Spring, Oct. 23 *; Lost Palm Canyon, Oct. 22.

House wrens probably pass through the Monument in large numbers, but, because they hide in dense cover whenever possible and because they are usually not vocal, they are seldom reported. The species winters in the lower levels of the Colorado Desert and also in coastal southern California. Consequently it may yet be found in the Monument in winter. However, the cold of the higher deserts probably forces these wrens out at least during the midwinter period. The migrants have been detected in the spring from March 31 to April 26 and in the fall from August 25 to October 23.

The birds appear singly, often about oases, and they have been seen skulking in thick chrysothamnus bushes, palms, and brush along the bases of rock outcrops.

A male taken on April 7 had testes that were only slightly enlarged (2 mm. long) and no deposits of fat. The five migrants taken in the fall likewise showed little or no fat; all were immatures that had completed the postjuvenal molt. Evidently they do not depend greatly on fat reserves in migration. One that weighed only 5.3 grams on August 29 was in dangerously weakened condition. These wrens probably conserve and augment their water supply by remaining in the shade of heavy cover and by foraging on insects and spiders.

The house wrens passing through the Monument are of the widespread western race *Troglodytes aedon parkmanii.*

Bewick Wren. *Thryomanes bewickii*

Description. A slender, fairly small (8—10 gram), long-tailed wren with conspicuous white superciliary line. Back brown; under parts light gray, darker on the flanks; tail black with white tips on four lateral pairs of feathers, the central pairs barred black and brown; under tail coverts barred black and white.

Range. Breeds from southwestern British Columbia, Nevada, southern Utah, southwestern Wyoming, Colorado, southwestern Nebraska, southern Iowa, the southern Great Lakes area, and central Pennsylvania south to southern Baja California, Oaxaca, Texas, and the northern parts of Gulf states. Largely resident.

Occurrence in Monument. Common resident. Recorded from: Morongo Pass, Apr. 28; Little Morongo Canyon, Mar. 31, Apr. 2 *; Black Rock Spring, Apr. 3, 4, Aug. 29, 30, 31, Sept. 1 *, 2, 3 *, 4 *; Upper Covington Flat, Aug. 28 *; Lower Covington Flat, May 19, Aug. 22, 23, 24 *, 25 *, 26 *, 27 *, 28 *; Smithwater Canyon, May 19; Quail Spring, Jan. 22 *, 23, 26, 27, Feb. 10, 11, 12 *, 13, Apr. 11, 23, May 17 *, 19, 20, July 3, Sept. 8 *; Stubby Spring, Feb. 10, Apr. 9, Sept. 5, 6, 7 *, 9 *, Oct. 31 *; Lost Horse Valley, Feb. 10, Oct. 31 *; Key's View, July 15; Indian Cove, Apr. 7, 8, May 13 *, 14; Fortynine Palms, Apr. 7 *, Oct. 12; Twentynine Palms, Mar. 21; Pinyon Wells, Oct. 9, 10, 12 *, 13 *, 15 *, 16 *, 17 *; Live Oak Tank, Apr. 6—8; Eagle Mountain, May 16 *, Oct. 18 *, 19 *, 20 *, 21 *; Cottonwood Spring, Feb. 7—11, Apr. 29, May 14—19, 17 *, Sept. 13, 14 *, 15, Oct. 14, 21—23 *; Lost Palm Canyon, Oct. 16, 22 *.

The Bewick wren is the most abundant member of its family in the area. It is not found, however, in the open desert flats; its virtual absence at Twentynine Palms is indicative of this. There is only one record of it there, possibly that of a vagrant, on March 21. Bewick wrens require low brush tangles. These are best afforded in the piñon belt, but the species occurs also in the canyons in catclaw, desert willows, chollas, and the lower yuccas. The birds use large Joshua trees chiefly for song posts, provided there are low retreat tangles associated, such as downed yucca logs, burro brush, and chrysothamnus thickets.

The abundance of the species in the upper vegetation belts is attested by estimates made of as many as 25 seen in a day, all of course as pairs or single individuals, as on Eagle Mountain on October 20. Again at Stubby Spring a total of 6 was detected along a mile of trail on September 6, and at Quail Spring on January 23 at least 10 were heard or seen in the general vicinity of the spring.

In the tangles and close to or on the ground Bewick wrens flit and hop through the branches or litter in search of small animal food, but they do not search in terminal foliage. In somewhat more open situations we have seen them forage among the spines of yuccas and along the ground, moving from one tuft of grass to another, or to low shrubs that intervene. Below Cottonwood Spring, at 2500 feet, one flew to a vertical granite wall and clung without use of its tail. Yet most of their time is spent in or under cover where details of activity are not easily followed, although the location of the bird may be readily revealed by the continual sidewise flicking of the upheld tail.

Shade of bushes and of rocks and ledges provides shelter from the sun, and the moist insects and spiders which are eaten ease the problem of moisture and water losses during summer heat. We have not seen these wrens resort to drinking water. The species as a whole must have a wide range of tolerance for temperature and moisture in its range from the dank, cool undergrowth of the Puget Sound area to the deserts. Yet it apparently meets the desert extreme chiefly by behavioral adjustments in the use of protective cover and moist foods.

Bewick wrens sing in the nonbreeding season as well as in connection with nesting. Especially is this noted in the fall, following or in the late stages of the molt. Thus occasional full song has been recorded on September 1, October 9, 12, 14, 15, 17, 19, and February 11. But on October 14 one was engaged in continuous song for 15 minutes. A bird collected on September 4 was giving a primitive song, not the full stereotyped adult performance; the bird was taken and proved to be by skull determination an immature male and nearly finished with its postjuvenal molt.

By late March, Bewick wrens are singing vigorously, and males with fully enlarged, functional gonads (testes 6–9 mm.) have been taken on April 2, May 13, 16, and 20. By mid-May well-grown young have been seen out of the nest, as on May 16 on Eagle Mountain. Such young must have represented eggs laid no later than the first week of April.

Bewick wrens, unlike some of the other resident passerine birds of the upper vegetation belts, show evidence of a high annual productivity and population turnover. Among fall-taken specimens in the period from August 23 to October 23, 32 were placed as to age by skull examination. Of these, 18 were adults and 14, or 44 per cent, were immatures hatched in the preceding spring. We found no significant differences in weights of the two age groups and no differences or trends associated with geographic locality or month. Bewick wrens, being essentially resident, deposit no conspicuous masses of subcutaneous fat. Accordingly weights for all fall-taken birds were grouped by sex alone and the following significant averages derived: 27 males, 9.25 grams (standard deviation 0.42; extremes 8.5–10.0); 10 females 8.64 (8.2–9.5) grams.

The postjuvenal molt in Bewick wrens is apparently complete; our material from the Monument shows that at least the outer four or five primaries are involved in many instances. By the last week in August most individuals, immatures and adults alike, have nearly finished the molt and only bits of sheaths remain on the outer primaries and on feathers of the head and neck. Indeed, two adults on August 26 and 28 had entirely finished. On the other hand, one immature had lagged far behind and was just beginning the molt of the breast and back on August 24.

The study of geographic variation of Bewick wrens of the area has brought out very clearly the large amount of individual variation in color in this species. In almost any large sample brown and gray variants may be sorted out, with several stages between extremes. Failure to appreciate this situation, especially among desert wrens, has led to dubious published occurrences of races based on single specimens; some of these may even have moved from their breeding grounds. Large series of birds in fresh fall plumage are needed to evaluate racial affinities or clinal systems.

Our fresh material from the Monument totals 39. The series represents a western group in the area from Black Rock Spring to Stubby Spring, a middle group from Pinyon Wells, and an eastern group from Cottonwood Spring and Eagle Mountain. Among these we see no clear-cut trend in coloration. Each has fairly rich brown individuals and paler, grayer-backed birds. The more extreme of the latter could well be matched in series of *Thryomanes bewickii eremophilus* from the Providence Mountains and from southern Arizona. The browner and intermediate colored birds are as rich and dark-colored as *T. b. correctus* from coastal southern California. The light gray individuals are more extreme than one finds in a series of *correctus,* but they are so few, amounting to no more than 5 of the 39, and scattered from Lower Covington Flat to Eagle Mountain, that the whole population should, we believe, be called *correctus.* If one had collected one or two of these gray birds and no others he would perhaps without hesitation have assigned them to *eremophilus.* Thus the population is one which is fundamentally like that of the coastal chaparral areas but which shows a slight trend, in the proportion of the polymorphic variants, toward *eremophilus* of the more easterly desert mountains. But no clear gradient in this direction can be seen within the Monument.

The question may be raised of course whether the distinctly gray individuals may be migrants from the breeding range of *eremophilus* to the north, since some winter movement does take place in the interior in the range of this species. Against such a possibility is the fact that one of the gray individuals was taken at Lower Covington Flat as early as August 25 while still in late molt, a circumstance in the Bewick wren not likely to represent a migrant. Also, others taken at later dates were among a general population of dark birds, all seemingly territorial and many doing some fall singing.

Cactus Wren. *Campylorhynchus brunneicapillus*

Description. A large (38–42 gram) wren. Crown brown; back gray-brown with broken white streaks bordered with black; throat and breast heavily spotted with black on white; flanks and belly with smaller and sparser spotting, the ground color becoming buffy posteriorly. Outer wing feathers blackish with rows of light spots on inner and outer webs. Lateral tail feathers black with subterminal white bar, the outer three pairs with one to five additional, partial cross bars or spots.

Range. Southern California, southern Nevada, southwestern Utah, western and southern Arizona, southern New Mexico, and central Texas south to southern Baja California, Michoacán, state of México, and Hidalgo.

Occurrence in Monument. Sparse but widespread resident. Recorded from: Morongo Pass, Apr. 28; Yucca Valley, Apr. 22; 3 mi. S Warren's Well, Aug. 31 *; Upper Covington Flat, Apr. 9, May 29 [1], June 2 [1], 25 [1], Aug. 24 *, 28; Lower Covington Flat, Feb. 17 [1], Apr. 8, 9, 14 [1], May 19, June 11 [1], 25 [1], Aug. 24, 25, 26 *; 6 mi. SE Joshua Tree P. O., Sept. 8 *; Quail Spring,

Fig. 48. Cactus wren on top of cholla cactus.

Jan. 19 *, 23, 27, Feb. 10, 11, 13, Apr. 22, May 19–21, July 1, 4, Sept. 8; Stubby Spring, Apr. 9; Lost Horse Valley, Nov. 1 *, 2; Indian Cove, May 13, 14, 15, Oct. 16; Pinyon Wells, Oct. 12 *, 13, 16 *; Split Rock Tank, Apr. 26, Oct. 29–31; Pinto Basin, Apr. 5, 6, July 14; Eagle Mountain, Oct. 18–21; Cottonwood Spring, Apr. 30, May 14–19, July 7, 8, Sept. 13, 14, 15 *, Oct. 21–24; 2 mi. SE Cottonwood Spring, Oct. 22 *; Lost Palm Canyon, Sept. 15, Oct. 22 *, 23.

Cactus wrens in the Joshua Tree Monument are shy birds, difficult to watch, yet often revealing their presence by their large, domed nests placed conspicuously in cholla cactuses or in spiny shrubs and yuccas. Nests are maintained throughout the year for dormitories, as Anderson and Anderson (1957, 1958) have well shown in their extensive study of this species near Tucson, Arizona. Remnants of old nests may persist for many months, but the rather loose structures, made chiefly of grasses externally, quickly fall into disrepair unless the wrens give them constant attention. A typically situated nest was one found at Quail Spring on July 4. It was 6½ feet up in the topmost branches of a cholla, and a group of three birds flushed from it at dusk at

7:15 P.M. Old nests were seen in chollas in Pinto Basin in April, 1951, but apparently they were all deserted in this particular year. In Morongo Valley on April 28, a nest $3\frac{1}{2}$ feet aboveground in staghorn cactus was noted; it was made of grasses, but most of the dome of the structure was gone. On Lower Covington Flat on April 9 a nest 3 feet up in a cholla (fig. 49) held 4 eggs. The nest pocket was about 3 inches below the sill of the entryway.

Cholla cactuses thus clearly are favored for nest emplacements, for these plants not only guard the nests from attack by some types of ground predators but afford dense retreats for the wrens themselves. Although a cactus wren can fly well and will move in flight up to 50 yards between bits of protective cover, it prefers to stay concealed. It will often creep through yuccas, chollas, and smoke-tree tangles and work about the ground beneath them in search for insects. Occasionally, as is true of verdins, they may get seriously caught in the cholla spines (see A. H. Miller, 1936). The chollas at times also provide fruit attractive to the wrens. Thus good scatterings of chollas or dense low Joshua trees almost always form the headquarters for

Fig. 49. Entrance passage of cactus wren nest well protected and supported in the branches of a cholla cactus; Lower Covington Flat, April 9, 1960.

a family of cactus wrens in this part of the desert. There is little use to look for them unless these facilities are available. However, the species is at times seen in other plants such as desert willows, and, at the oases where there are palms, they will hide and also forage in the mat of dead hanging palm leaves. At Lost Palm Canyon we saw them 25 to 40 feet aboveground in the thick palm tangles.

Cactus wrens regularly associate in family groups from the late spring through the winter. Only as the nesting season starts does the adult pair of the family chase off the others and remain in sole possession of a territory, as the Andersons have shown. Thus we have seen groups of 3 wrens on July 4 at Quail Spring, of 5 in October at Pinyon Wells, and of 4 at Quail Spring on January 19. Even on April 28, 3 were seen together, possibly representing young quite recently emerged from an early nest.

Because cactus wrens are insect-feeders and resort also to cactus fruit, they apparently are confronted with no serious problems of water intake. They can withstand the high desert temperatures readily, yet actually escape the most severe temperatures of summer days by limiting midday activity and by using the shade of the denser plants, which they favor. The nests provide shelter but chiefly or entirely, as far as adults are concerned, for the night period. In fact the dormitory nests may be a critical factor in survival in mid-winter when freezing temperatures and cold winds are common in the desert nights.

In spite of the adaptations and behavioral adjustments of this species for desert existence, there is likelihood that at times its food resources are so depleted that the wrens abandon an area. Such seemed to be true in Pinto Basin in 1951, and we believe that the situation may be like that postulated for the Le Conte thrasher (p. 177), in which these desert species may have the capacity to shift locations of residence several miles to areas where the spotty seasonal rainfall and the dependent insect food has developed adequately in a given year.

Molt is conspicuous in specimens taken between August 24 and September 15. Late traces of molt were seen in a bird of October 12, but specimens of the 16th and 22nd were finished with the molt. A juvenile of September 15 had only just started the molt, but another bird of the year of August 24 had replaced most of the inner primaries and much of the body was in fresh plumage. Adults of August 31 and September 8 were halfway through the molt of the primaries.

The cactus wrens of the California deserts belong to the race *Campylorhynchus brunneicapillus couesi*.

Long-billed Marsh Wren. *Telmatodytes palustris*

Description. A small (9–11 gram) wren with short tail that lacks white spots; center of back striped black and white, and white line over eye. Upper parts other than center of

back brown, darker on crown and nape; under parts whitish, with gray or buff on breast and flanks; tail barred black and brown or buff.
Range. Breeds from central British Columbia and northern Alberta southeast to southwestern Quebec and eastern New Brunswick and south to northern Baja California, central New Mexico, the Gulf coast, and central Florida; also central México. Winters over most of breeding range and south to southern Baja California, the state of México, Veracruz, and southern Florida.
Occurrence in Monument. Rare vagrant or local resident. Recorded from: Twentynine Palms, Jan. 26, Feb. 5, Mar. 8, 21, Apr. 5, Oct. 25, Nov. 7.

This marsh wren, which even in winter and in migration seeks sedges and cattails of marshes, may be expected occasionally in the Monument as a vagrant or migrant, especially at oases. Apparently when marsh vegetation is allowed to develop, it may colonize, or attempt to do so, in such an area and stay for the winter or even undertake nesting. Miss Carter found it at Twentynine Palms in the tules and sedges of the oasis as it existed there in 1934 and 1935. She states that it was common and gives dates from October 25 to April 5, with singing recorded in winter and early spring. It is not actually clear how many individuals were present, but several males must have been there. She thought it was nesting in early February, and partial nests, of palm fibers, were found; combings of dog hair were used in the linings. She saw no young, nor does she mention occurrences of the species in late April or May.

Since males of this wren build many nests in the early spring season as a part of their attraction of mates, it may well be that the nests observed were never used for breeding and that the species moved out in April. Singing in the winter and long before nesting is frequent also.

One later occurrence, without details, reported for March 21, 1945, indicates that the oasis was still attractive to the long-billed marsh wren in that year.

Canyon Wren. *Catherpes mexicanus*

Description. A moderate-sized (9–12 gram) wren with chestnut brown belly and flanks contrasting with white chin and throat. Upper parts grayish brown anteriorly, becoming rich brown posteriorly, and finely dotted with light gray and black; wings and tail chestnut brown barred with black, the fine black bars of the tail widely spaced. Bill long (2 cm.), moderately curved.
Range. Resident from central southern British Columbia, northern Idaho, southeastern Montana, southwestern South Dakota, and central Texas south to southern Baja California and Chiapas.
Occurrence in Monument. Resident in small numbers. Recorded from: Black Rock Spring, July 6; Lower Covington Flat, Aug. 23, 26, 27; Smithwater Canyon, Apr. 10, Aug. 26; Quail Spring, Jan. 23, 26, Apr. 23, 25; Key's Ranch, Feb. 24 *; Stubby Spring, Sept. 5, 6 *, 9 *; Indian Cove, Apr. 7, 8, May 13, 14, Oct. 16; Fortynine Palms, Apr. 7 *; Pinyon Wells, Oct. 9; Split Rock Tank, Oct. 29; Live Oak Tank, Apr. 6–8; Virginia Dale Mine, Apr. 24; Eagle Mountain, Oct. 19, 20 *, 21 *; Cottonwood Spring, Feb. 7–11, Apr. 26, May 14, Sept. 13, 14, 15, Oct. 14, 21–24; Lost Palm Canyon, Sept. 14 *, Oct. 22.

This wren is present in the Monument wherever cliffs (fig. 51) and large tangles of boulders provide them shaded rock faces and deep crevices and crypts in which they forage and nest. As is typical of the species, the population is sparse, a condition probably dictated by the scattered occurrence of suitable canyon and rock formations. In the Monument we never saw more than one pair in a given canyon that was suitable for the species. Male canyon wrens are belligerent in response to imitations of their song, which consists of a series of clear whistling notes of descending pitch. Their response is to come close and to sing competitively or to give their location note. Doubtless this reaction leads to challenges between wrens and to the wide spacing of pairs.

On the north side of Eagle Mountain in October we found canyon wrens on the steep, shady, and broken cliffs and took two males on successive days but not in the same section of this extensive mountain face. At this time of the year parts of these slopes were in continuous shadow. Occasionally one finds a bird in more open, small, rocky outcrops, as was true on August 27 at Lower Covington Flat, but such places are more typical of the habitat of rock wrens and do not constitute the permanent stations of canyon wrens. They probably are used only briefly in passing between locations that are more favorable.

Canyon wrens are entirely insectivorous, and the habitat they select provides them protective shade from intense midsummer heat and exposure. Thus their immediate habitat is not desertlike and their moisture problem would not seem to be serious. Not infrequently the canyons they occupy have residual tanks of water, if not springs, which they could use, although in some places these must be lacking in summer.

Canyon wrens occasionally sing spontaneously even in the fall, as we noted at Indian Cove on October 16, at 5:30 A.M. But in spring, song is often heard.

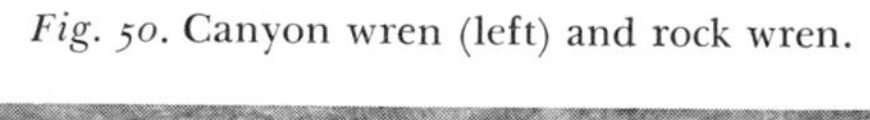

Fig. 50. Canyon wren (left) and rock wren.

Fig. 51. Canyon wren habitat in cliffs above yucca forest in Lost Horse Valley.

Male wrens taken on April 7 at Fortynine Palms were not yet in full breeding state (testis only 3 mm.). However, by May nesting is in progress. On May 13 at Indian Cove a nest, inaccessible on a rocky ledge, was noted, but the exact stage of the nesting cycle could not be ascertained.

Canyon wrens of the Monument belong to the widespread northern race, *Catherpes mexicanus conspersus.*

Rock Wren. *Salpinctes obsoletus*

Description. A moderate-sized (14—17 gram) wren, gray above with fine black and white flecks, buffy on the rump. Below whitish to light gray, obscurely streaked, the flanks pale buff; tail gray basally, finely barred with buff and with subterminal dark patches and light buff tips; edges of tail coarsely barred with black and buff; wings extremely finely barred with gray and buff. Bill long ($1\frac{1}{2}$ cm.) and moderately curved.
Range. Breeds from central southern British Columbia, southern Alberta, southwestern Saskatchewan, and western North Dakota south to southern Baja California, Guadalupe and Revilla Gigedo islands, and through mainland of México and Central America to northwestern Costa Rica. Winters from Washington southward; southern populations permanently resident.
Occurrence in Monument. Common resident. Recorded from Little Morongo Canyon, Mar. 31; Black Rock Spring, Oct. 16; Upper Covington Flat, Aug. 24, 28 *; Lower Covington Flat, Apr. 8, 10 [1], June 26 [1], Aug. 23, 24 *, 25, 26, 27 *, Sept. 30 [1], Nov. 4 [1]; Smithwater Canyon, Apr. 10; Quail Spring, Jan. 19, 23, 27 *, Feb. 10, 11, 13, Apr. 25, May 19—21, July 1, Sept. 10 *; Stubby Spring, Apr. 9, Sept. 5, 6 *, 7 *, 10, 11 *; Barker's Dam, Feb. 18; Indian Cove, Apr. 7, 8, 27, Sept. 5, Oct. 16; Fortynine Palms, Apr. 7, Oct. 12; Twentynine Palms, Nov. 6; Pinyon Wells, Oct. 10, 12, 13, 17; Split Rock Tank, Oct. 29; Live Oak Tank, Apr.

6–8; Pinto Basin, Apr. 6; Virginia Dale Mine, Apr. 23; Eagle Mountain, Oct. 20, 21; Cottonwood Spring, Feb. 7–11, Apr. 5, 25, Sept. 6, 13, 14, 15, Oct. 22; Lost Palm Canyon, Oct. 22.

The open, rocky habitat of this wren is much more abundant and widespread than are the shaded cliffs required by the canyon wren. Accordingly, rock wrens are more abundant and may be seen readily except on the open flats. Even here occasional individuals may occur. Thus one was singing from a low rocky hill north of the sand dune in the middle of Pinto Basin on April 6, and one visited the buildings at Twentynine Palms on November 6. Indicative of abundance is the fact that one or a pair was detected on rocky slopes about every ¼ mile of a 1½ mile section of Smithwater Canyon on August 26. In the same canyon, canyon wrens were found only once.

Rock wrens expose themselves to sun and wind much more than canyon wrens and seem to tolerate warmer rock outcrops and experience higher average air temperatures. Yet it should be realized that even in the exposed or shallow rock piles they can and do find shade in midday. Like canyon wrens, they are entirely insectivorous and thus have a moist diet.

Rock wrens are profuse singers. We have specifically noted song on February 18 in the nonbreeding season, but song is common from late March to mid-May. We heard them singing on April 7 in the middle of the night by moonlight.

Immatures from the preceding nesting season are finished with the postjuvenal molt in late August and the first week of September.

The race represented in the Monument is *Salpinctes obsoletus obsoletus.*

MOCKINGBIRDS AND THRASHERS

Mockingbird. *Mimus polyglottos*

Description. Slender, gray-bodied, about 10 inches long. Wings black with large white "windows" at bases of primaries and secondaries; the wing coverts edged with white; tail black, the outer three pairs of feathers white or with longitudinal white areas; belly white. Eye yellowish.

Range. Northern interior California, southern Great Basin, South Dakota, central Iowa, central Illinois, the Ohio Valley, and southern Pennsylvania south to the Isthmus of Tehuantepec, Jamaica, and the Virgin Islands.

Occurrence in Monument. Fairly common resident. Recorded from: Little Morongo Canyon, Mar. 30, Apr. 3; Black Rock Spring, Sept. 1 *; Upper Covington Flat, Apr. 9; Lower Covington Flat, Apr. 7, May 15[1], 19; Quail Spring, May 21, July 1, 2, 4; Split Rock, May 29[1]; Indian Cove, Apr. 27; Twentynine Palms, Nov. 18, Jan. 18, Mar., Apr. 22, 28, May 24; Pinyon Wells, Apr. 10; Live Oak Tank, Apr. 6–8; Pinto Basin, Apr. 11, May 30[1], July 15; Cottonwood Spring, Sept. 13, Apr. 11, July 14.

Mockingbirds are somewhat less abundant in the Monument than is apparent, for a mocker is readily noticed when present. For example, Cottonwood Spring is a well-studied location and yet many times the species is not recorded in the course of a visit there. Probably occasional pairs reside in this section but are not established at the spring itself. Miss Carter in her stay at Twentynine Palms only occasionally recorded the species. One that seemed to be resident there in a distant mesquite patch did not come close to the buildings. Once she recorded as many as four individuals. Yet in April, 1960, mockers were much in evidence at several stations, giving frequent songs by day as well as in the night. There may have been extra concentrations of them in this year when the desert was in flower, as at Pinyon Wells and in Pinto Basin.

Mockers were evidently established at Quail Spring, in 1945 and 1946, at least, where they were active throughout the day in July and were noted in the heavy catclaw thickets of the valley bottom in May. The species seems not to occur in the piñon belt except as stragglers, but open stands of Joshua trees, as in Covington Flat, and brushy washes are inhabited at the bases of the steeper mountains as well as on the flats.

On April 11 a nest was found in a smoke tree in a wash that runs through the dunes of Pinto Basin. It was $3\frac{1}{2}$ feet up and two-thirds shaded by branch work. The nest had a stick base and a lining of rootlets; outside dimensions were 8 x 6 inches deep, and inside, 4 x $2\frac{3}{4}$ inches deep. The nest held four fresh eggs and the parent flushed from it at 50 feet, not returning. In this basin a young bird recently out of the nest was taken on May 30. Song has been heard in March and April chiefly.

Mockingbirds in general are heat tolerant. Their food of insects and berries would seem to afford them adequate moisture for desert existence.

On September 1 a vagrant, independent juvenile was taken at Black Rock Spring, where apparently it was coming for water. It showed no commencement of the postjuvenal molt.

The race of mockingbird represented is the widespread western form, *Mimus polyglottos leucopterus.*

Brown Thrasher. *Toxostoma rufum*

Description. A thrasher with prominently streaked under parts, no white in tail, and yellow eye. Dorsal surface and tail rich brown; wing coverts with two white bars or rows of spots. Bill only slightly curved.

Range. Breeds in southern Canada from Alberta to Quebec, and south to eastern Colorado, the Gulf coast of Texas, and southern Louisiana. Leaves northern part of breeding range in winter; casual west to California.

Occurrence in Monument. Vagrant. One record: Cottonwood Spring, Oct. 22 *.

At Cottonwood Spring on October 22, 1945, a thrasher of this species was "working slowly through the tops of the cottonwoods, calling loudly." It then "descended to some nearby catclaw where it could be heard scratching among dry leaves. After several minutes, the bird flew down canyon 150 yards to some small cottonwoods and then retraced its route on the ground, beneath the brush. All the while it uttered calls as if hoping for an answer" (Russell, 1947).

Fig. 52. A mockingbird in flight, showing its conspicuous white wing and tail markings.

The bird proved to be a male belonging to the western race of the species, *Toxostoma rufum longicauda.* Judging from plumage, this thrasher was a bird of the year. Immature individuals frequently are the ones involved in such vagrant records.

Bendire Thrasher. *Toxostoma bendirei*

Description. A gray-brown thrasher with fairly short, slightly curved bill, yellow eyes, and white tips on ends of lateral tail feathers. Throat and breast lightly streaked with dark brown.
Range. Breeds in southeastern California, southern Nevada, southern Utah, Arizona, southwestern New Mexico, and Sonora. Winters chiefly in Sonora and Sinaloa.
Occurrence in Monument. Occasional summer resident. Recorded from: Warren's Well, May *; Lost Horse Valley, Apr. 19.

This thrasher breeds in the Mohave Desert, but very locally, and the numbers are distinctly variable from year to year. It occurs in the Joshua tree belt, nesting often in cholla cactuses (Pierce, 1921; Johnson, Bryant, and Miller, 1948). Our parties have not found Bendire thrashers in the Monument, but Heller (1901) in May of 1896 found them to be fairly common at Warren's

Well at the western edge of the Monument and took specimens there. On April 19, C. A. Harwell and his party watched a singing Bendire thrasher for a half-hour in Lost Horse Valley.

The Bendire thrashers of the southwestern United States belong to the race *Toxostoma bendirei bendirei.*

California Thrasher. *Toxostoma redivivum*

Description. A large, brown thrasher, 12 inches long, with long curved bill. Back uniform brown, the tail only slightly darker brown; throat whitish, breast gray, and belly and under tail coverts buff, the flanks browner.

Range. Resident of chaparral from Humboldt, Trinity, and Shasta counties, California, south to northwestern Baja California.

Occurrence in Monument. Resident. Population sparse. Recorded from: Little Morongo Canyon, Apr. 1 *, 2, 3; Black Rock Spring, Apr. 4, Aug. 28, 29, Sept. 2 *, 3, 4 *; Upper Covington Flat, Apr. 9 [1]; Lower Covington Flat, Aug. 27 *; Quail Spring, Feb. 11, May 20 *, 21, 22, July 1; Stubby Spring, Sept. 5, 7, 11 *; Lost Horse Valley, Dec. 4 *; Key's View, May 14, July 15; Indian Cove, May 14 *; Pinyon Wells, Oct. 12, 13 *, 14, 15 *.

One of the distinct surprises of our field work in the Monument was finding the California thrasher resident in the rather sparse chaparral of the upper levels. Before 1945 we had thought (Grinnell and Miller, 1944:351) that it was limited to the chaparral of the coastal slopes and to that of the eastern borders of the high mountains of southern California. However, a sparse population does occur in the western half of the Monument, centering in the scrub oak and manzanita thickets of the Little San Bernardino Mountains and extending east to Pinyon Wells. This thrasher ranges locally down canyons below the chaparral in a few places, as at Quail Spring and Indian Cove, and there occupies valley bottom thickets of catclaw and desert willow. In the junipers and willow thickets of Little Morongo Canyon thrashers were also below the chaparral. Even in the densest cover the species occupies, the chaparral patches are rarely more than 60 feet in diameter, and 30 to 50 feet of open terrain may intervene between bush clumps. Low piñons intermixed with chaparral are used to some extent also.

California thrashers dig and whisk the bill in leaf litter and humous layers beneath or at the edge of chaparral as a standard method of obtaining food. In the relatively open habitat which this species encounters in the Monument, it is often seen running between foraging places, the tail tilted up, as the bird retreats ahead of a person. California thrashers generally are shy but will respond to squeaks and then come toward one or mount a bush top briefly to look about and give their alarm note, a throaty *chuck*. On May 21 at Quail Spring a pair of thrashers was agitated and scolding near some scrub jays; they then came readily to squeaking.

A specimen taken at Pinyon Wells on October 15 had a complete camel cricket *(Stenopelmatus)* in the stomach. Such a large ground-dwelling insect probably typifies the food source that these thrashers draw upon. Such moist food, and possibly at times berries, makes unnecessary the seeking of water, although elsewhere this species of thrasher has often been seen drinking where water is available. Very possibly the lack of real dependence on water because of the nature of the food has made it possible for this thrasher to penetrate the marginal type of chaparral of the Monument, whereas the chaparral-dwelling coastal brown towhees, so regularly associated with California thrashers, seem unable to do so, partly or solely because of the water factor (see p. 252).

California thrashers are well known to break into song at almost any period of the year. In the Monument we have specifically recorded song on September 7 and 11, October 15, April 3 and 4, and May 14. The autumnal singing, like that in the spring, may be brief or not full voiced, and yet on other occasions it may be loud and persistent. On September 7 and 11 two different birds were singing steadily and in full voice near Stubby Spring in oak and juniper tops. One on September 11 had in its sound repertoire a perfect imitation of the "chickadee" call of the mountain chickadee, a species which winters at this location and which had just arrived in the area in the same week. This thrasher proved to be an adult bird with testis about 4 mm. long and thus not fully quiescent but yet far below the breeding level (8 to 10 mm.). In the milder climate of the coastal slopes, some California thrashers may breed in the fall as well as in spring (Sargent, 1940), but thus far there

Fig. 53. California thrasher (right) and sage thrasher (left).

is no evidence of this in Joshua Tree Monument and singing evidently may be quite independent of breeding condition and may be engaged in by females. An adult female taken at Pinyon Wells was inactive sexually on October 15, and an adult male on August 27 at Lower Covington Flat had a testis 3 mm. long, close to or at minimum size.

Evidence of nesting was obtained at Little Morongo Canyon on April 1. Here a female with a dry postbreeding brood patch and remnants of empty ovarian follicles was taken on April 1. This would suggest egg laying about March 10. The ovary of this female showed new eggs somewhat enlarged (2 mm.), and probably another set would have followed. A second female on this date had no brood patch and no empty follicles and thus had not bred in the current year but, like the other female, might soon have produced a set (ovum 2 mm.). A male taken at Indian Cove on May 14 had a testis 6 mm. long and was thus probably in a stage of postbreeding regression, and a female at Quail Spring on May 20 had a postbreeding brood patch. Thus the scattered evidence indicates egg-laying in the Monument in March and April, and this accords with the principal laying period in coastal southern California (Willett, 1933:129).

The 12 specimens of California thrashers from the Monument plus two from the desert slopes of the Santa Rosa Mountains (3 mi. NW Rabbit Peak) to the south all indicate by degree of wear of the plumage an annual molt in July and August. Late stages of this molt are indicated by an adult male of August 27, in which the outer two primaries are still in sheaths but the body feathers are all new, and an adult female on September 2 in which the molt is finished.

The population of thrashers isolated in the Little San Bernardino Mountains might be expected to show differentiation from birds of coastal southern California parallel to that found in plain titmice, bushtits, scrub jays, and mountain quail. Careful comparison of the seven fresh-plumaged specimens we have taken with comparable material from the adjacent coast districts convinces us that there are no consistent differences in coloration or other features. Racial evolution in thrashers of this area seems therefore not to have taken place. The birds of the Monument should therefore be designated as *Toxostoma redivivum redivivum.*

Le Conte Thrasher. *Toxostoma lecontei*

Description. A slim, light ashy gray thrasher (10 inches long), with long curved bill. Tail feathers black, contrasting with body plumage, the tips brown; under parts pale gray to buffy gray; belly white. Iris dark in adults.

Range. Resident of desert regions of interior southern California, southern Nevada, southwestern Utah, and western and central Arizona south to northeastern and central Baja California and northwestern Sonora.

Occurrence in Monument. Sparse resident, locally. Recorded from: 6 mi. SE Joshua Tree P. O., Sept. 8; Quail Spring, May 19; near Key's Ranch, Apr. 21; Twentynine Palms, Mar. 21, 23, Apr. 15, July 12, 13; Pinto Basin, Apr. 11 *, July 15, Aug. 22 [1], Sept. 13; Cottonwood Spring, May 17 *, 19, July 7.

Fig. 54. Le Conte thrasher running over sandy flat in Pinto Basin.

Le Conte thrashers occur in flat open desert or in large washes where large shrubs or chollas are widely scattered. The soil over which they run swiftly is usually fine alluvium or sand. The creosote bush association is typically occupied, and the scanty litter of leaves under this plant and about the more diversified low shrubs of the washes (fig. 55) constitute the foraging ground, where they whisk and dig into the soil, principally for insects. In periods of lush annual flowers and attendant insect abundance, these thrashers take caterpillars (fig. 4). Thus in April of 1960 in Pinto Basin, a male was collected which had a large sphinx moth caterpillar in the stomach and whose breast plumage was stained yellow either from caterpillar juice or pollen of the flowers.

The Le Conte thrashers occur in the Monument in only a few areas; they are by no means found in all sections that would seem to be usable. Stout, dense bushes are required for holding and protecting their bulky nests, cholla cactuses being ideal, and in a few places the absence of such bushes may render the locality unsuitable for them. Yet many open sandy flats in which chollas are present have no thrashers. Altitude as such seems to be no barrier if the terrain is fairly flat and open.

Not infrequently Le Conte thrashers may be discovered by first finding their nests, which are large stick structures, characteristically lined with a felted mat of cottony seed heads, most often made of the seed clusters of the creosote bush. Nests may persist for several years, even after the thrashers have disappeared. But careful search among a scattering of nests will usually

Fig. 55. A wash bordered by smoke trees in Pinto Basin, habitat of Le Conte thrashers.

reveal an occasional thrasher, running in the distance or shyly skulking behind bushes.

A nest in Pinto Basin was placed 3 feet up in a smoke tree against the main trunk and held firmly in the spiny twigs, where it was about one-half shaded. Its heavy framework was 8 x 8 inches deep outside, and the lined cup was 4½ x 3 inches deep.

The breeding season is early, perhaps triggered by the light and erratic winter rains and the subsequent rapid appearance of annual green plants. At Twentynine Palms on March 21, C. A. Harwell recorded a bird sitting closely on a nest in a low shrub, but no eggs had yet been laid. North of here in the sand dunes, Miss Carter saw a nest in a cholla cactus on April 15 which held young. On April 21 Harwell reported a nest under construction at Key's Ranch, and at Cottonwood Spring on May 17 a fully grown young of the species was taken, in an association of *Yucca mohavensis,* junipers, and chollas in an open bajada. Males were singing in this area (testis of specimen 7 mm.). But two days later near Quail Spring a nest in a cholla in a gently sloping wash contained 3 young estimated to be 5 days old. In Pinto Basin on April 11, 1960, males were singing, and a male (testis 10 mm.) visited a completed new nest, then mounted above to the 10-foot bare tip of a smoke tree to resume singing. The nesting season thus evidently is prolonged, perhaps is variable from year to year in timing, and doubtless extends late into spring at times with replacement nests or second-brood efforts.

Despite the difficulty at times of locating adults in places where there are nests, we are convinced that occasionally areas once occupied by Le Conte thrashers become deserted. When such locations are visited in the middle of the nesting season and no fresh nests are found nor adults detected by sight or sound on prolonged search, this conclusion seems inescapable. This was so at scattered old nests among smoke trees and chollas in two places a mile apart in Pinto Basin on April 5 and 6, 1951; the area then was evidently unoccupied by thrashers, whereas we had seen them near these nests in July of 1946, in September of 1950, and again in April of 1960. Before April of 1951 there had been almost no rain in this section for three years and the desert vegetation in 1951 was showing the effect of this extreme drought. We believe that species such as the Le Conte thrasher, which exist close to the level of tolerance even for desert species, have to move out at times. The Le Conte thrasher of course requires no drinking water, but in prolonged droughts annual plants and leaf litter below bushes must yield fewer insects than normal, and cactus fruits as possible sources of water do not materialize. Then the birds must move or die. Rains in the deserts are spotty and local, and a Le Conte thrasher "colony" could move a few miles to a new site where rains had fallen. We suspect that this kind of shifting goes on and that there is perhaps a premium on a special ability to do so among organisms that live under desert extremes.

Le Conte thrashers of the Monument belong to the race *Toxostoma lecontei lecontei.*

Sage Thrasher. *Oreoscoptes montanus*

Description. A small thrasher (8 inches long), with short, almost straight bill and heavily streaked under parts. Back gray, obscurely streaked; wings dark gray, the coverts and secondaries edged with white; tail dark gray, the lateral four pairs of feathers tipped with white. Eyes pale yellow (fig. 53).

Range. Breeds in Great Basin and Rocky Mountain brushlands from southern British Columbia, Montana, and southwestern Saskatchewan south to upper levels of Mohave Desert and northern New Mexico. Winters from central California, southern Nevada, northern Arizona, southern New Mexico, and central Texas south to southern Baja California and Guanajuato.

Occurrence in Monument. Fairly common winter visitant. Recorded from: Upper Covington Flat, Mar. 19[1]; Lost Horse Valley, Apr. 3, 6; Twentynine Palms, Dec. 11, Jan. 7, 29, Feb., Mar. 21, 23, 24; Pinto Basin, Sept. 13.

Sage thrashers forage in open desert terrain, although they utilize as retreats the low shrubs and bushes present there. They often are in small, loose flocks of 6 to 8 individuals. Miss Carter found them variable in numbers in two successive winters. They were "quite numerous" in 1934, frequenting mesquite at the edge of the oasis at Twentynine Palms, but the next year they were scarce.

The earliest date recorded is that of September 13, when one was taken in Pinto Basin; it was an immature female seen perched low down in a large creosote bush on the crest of a sand dune.

This species of thrasher breeds not far away in the Mohave Desert in the vicinity of Victorville. We have no evidence that it breeds in the Monument, and the behavior of the birds seen there suggests that of migrants or winter visitants. The species is primarily one of the sagebrush belt to the north where temperatures are somewhat lower in summer than in the Monument and aridity is less severe. However, winter residence in the Joshua tree belt should present no special problems for this insect-eating species.

THRUSHES, GNATCATCHERS, AND ALLIES

Robin. *Turdus migratorius*

Description. Above gray, the head, wings, and tail blackish; below brick red; throat streaked black on white; tail with very small lateral white tips. Female usually duller than male.
Range. Breeds from northern tree line of North America south to southern California, Guerrero, Oaxaca, and the Gulf coast of the United States. Winters from southern Canada south to Guatemala, Yucatán, and southern Florida.
Occurrence in Monument. Fairly common migrant and winter visitant. Recorded from: Little Morongo Canyon, Mar. 30, 31, Apr. 1 *; Upper Covington Flat, Nov. 18 [1]; Lower Covington Flat, Mar. 24 [1]; Smithwater Canyon, Feb. 16 [1], Mar. 10 [1]; Quail Spring, Feb. 17, Apr. 25; Stubby Spring, Oct. 15; Barker's Dam, Feb. 18; Twentynine Palms, Oct. 29, 30, Nov. 5, 6, 7, 14, Dec. 8, 9, Jan. 26, 30, Feb. 25, Apr. 12; Pinyon Wells, Oct. 12, 13; Live Oak Tank, Apr. 6–8; Cottonwood Spring, Oct. 21, 23 *, Feb. 7, Apr. 5; Lost Palm Canyon, Oct. 22.

Robins are widely scattered in the Monument area in winter, but seldom are more than two or three seen at a time. Some must be en route in migration in October and in late March and early April, but only once has a conspicuous concentration been noted, namely on March 31 in Little Morongo Canyon, when about 25 were present along the small stream. Extreme dates known to us are October 15 and April 25.

Robins come to water at the oases, as we have recorded specifically on April 25 at Quail Spring and on October 21 at Cottonwood Spring. There are few places where they can feed, as they prefer, on moist ground or turf, such places being at a well outflow at Twentynine Palms and in small marshy borders of Little Morongo Creek. Otherwise robins probably use berries as food or glean insects from dry ground surfaces while in the desert area. The period of occurrence in the Monument excludes the warm part of the year.

The widespread western race of robin, *Turdus migratorius propinquus,* is the form that commonly winters in Californian deserts, coming from breeding grounds to the north in the Rocky Mountain and inner coastal districts or from the nearby Great Basin areas and Californian mountains. Specimens from the Monument generally represent this form. However, a female, taken on October 23, 1945, at Cottonwood Spring, was very dark-colored and seemed certainly to be *T. m. caurinus.* As such it represents a considerable southward extension of the known winter range of this subspecies which breeds in southeastern Alaska and south along the coast to northern Oregon. Previous winter records of *caurinus* are known south only to the San Francisco Bay region (Grinnell and Miller, 1944) and the Santa Cruz area in central California (specimen, Santa Cruz, December 20, 1939, in the Museum of Vertebrate Zoology).

Hermit Thrush. *Hylocichla guttata*

Description. A slender (20–29 gram), long-legged thrush, brown or gray-brown above, the rump brighter rusty brown than back and tail; light eye ring; under parts spotted.
Range. Breeds from central Alaska east in timber belt to southern Labrador and south in the mountains to southern California, southeastern Arizona, and southern New Mexico. Also south to northern Great Lakes region and southern New York and in Appalachian Mountains to western Maryland. Winters from southern British Columbia, southern Nevada, central Colorado, northern Texas, the Ohio Valley, and Massachusetts south to southern Baja California, Guatemala, the Gulf coast of the United States, and southern Florida.
Occurrence in Monument. Fairly common fall and spring migrant. Winter resident in small numbers. Recorded from: Little Morongo Canyon, Apr. 1 *, 2 *; Black Rock Spring, Apr. 3; Upper Covington Flat, Oct. 14 [1], 27 [1]; Lower Covington Flat, Apr. 30 [1]; Smithwater Canyon, Apr. 8 *, 23 [1], 30 [1]; Quail Spring, Jan. 26; Stubby Spring, Sept. 7 *, Oct. 15; Twentynine Palms, Mar. 21, 23, Apr. 2, 3, 16, May 5; Pinyon Wells, Oct. 9 *, 12, 13 *, 14 *, 17; Eagle Mountain, Oct. 19, 21 *; Cottonwood Spring, Oct. 13, 22 *, Apr. 26.

The fall migration extends from September 7 into October, some of the occurrences in the middle of that month probably reflecting birds settled on winter quarters in the piñon and oak scrub. In the spring, migrants are in evidence chiefly from March 21 to April 30, with one late occurrence on May 5. Even when hermit thrushes are most certainly in migration, they are seen by day in relatively good plant cover. Thus at Twentynine Palms, where they seemingly did not winter during Miss Carter's sojourn there, the thrushes when seen in spring migration were found in the underbrush near the water. Spring migrants in Little Morongo Canyon were in baccharis bushes beneath tall willows by the stream, and similarly in Smithwater Canyon a bird was in a willow clump. An early fall migrant on September 7 was moving under cover of junipers and oaks close to the ground. Unlike the Swainson thrush, the hermit thrushes have not been seen in obvious diurnal migratory flights and have not been heard at night, but perhaps this species does not call in nocturnal flight.

In mid-October in the piñons and oak brush and in shaded areas about rocks and steep canyon walls at Pinyon Wells and on Eagle Mountain, hermit thrushes probably were settled for winter. As usual they were solitary, but they were spaced out in places with good leaf litter, partly under brush canopy, where they typically forage. As many as 6 to 10 were noted on a single morning.

On October 15 at Stubby Spring one was watched eating honeybees. At Quail Spring on January 26 two different individuals were in a catclaw

thicket foraging by flipping over leaves on the ground in search for insects.

This species, by reason of its migration schedule, usually arrives after the severe summer heat has broken, and it departs in spring, earlier than the Swainson thrush, and before temperatures are again high. Winter residence is in the upper brush and woodland belt. Thus severe drought and heat are largely avoided.

The races of hermit thrushes represented in the sample of specimens taken are two. *Hylocichla guttata slevini,* the breeding form of coastal northern California and the Cascade Range, was taken on October 13 at Pinyon Wells, on April 23 and 30 in Smithwater Canyon, and on April 30 at Lower Covington Flat. All others are of the race *H. g. guttata* which breeds in the interior of British Columbia and in south-central Alaska. Birds collected on September 7 and on October 9, 13, and 21 were fat and thus may still have been in migration. One taken on October 22 had no fat. Fat was not noted in the early April birds, two of which were males with testes 2 and 2½ mm. in length and thus showing little progress in recrudescence.

Swainson Thrush. *Hylocichla ustulata*

Other name. Known widely in the West, under the name of one of the subspecies, as russet-backed thrush.

Description. A slender, long-legged thrush (24—30 grams), uniform russet-brown or olive-brown (dependent on race) above, the rump and tail not brighter than the back; buffy eye ring; under parts spotted.

Range. Breeds from northern limit of the boreal forests of Alaska and Canada south to southern California, southern Colorado, the Great Lakes area, and West Virginia. Winters from southern México south to Paraguay and northwestern Argentina.

Occurrence in Monument. Common spring and fall migrant. Recorded from: Upper Covington Flat, May 13[1], 14[1]; Lower Covington Flat, May 14[1]; Smithwater Canyon, May 13[1], 14[1]; Quail Spring, Apr. 25, 26, May 19—21; Stubby Spring, Sept. 5, 9 *, 12; Indian Cove, May 13; Twentynine Palms, May 7, 14, 17, 30; Cottonwood Spring, Apr. 29, May 6, 9, 12, 13, 14, 17 *, June 3.

Swainson thrushes regularly pass through the desert in considerable numbers, but they always move solitarily. They often fly long distances through the desert scrub, but, when they stop, they usually seek as much concealment as possible in thick cover such as that afforded by junipers. About springs as many as five have been seen at one time scattered through the vegetation and on the ground. Often just before dawn they can be heard calling in flight, indicating that migratory flights are carried on at night; yet migratory movement is considerably in evidence by day also. On May 17 near Cottonwood Spring a Swainson thrush was seen flying eastward, over rolling terrain and along washes; this direction may not have been representative of the general migratory route taken.

The period of spring migration extends from April 25 to May 30. In the fall we have found these thrushes only between September 5 and 12, but probably they occur somewhat later also; the most severe heat and drought

Fig. 56. Townsend solitaire (above), hermit thrush (below), and Swainson thrush (left).

of August, 1950, was apparently avoided by the late arrival of Swainson thrushes in the area.

The birds probably visit springs whenever they pass near them, but it seems unlikely that they are much dependent on them for water. Their insect food should contribute considerable water, as it does to other insectivorous types. Migrants are normally well supplied with fat reserves.

On May 9, 1946, at Cottonwood Spring, a Swainson thrush was found that had just been killed by a rattlesnake (*Crotalus mitchelli*). The bird had evidently come to the water and had been struck and mouthed; the snake was waiting beside it but slipped away without rattling. On April 29, 1952, at the same locality a Swainson thrush was seen tussling with a lizard 4 inches long which it seemed to be trying to eat.

All specimens taken belong to the coastal race *Hylocichla ustulata ustulata*. It is probable that the form breeding in the Great Basin passes through the Monument also. The specimen of September 9 was an immature that had completed the postjuvenal molt. The specimen of May 17, a female, was exceptional in that it had no fat and weighed but 21 grams; it apparently was in rather poor condition.

Western Bluebird. *Sialia mexicanus*

Description. A bluebird marked with rust-brown areas on breast and back. Male with head, neck, lower back, wings, and throat deep, bright blue; chestnut-brown areas in lateral and sometimes in middle parts of back; breast and flanks rust-red; belly dull blue to white. Female with blue and brown areas much duller than in male, the throat often nearly gray.

Range. Breeds from southern British Columbia and central Montana south to northern Baja California, Morelos, and western Veracruz. Partly migratory in north and locally, moving to deserts and lower altitudes for winter.

Occurrence in Monument. Common winter resident. Recorded from: Morongo Valley, Apr. 11, 14; Little Morongo Canyon, Mar. 31; Upper Covington Flat, Apr. 8 [1], Aug. 28 *; Lower Covington Flat, Apr. 9 [1], 23 [1]; Quail Spring, Oct. 15, Dec. 4 *, Jan. 21, 25, 27, Feb. 13 *; Barker's Dam, Feb. 18; Twentynine Palms, Nov. 10, Jan. 18, 22, 27, Mar. 1, 20–29.

Western bluebirds visit the Monument from the end of August through March and may be found both in the piñon belt and in the lower, open desert. Our records are all for areas from Twentynine Palms westward. Doubtless there are more easterly occurrences, and yet it would appear that most of the wintering population is in the western part of the Monument.

A prominent feature of winter activity in these bluebirds is the feeding on mistletoe berries. Miss Carter found them working on mistletoe (*Phoradendron californicum*) in the mesquites at Twentynine Palms in January, and we saw them in the same activity in mesquites in Little Morongo Canyon on March 31. Concentration at mistletoe was also noted in December. At other times these bluebirds forage on insects, dropping down to the desert floor. At Quail Spring they were seen drinking at the tank on three occa-

Plate III. Male bluebirds and goldfinches: Mountain bluebird

Western bluebird

Lawrence goldfinch

American goldfinch (winter plumage)

Lesser goldfinch

GMC

sions in January and February. Moist food and water is thus taken in the winter period of residence.

Bluebirds, like phainopeplas (see p. 192), are agents in the dispersal and implantion of the sticky mistletoe seeds, and thus a mutually beneficial relationship between the bird and the plant species exists.

Western bluebirds are usually seen in flocks in winter. These are normally in the order of 3 to 10 individuals, but in January at Twentynine Palms groups, with some mountain bluebirds intermixed, were as large as 25 and 50.

An adult male taken on August 28 and presumably a very recent arrival, possibly from the nearby San Bernardino Mountains, was in the middle of the annual molt. Old primaries 9 and 10 were still retained, and most of the head and neck was still covered with old worn feathers.

Specimens taken on August 28, February 13, and April 9 and 23 have been of the race *Sialia mexicanus occidentalis,* the form breeding to the west and northwest of the Monument. However, van Rossem took a specimen at Quail Spring on December 4 which is *S. m. bairdi,* the Rocky Mountain race, and one from Upper Covington Flat on April 8 is also of this form. Winter visitants of both eastern and western breeding populations are thus to be expected.

Mountain Bluebird. *Sialia currucoides*

Description. A bluebird without any rich brown areas in its plumage. Male light azure blue above and below, except belly, which is white. Female dull brownish with pale, dull blue rump, tail, and wings.

Range. Breeds in western North America from central Alaska east to southwestern Manitoba and south in mountains to San Bernardino Mountains of southern California and to central Arizona and southern New Mexico. Winters from southern British Columbia, western Montana, and Kansas south to northern Baja California, Michoacán, Guanajuato, and coastal Texas.

Occurrence in Monument. Winter visitant in moderate numbers. Recorded from: Upper Covington Flat, Mar. 10[1]; Quail Spring, Jan. 27; Twentynine Palms, Nov. 16, 18, Jan. 22, 25, 28, Feb. 9, Mar. 7, 10, 13, 24.

This bluebird frequents open desert in winter, and our principal information on it is derived from Miss Carter's observations at Twentynine Palms. The birds arrive late and depart early, the extreme dates being November 16 and March 24. Mountain bluebirds often will join in loose flocks with western bluebirds. Miss Carter saw such an aggregation catching insects in the open desert on February 9. The greatest numbers were reported in January; she noted about a dozen on January 25. At Quail Spring on January 27 three came together to the catclaw by the water tank, probably attracted to the water source, and then flew off together.

This arctic and alpine species, it is noteworthy, frequents the desert only in the coldest months. Its insect food, taken chiefly from the open ground, should afford it good moisture.

Townsend Solitaire. *Myadestes townsendi*

Description. A slender, long-tailed member of the thrush group with gray body, unspotted in the adult. Tail with outer part of outermost feathers white; wing with dull yellowish patch at base of flight feathers, exposed as a light patch in flight; secondaries edged subterminally with buff in form of a band, and inner secondaries white-tipped; eye ring white (fig. 56).
Range. Breeds in western North America from central eastern Alaska, southern Yukon, and southwestern Mackenzie south to southern California and Durango. Winters from southern British Columbia and western Nebraska south in lowlands to northern México.
Occurrence in Monument. Winter visitant in small numbers. Recorded from: Lower Covington Flat, Oct. 14[1], 15[1]; Quail Spring, Jan. 22, 23 *, 25, 26, 27; Stubby Spring, Sept. 5, 8 *, 11 *, Dec. 4–6, Apr. 9 *; Pinyon Wells, Oct. 13, 14 *, 17; Eagle Mountain, Oct. 19; Cottonwood Spring, July 11[1], Oct. 23 *; Lost Palm Canyon, Oct. 23.

The period of winter residence extends from September 5 to mid-April, and the birds seem always to settle in the piñon belt or in canyons. The juniper areas are especially attractive, as solitaires regularly eat juniper berries. There was a good supply of these available in October, for example, at Pinyon Wells.

In keeping with their name, these birds are usually seen singly, but on October 13 at 5000 feet, above Pinyon Wells, three were seen, one chasing off two others. Usually they are silent, but on October 17 one was giving its single, rhythmic, clear whistle in a shady canyon at 6:30 A.M. On Eagle Mountain one came to a disturbance caused by imitating horned owl notes.

We have seen Townsend solitaires in willows, catclaws, and paloverdes near the oases and at other points in scrub oak bushes and mountain mahogany. The species thus seldom appears in truly desert vegetation except in canyons and near watering places. Berries and the insect food utilized should afford good moisture for it during its winter sojourn in the higher desert areas.

The specimen in the Long Beach State College Collection obtained on July 11 at Cottonwood Spring obviously represents a nonbreeding bird that failed to migrate from the wintering grounds.

The birds taken in the Monument are of the nominate race, *Myadestes townsendi townsendi.*

Blue-gray Gnatcatcher. *Polioptila caerulea*

Description. Size very small (5 grams). Blue-gray above and white to very pale gray below; tail black centrally, the outer tail feather white, next to outer with terminal third white, and adjacent feather medially with white tip. Males in spring plumage with black frontal area extending back as narrow line above eye.
Range. Breeds from northern California, central Nevada, Utah, and Colorado east to southern Great Lakes area and New Jersey and south to southern Baja California, Guatemala, the Gulf coast, and the Bahama Islands. Winters from southern United States southward, extending to Cuba.
Occurrence in Monument. Common summer resident. Recorded from: Morongo Pass, Apr. 28; Little Morongo Canyon, Mar. 31; Black Rock Spring, Aug. 29, 30, 31, Sept. 4 *; 3 mi. S Warren's Well, Aug. 31; Upper Covington Flat, Apr. 9, Aug. 28; Lower Covington

Flat, May 19, Aug. 23 *, 24, 25, 26, 27, 28, 29; Smithwater Canyon, May 13 [1]; Quail Spring, Apr. 10, May 21, June 3, July 1, Sept. 8; Stubby Spring, Sept. 5, 6 *, 7; Key's View, May 14, July 15; Indian Cove, Apr. 7, 8, May 12, 14; Twentynine Palms, Mar. 18, July 12, 13; Pinyon Wells, Apr. 10, Oct. 12—18; Pinto Basin, Apr. 6; Eagle Mountain, May 16; Cottonwood Spring, Apr. 5, May 14—19, 16, July 7, Sept. 13, 14, 15.

This gnatcatcher favors the upper levels of the Monument and nests in the piñon-juniper and chaparral habitats. However, it ranges downward to overlap the breeding range of the black-tailed gnatcatcher which is typical of the low-desert washes. Blue-gray gnatcatchers in the periods of migration appear at low-level stations, such as Pinto Basin and Twentynine Palms; it is unlikely that they nest there in the creosote-bush and mesquite habitats which are typical of the other species. Postbreeding dispersal of blue-gray gnatcatchers occurs early, and even some of our records in July may represent this and the beginning of migration, especially in areas apart from the normal nesting habitat. The species might well spend the winter at such places as Twentynine Palms, for it winters in the floor of the Coachella Valley and in low desert areas to the east of the Monument. However, the one date Miss Carter specifies at Twentynine Palms, namely March 18, could represent migration rather than wintering.

In the breeding season, blue-gray gnatcatchers are fairly conspicuous in the piñons and junipers, as at Key's View, Stubby Spring, Quail Spring, Lower

Fig. 57. Black-tailed gnatcatcher (above), blue-gray gnatcatcher (right), and ruby-crowned kinglet (left).

Covington Flat, and Pinyon Wells. The sustained whining song of the males was noted on April 7, 9, and 10, and May 21. On the latter date at Quail Spring, in the heavy catclaw thicket of the bottomland, a nest was found 10 feet up. It was still being built by the pair.

On April 7 in the desert willows at Indian Cove, below the canyon mouth, a pair had a nest 7½ feet up in a three-way upright crotch. The neat, felted cup was 2½ x 2½ inches deep outside, and 1½ x 1½ inches deep inside. The material consisted of bark and dark gray fibers held together with cobwebs, and it had no external lichen cover. The nest held one egg, and the female settled on it while we were within 15 feet, possibly anxious to lay. The shade temperature at the time was only 90° F. at 11:15 A.M.; the nest was half-shaded and the pit of the cup completely so. The male sang vigorously in the same tree and nearby bushes in sun temperature of 94° F.

On May 16 birds were spaced out on territories in the juniper-piñon scrub on the crest of Eagle Mountain. On August 28 at Lower Covington Flat one gnatcatcher was seen fluttering its wings and begging in the presence of another. This probably indicated a dependent juvenile, as courtship feeding seems unlikely at that date. An immature was collected at this station on August 23 which was halfway through the postjuvenal molt.

Gnatcatchers are insect feeders exclusively. Despite their small size, which aggravates their problem of water loss, the constant insect food enables them to stay in water balance without normal recourse to springs. However, one was seen to drink at a pool below Lower Covington Flat on August 24, at 4:00 P.M. on a hot afternoon. Gnatcatchers of both species are much exposed in their habitats and often, as indicated, are active in strong midday sunlight and in high temperatures. The average temperatures met by this species are not quite so severely high as those experienced by the black-tailed gnatcatcher, but they must nonetheless be formidable; slightly more shade areas are available in the habitat of the blue-gray. When, as often happens, air temperatures exceed the body temperature, water loss in evaporative cooling should be rapid and a premium must be placed on success in ingesting moist food with little interruption.

The blue-gray gnatcatchers are of the widespread western race *Polioptila caerulea amoenissima.*

Black-tailed Gnatcatcher. *Polioptila melanura*

Description. Size very small (5 grams). Plumbeous gray above and light gray (local race) to white below; tail black, the outer pair of feathers with white lateral web and white tip, the next pair with white tip. Male in spring plumage with lores, top of head, and postocular region black, but the eye ring white.

Range. Southern California, southern Nevada, central Arizona, southern New Mexico, and the lower Río Grande Valley in Texas south to southern Baja California, Durango, San Luis Potosí, and Tamaulipas.

Occurrence in Monument. Common resident. Recorded from: Little Morongo Canyon, Mar. 31 *—Apr. 3; Quail Spring, Jan. 23 *, Feb. 10, 11, 13, 17, May 19—21, July 1; Indian

Cove, May 13, 14; Twentynine Palms, July 12, 13; Pinyon Wells, Oct. 13 *—18; Live Oak Tank, Apr. 6—8; Virginia Dale Mine, July 12, 14; Cottonwood Spring, Feb. 7—11, May 14—17 *, July 7, 14, Sept. 13, 14, 15 *, Oct. 22 *, 23; Pinto Wash Well, Oct. 16[1].

The black-tailed gnatcatcher is primarily a bird of the catclaw (*Acacia greggii*) thickets in the Monument, but it also uses the limited patches of mesquite and occasionally is seen in smoke trees. Adjacent scrub plants are at times frequented, but the center of activity is in these larger plants below the piñon belt. It is noteworthy, however, that the species follows the catclaw upward where it interdigitates with piñons. This circumstance and the use of catclaw at the upper levels by the blue-gray gnatcatcher by extension downward from the piñon woodland means that the two species of gnatcatchers are not uncommonly in contact. Thus both evidently breed in the thickets just below Quail Spring and at Indian Cove.

Unlike the blue-gray gnatcatcher, this species stays through the winter in normal numbers, as specifically noted on a number of occasions at Quail Spring and Cottonwood Spring in January and February.

Nesting was observed at Indian Cove on May 13, when a pair was building in a mistletoe clump in a catclaw. The nest was slung in hanging dead twigs of mistletoe. This sort of location is frequent in this species; the dense mistletoe twigs, even when dead, afford some concealment and better shade than would the open, almost leafless branches of the catclaw. That this nest was later than some is shown by the occurrence of fledged but dependent young on May 17 at Cottonwood Spring, birds hatched from eggs probably laid in early April. However, similar dependent young were seen at this station on July 14, so the breeding season must be extended. Males on March 31 and May 17 had testes 3 and 4 mm. long, respectively, the latter at functional level.

As noted in connection with the blue-gray gnatcatcher, these small species must take small moist insects at frequent intervals in order to keep pace with their water loss, which is severe due to exposure to heavy sun and high temperatures, particularly in a bird of small body mass. In the low-desert habitat of this species shade is minimal, but we think these gnatcatchers tend to seek it in midday, as we often see them in and about the dense mistletoe clumps of the catclaws and mesquites. These clumps can afford significant protection from sun and wind, a protection which the small verdin achieves in part by its enclosed nests. Among the insect-feeders of the desert, the black-tailed gnatcatcher, because of its size and its lowland, hot, and exposed habitat, must, if lacking special adjustments and temperature tolerances, live dangerously close to critical limits of the environment.

The annual molt is completed rather late in this gnatcatcher. An adult female on September 15 had not yet dropped outer primaries 7 to 10 and the body was in heavy molt. An adult on October 13 still had sheaths at the bases of the outer primaries and pinfeathers on the body, especially about the head.

Black-tailed gnatcatchers of the Monument are of the race *Polioptila melanura lucida* of the desert province. They show no approach whatsoever in color to the much darker race of the semidesert scrub of coastal southern California.

Golden-crowned Kinglet. *Regulus satrapa*

Description. Tiny (6 gram), slightly built bird, the crown gold with black bordering line and white superciliary stripe. Back green; under parts gray, darker and greenish on flanks; wing coverts with two light buff bars.
Range. Breeds from southern Alaska east across tree belt to Quebec and south to mountains of southern California, eastern Tennessee, and western North Carolina; also resident in highlands of southern México and Guatemala. In winter from southeastern Alaska, northern United States, and Newfoundland south to lowlands of coastal southern California, Arizona, and New Mexico and to northern Coahuila, Tamaulipas, the Gulf coast, and northern Florida.
Occurrence in Monument. Two records: Upper Covington Flat, Nov. 18, 1961 [1]; Stubby Spring, Apr. 9, 1951 *.

This kinglet in winter adheres typically to dense tree growth and thus, especially at the lowland southern periphery of its winter range, it avoids the deserts. Unexpectedly a straggler was encountered in a solitary piñon next to a large willow at Stubby Spring on April 9. While wounded, it called vigorously in such a manner as would have elicited response from other members of this very social species had they been present. None other was seen or heard, showing that the bird was indeed alone, a most unusual circumstance for this kind of kinglet. The bird was fat and in good condition and thus showed no evidence of difficulty resulting from its unusual exposure in the open desert.

Ruby-crowned Kinglet. *Regulus calendula*

Description. Tiny (6 gram), slightly built species, the head gray, without black; ruby crown in males, visible only when feathers spread. Back olive-green; under parts pale yellowish or greenish gray; whitish eye ring and wing bars conspicuous (fig. 57).
Range. Breeds from northwestern Alaska east across tree belt of Canada to southern Labrador and south, in mountains, to southern California, southern Arizona, and central New Mexico, and to northern Michigan, northern New York, and northern Maine; also Guadalupe Island, off Baja California. Winters from southern British Columbia, Idaho, Nebraska, the southern Great Lakes region, and New Jersey south to southern Baja California, Guatemala, the Gulf coast, and Florida.
Occurrence in Monument. Common winter resident. Recorded from Little Morongo Canyon, Mar. 31, Apr. 1, 2; Black Rock Spring, Apr. 3, Oct. 16; Lower Covington Flat, Apr. 7, 8, Oct. 23 [1]; Smithwater Canyon, Mar. 24 [1]; Quail Spring, Jan. 22, 23 *, 26, Apr. 22; Stubby Spring, Nov. 1, Dec. 4–6, Apr. 9; Lost Horse Valley, Nov. 1, Feb. 10; Key's View, Oct. 31; Barker's Dam, Feb. 18; Indian Cove, Apr. 8; Twentynine Palms, Oct. 20, 26, 27, 30, Jan. 10, Feb. 25, Apr. 3, 9; Pinyon Wells, Oct. 10, 12, 14 *, 16, 17; Split Rock Tank, Oct. 29—31; Live Oak Tank, Apr. 6–8; Squaw Tank, Mar. 17 [1]; Eagle Mountain, Oct. 19; Cottonwood Spring, Oct. 13, 23, Feb. 7—11; Lost Palm Canyon, Oct. 23.

This kinglet in winter favors low, open growth as well as the foliage of larger trees and thus finds conditions generally favorable in the desert at this season. Its arrival in October is late enough and its departure in spring early enough to escape the desiccating heat period. Birds forage singly, as typical of this species, and use all manner of desert bushes and trees four feet or more in height, even though widely spaced. We have seen ruby-crowned kinglets in catclaws, baccharis, desert willows, junipers, scrub oaks, piñons, willows, and cottonwoods, but more commonly in the conifers and in the willows and cottonwoods about the springs than in other cover.

These delicate little insect feeders hover in the foliage and dart out from it, capturing both moving and stationary prey. Individuals may concentrate at favorable food areas, for example a willow coming into flower and leaf (April 9) in the spring, and two to five individuals have been seen working individually through such a tree, challenging one another on close approach with their grating chatter.

Our earliest and latest dates for the Monument are October 10 and April 22. In mid-October at Pinyon Wells kinglets were particularly numerous, and we think that some of them were in migration, when, on October 17, birds could be detected every few minutes in the piñon belt. However, the remaining winter residents are numerous, even at the high levels in periods of nightly freezing temperatures. Occasional singing is heard, as on October 23 and April 2 and 7, in the first case on the part of a male at sunrise while he simultaneously exposed his red crown.

One of the two birds taken on October 14 just after arrival appeared to be dangerously lightweight at 4.8 grams. However, it did not necessarily have to migrate farther and it might have been able to survive once in this wintering area.

The ruby-crowned kinglets wintering in the Monument are of the race *Regulus calendula cineraceus,* the form breeding in the mountainous areas of western North America.

Water Pipit. *Anthus spinoletta*

Description. Size of a white-crowned sparrow, but with slender bill and white outer tail feathers; shows conspicuous action of bobbing the tail and hindquarters while walking. Above brown, obscurely streaked, the prevailing color becoming grayish in spring plumage; below buff, with varying amount of streaking on breast and flanks, the buff becoming richer in spring.

Range. Breeds in tundra and high mountains of Northern Hemisphere, extending south in mountains of western North America to northern Oregon, northern Arizona, and northern New Mexico. In North America, winters from southern British Columbia, central Nevada, southern Utah, Texas, Arkansas, Tennessee, and the Delaware Valley south to Guatemala and Florida.

Occurrence in Monument. Occasional winter visitant. Recorded from: Twentynine Palms, Oct. 31, Nov. 11, Jan. 18, Feb. 28; Pinto Wash Well, Oct. 16[1].

This species does not remain long in migration or in winter except in areas where there is some grass or moisture in the open ground it frequents. Miss Carter mentions a flock only once at Twentynine Palms, on

October 31. On November 11 and February 28 single individuals were noted.

The specimen in the collection of Long Beach State College is of the race *Anthus spinoletta pacificus.*

Cedar Waxwing. *Bombycilla cedrorum*

Description. Size of a large sparrow (32–34 gram) with conspicuous crest and generally sleek appearance. Above brownish fawn color, grayer posteriorly and on wings; below similar, but blackish on chin, grading to light yellow posteriorly; head marked by black mask from base of upper mandible through the eye; tip of tail yellow. Adults with varying number of small waxy flecks of red on ends of secondaries. Juvenal plumage with coarsely streaked breast.
Range. Breeds from southeastern Alaska and north-central British Columbia east to Newfoundland and south to northern California, northern Utah, Colorado, Missouri, and northern Georgia. Winters from southern British Columbia, northern Idaho, northwestern Colorado, Missouri, southern Great Lakes region, and Massachusetts south to Panamá, the Gulf coast, and central Florida; casually south to northern South America.
Occurrence in Monument. Spring and fall migrant. Occasional in winter. Recorded from: Stubby Spring, Dec. 4–6; Upper Covington Flat, June 2 [1]; Twentynine Palms, Oct. 17–31, Nov. 4, Mar. 1, 3, 8, Apr. 14–16, May 7, 12, 17, 24–27; Pinyon Wells, Oct. 9, 12, 14, 17; Eagle Mountain, Oct. 19, 20; Cottonwood Spring, Sept. 15 *, Oct. 14, 21, Apr. 29, May 14–18 *, June 3.

This species nests in the north relatively late and migrates late accordingly in both fall and spring. In migration and in winter there is much irregularity in movements and local vagrancy. In the Monument most occurrences are in the migration seasons, from September 15 to November 4, and from April 14 to June 3. Occurrences in January and early March may be regarded as winter records. In October single birds or small groups were detected, usually in flight, above the piñon woodland at Pinyon Wells and on Eagle Mountain. In the spring migration they have been found chiefly near the oases, where they may linger about the water and feed on the buds of cottonwoods and willows.

Flocks noted are usually small, although at Twentynine Palms Miss Carter saw groups of 12 and one of at least 17 in March.

A specimen taken on September 15 was in full juvenal plumage with postjuvenal body molt just starting. Migration in this plumage is not rare in this species. A male taken on May 18 had a testis 3 mm. long, which is well below functional state.

Waxwings move about a great deal in search of berries and buds, on which they feed almost exclusively. They engage in long exploratory flights with ease. Thus desert crossing or vagrancy about arid areas is not especially hazardous to such a species. Waxwings can readily leave if moist foods are not found. We have noted, however, that when birds locate an oasis, as at Cottonwood Spring, they commonly use it for drinking. At this place we saw them at water, drinking, on September 15, October 21, and May 14.

PHAINOPEPLAS AND SHRIKES

Phainopepla. *Phainopepla nitens*

Description. A slender, long-tailed, flycatcher-like species, about 7½ inches long (24–30 grams), with long, pointed crest and conspicuous white "windows" in the wing formed by white at the bases of the primaries. Plumage of adult male otherwise glossy black. Females and juveniles dark gray, the wing coverts, secondaries, and under tail coverts edged with white. Eye red.
Range. Arid woodlands and deserts from central California, southern Nevada, and southern Utah east to western Texas and south to Baja California and over the Mexican Plateau to Puebla and Veracruz.
Occurrence in Monument. Common resident, locally. Numbers at times augmented by migrants and winter visitants. Recorded from: Morongo Pass, Apr. 28; Little Morongo Canyon, Mar. 30, Apr. 1, 2 *, 3 *; Black Rock Spring, Aug. 31 *, Sept. 3 *; Lower Covington Flat, Apr. 15 [1], May 28 [1], Aug. 28; Smithwater Canyon, May 19; Quail Spring, Feb. 11, May 21; Stubby Spring, Sept. 10, 11 *, Nov. 1; Key's View, May 14; Indian Cove, Apr. 7, 8, 27, May 12, 14; Fortynine Palms, Apr. 7; Twentynine Palms, Apr. 29, May 7, 23, Oct. 11, 17, Nov. 7, Dec. 11; Pinyon Wells, Apr. 10, Oct. 16 *; Split Rock Tank, Oct. 30; Cottonwood Spring, Feb. 7–11, Apr. 5 *, 11, 25, 26 *, 29, 30, May 14–19, June 22, Oct. 21–24; Pinto Wash Well, Oct. 16 [1].

Phainopeplas are among the most spectacular birds of the Monument and are easily seen because of their flycatching forays out from the tops of tall shrubs and trees and their frequent challenges of one another while on the wing. Although they may be seen at all times of the year, the species is partly migratory and also may move about locally to favored foraging areas. In winter they tend to concentrate in the mesquites of the lower stations, and their numbers then are almost certainly added to by migrants from northern and western sections of the breeding range. At Twentynine Palms, for example, Miss Carter saw them almost daily through the winter of 1933–1934 but states that by May 7 only a few pairs remained, "which suggests that some individuals migrate." In the next year they became plentiful by November 7 and were numerous in December and on into the early spring.

Phainopeplas have two divergent foraging procedures. First, they take insects in flight with erratic maneuvers, sometimes for many yards away from the initial perch. The birds are light on the wing and turn and cruise about with little apparent effort. Second, they take mistletoe berries in quantity, especially from the clumps of this plant in mesquite thickets. For example, on April 2 and 3, birds which were taken in a mesquite patch proved to have mistletoe berries in the stomach, about 5 berries constituting the contents in one case. On April 5 two were seen plucking berries in a mistletoe clump (fig. 59). Again such action was noted on April 8 and 26.

Cowles (1936: 352–356) has reported that the fruiting season of desert mesquite mistletoe (*Phoradendron californicum*) extends from November to April and affords a succulent nutritive diet, especially for phainopeplas. This bird species is the prime agent for dispersal of this parasitic plant, the

seeds, passed in the feces, accumulating on plant branches and germinating and penetrating the bark to establish a new mistletoe cluster. Thus the dense spread of mistletoe clusters in mesquite may be thought of as "planted" largely by the phainopeplas and a supply on which later generations will largely depend for food during the winter season when insects are relatively scarce.

The two types of food used by phainopeplas assure a good moisture supply, and no resort to drinking water is necessary.

Apart from mesquites, we have seen phainopeplas in cottonwoods, desert willows, palo verdes, and catclaws, all but the desert willow supporting mistletoe. Lower shrubs, less than 6 feet in height, are seldom visited, and we have seen phainopeplas only as vagrants in the piñons.

As to numbers, a check at Cottonwood Spring in February showed 4 or 5 phainopeplas present about the oasis. At Indian Cove a loose group of 6 to 8 was seen on May 12. In Little Morongo Canyon on April 1, nesting pairs were spaced at intervals of 100 to 150 yards along a mile or more of canyon bottom. At Cottonwood Spring in late April an estimate was made of one per acre in the washes and flats near the spring.

Nesting has been found to be about two months earlier in the desert than in the coastal breeding range of the phainopepla (see A. H. Miller, 1933: 432–433). The principal laying period on the desert is in early April. Our further specific evidence on this matter from the Joshua Tree Monument bears this out. In Little Morongo Canyon, on April 1, two nests were found that were being built. They were discovered by seeing the birds fly from the sites in the mistletoe clumps in the mesquites. Both nests were on the underside of a clump on some broken twigs, in the shade. The nests were, at this stage, rough cups formed of bark and twig fragments with some cobweb binding. The nests were 7 feet aboveground, one of them over a trail. The phainopeplas stayed close by, the males singing and snarling frequently. On April 3 one of these males was seen making four different pursuits in defense of his territory against passing or invading males. Two males taken in this canyon on April 2 and 3 had testes 7 and 6 mm. long.

At Indian Cove on April 7, two nests, each with sets of two eggs, were found. One was saddled on a horizontal, forked mesquite limb, not among mistletoe, 9 feet up, and the other was 4 feet up in a horizontal fork of a desert willow (fig. 60). The nests were "brushed" onto the limbs with binding cobweb, and at the nest in the mesquite the male was seen attaching the cobweb with his bill as he took his turn incubating. The nests were both 4 x 2 inches deep outside and 2½ x 1 inch deep inside, made of gray plant strips and fiber shreds, cobweb bound, with felting inside of cottony seed heads, chiefly of the creosote bush. At the first nest, after a change-over in incubation, the female mounted to the tip of the tree directly overhead and

Fig. 58. Phainopeplas in the mesquites; male on upper lookout.

Fig. 59. A fruiting mistletoe clump in a mesquite; the mistletoe berries are eaten and dispersed especially by phainopeplas; Monument Headquarters, Twentynine Palms, December 23, 1960.

sat on lookout. A nest at Pinyon Wells was being built on a flat limb of a desert willow on April 10.

At Twentynine Palms on May 23, two fully grown young were seen being fed by their parents. These young would be from eggs laid in April almost certainly.

Nesting at Cottonwood Spring at times must be a little earlier. Here on April 26 a fledgling was taken, with tail only 1½ inches long; the bird must only recently have left its nest. This juvenile would be from an egg laid late in March. On the other hand, at the same place in another year, a female that had no brood patch as yet was taken on April 5, and a pair with an active nest was observed on April 29.

Adults in molt were taken on September 3 and 11. Each was far along in growth of new body plumage; old outer primaries 7—10 and 9—10 were still present, respectively. A bird of the year on August 31 had no more than just started the postjuvenal body molt.

Phainopeplas of this area are of the widespread northern race *Phainopepla nitens lepida.*

Fig. 60. Nest and eggs of the phainopepla in a desert willow at Indian Cove on April 7, 1960.

Loggerhead Shrike. *Lanius ludovicianus*

Description. A stocky, large-headed species, 8½ inches long. Above light gray, the scapulars and rump whitish; a black mask extends from bill through eye to ear area; below white; wings black with white "windows" at bases of primaries; secondaries tipped with white. Tail black with white tips on all but central pair, the white areas progressively larger laterally; tail distinctly graduated.

Range. Southern British Columbia, central Alberta, central Saskatchewan, southern Manitoba, southern Ontario, southern Quebec, and southwestern New Brunswick south to southern Baja California, Guerrero, Oaxaca, the Gulf coast, and southern Florida. Winters chiefly south of latitude 45° N.

Occurrence in Monument. Sparse but widespread resident. Recorded from: Morongo Valley, Dec. 31; Morongo Pass, Apr. 28; Black Rock Spring, Aug. 29; 3 mi. S Warren's Well, Aug. 31; Upper Covington Flat, Jan. 28 [1], Mar. 20 [1], Apr. 8 [1], 9 [1], Aug. 28; Lower Covington Flat, Jan. 22 [1], Feb. 11 [1], 17 [1], Mar. 4 *, 27 *, Apr. 1 [1], 15 [1], Aug. 24, 25 *, 26, 27 *, 29, Nov. 6 [1]; Quail Spring, Jan. 25, 27, Feb. 10, 11, 13 *, July 4, Dec. 4–6; Lost Horse Valley, Jan. 12 [1], Oct. 30–Nov. 1; Key's View, July 4; Indian Cove, May 13–14; Twentynine Palms. Jan. 14, Apr. 20, July 13, Oct. 29, 30; 4 mi. N Twentynine Palms, Mar. 1 [1]; Pinyon Wells, Oct. 14, 15, 16 *, 17 *; Split Rock Tank area, Oct. 29–30; Live Oak Tank, Apr. 6–8; Pinto Basin, Apr. 11, Feb. 18, Sept. 12; Virginia Dale Mine, Apr. 23, 24 *; Eagle Mountain, May 16, Oct. 18–21, 19, 20, 21 *; Cottonwood Spring, Feb. 7–11, May 1, 12, 14–17 *, Sept. 13, 14, 15.

The loggerhead shrike occurs in all parts of the Monument in flat lands and on canyon walls and mountain slopes. It is a species that requires open ground, free of dense cover, on which to forage, but even the piñon belt of the Monument is sufficiently open for it. The Joshua tree "forests" seem especially favorable, as they provide needed lookout perches from which the shrikes can scan the surrounding ground and drop down in power glides to catch food.

Each shrike requires a large area which it defends as a territory. In the deserts these territories are in the order of 30 to 40 acres (A. H. Miller, 1931: 148–159). As a correlate, numbers of shrikes are small. One or two only may be seen in a morning afoot. Shrikes visible along roadways have been estimated at 5 to 7 in 30 miles; such was the figure noted on January 24 between Quail Spring and Twentynine Palms.

Shrikes use the song, as do many species, to advertise their presence in a territory and their readiness to defend it. Since territories are maintained throughout the year, by each bird alone in the postbreeding period and by the pair in the nesting season, song may be heard at any time. It is likely to be less melodic and more interspersed with screeches in late summer, as was true on August 24 at Lower Covington Flat. We have recorded song also on August 27 and 29, October 14 and 20, February 11, and April 9 and 11.

Large ground-dwelling insects are the principal food of loggerhead shrikes, orthopterans which are moist and fairly soft being much favored. But also lizards, small rodents, and occasional birds are captured, and especially when they are unduly large are hung on thorns to enable tearing up or "butchering." On October 19 a shrike was worrying an acorn woodpecker on the top of Eagle Mountain, the woodpecker being a species too large for the shrike to overpower, but it is not rare to find shrikes attempting to attack a large species. On May 1 at Cottonwood Spring a shrike captured a young goldfinch.

The food of shrikes, whether insect or vertebrate, provides all the moisture needed, and we did not see them go to water to drink, even though they are often near open water, as at Cottonwood Spring and Quail Spring. Shrikes stand exposure to sun well; one was singing at 10 A.M. on August 24 in the sun when shade temperatures were in the high 90's. If in flying they become overheated, they pant and show evident heat stress; they then often seek shady perches beneath the dense spine clusters of Joshua trees.

Shrikes nest in March and April primarily in this part of the desert. In Pinto Basin on April 11 a pair had a nest 4 feet up in open, but strong, supporting twigs of a smoke tree. It was about half shaded and contained five young just ready to leave. One of the young (fig. 62) moved out onto the rim and into the surrounding twigs as we watched at close range. The nest was 8 x 6 inches deep externally and 5 x 3 inches deep inside. The cup inside the outer stick structure was felted with soft, cottony material.

On May 16 on Eagle Mountain, at 3800 feet, a family of dependent young was noted. They may well have represented an early April laying. Testes of a first-year male were small (2 mm.) on May 17, supposedly having regressed from an active period, although possibly this individual had not bred in its first year.

Fig. 61. Loggerhead shrike in a glide toward the ground in pursuit of prey.

The five specimens taken in the autumn are in the following molt stages and plumages: Adult ♂, August 25, primaries 5–10 old and body $\frac{4}{5}$ new; immature ♀, August 27, no molt of remiges, body $\frac{4}{5}$ new; immature ♂, October 16, partial primary molt, the outer 3 still in sheaths, lateral rectrices still in sheaths, traces only of body molt; adult ♂, October 17, outer 2 primaries and lateral rectrices still in sheaths, traces of body molt; immature ♂, October 21, traces of body molt. For details on molt sequence in this species, see A. H. Miller (1928, 1931).

The shrikes collected near Twentynine Palms and at Pinyon Wells, Virginia Dale Mine, Eagle Mountain, and Cottonwood Spring are typical examples of the resident race of the Sonoran deserts, *Lanius ludovicianus sonoriensis.* The two August specimens and 10 of 12 collected by the Long Beach State College group in the Covington Flat area in the northwestern segment of the Monument are typical *L. l. nevadensis* (A. H. Miller, 1931), which occupies the Joshua tree belt of the Mohave Desert. It appears then that the northwestern Joshua tree plateaus and basins are occupied by the Mohave Desert form and that *sonoriensis* replaces it eastwardly and to the south. Such an interpretation represents only a slight adjustment in the mapping of the racial boundary through this area postulated without adequate

Fig. 62. A young shrike near its nest in a smoke tree; Pinto Basin, April 11, 1960.

documentary specimens by Grinnell and Miller (1944). We have not taken any shrikes in the Monument which suggest migrants from the north of the dark race *gambeli*. Such, however, may well occur at times.

VIREOS

Four kinds of vireos have been found in the Monument; the warbling and solitary vireos are regular migrants; the Hutton vireo is a rare vagrant; and the fourth, the gray vireo, is a summer resident, nesting in the area. None is abundant or conspicuous, although the warbling vireo can be found readily in migration periods by persons scrutinizing the smaller birds about oases and wooded areas.

The vireos move deliberately and solitarily, as is typical of all members of their family, and they are dependent on insect food. We have not seen them coming to water, but we have found warbling vireos in weakened condition apparently as a consequence of desert conditions. The general problems of desert crossing for the migrant vireos would appear to be the same as those for the migratory warblers and are discussed at some length later (see pp. 202–204). The directions of movement of the solitary and warbling vireos likewise are similar to those of the warblers. Vireos are, however, seldom seen taking off in long flights but instead usually are moving along slowly through the branches of larger trees and shrubs of the open woodland. They are especially attracted to cottonwoods and willows at the oases.

KEY TO THE VIREOS OF THE MONUMENT

A. White superciliary line present, extending the length of the head; no white eye ring . . . . *Warbling vireo*
A′. No white superciliary line; white or faint gray eye ring present.
  B. Two conspicuous whitish bars on tips of wing coverts.
    C. White line from base of bill connecting with prominent white eye ring; under parts, except flanks, white . . . . *Solitary vireo*
    C′. Diffuse gray area in front of eye ring; under parts pale gray green . . *Hutton vireo*
  B′. No conspicuous wing bars; under parts white . . . . *Gray vireo*

Hutton Vireo. *Vireo huttoni*

Description. A moderately small (9–11 gram), fairly uniformly gray-green vireo with two prominent whitish or pale yellow-green wing bars on secondary coverts; eye ring whitish or pale yellow; area in front of eye gray but not forming a well-defined white line; lighter gray-green of under parts not sharply delimited from gray-green of back; wing and tail feathers dusky, edged with green.
Range. Southwestern British Columbia south along Pacific Coast, west of the Cascade–Sierra Nevada axis, to northwestern Baja California; also Cape district of Baja California; Rocky Mountains and mountains of western and southern México from central Arizona to Oaxaca; also mountains of Guatemala. Essentially resident. Occasionally vagrant outside breeding range.
Occurrence in Monument. One substantiated record: north side Eagle Mountain, 4750 feet, Oct. 20, 1945 *.

The record specimen taken was an adult female, a member of the coastal race *Vireo huttoni huttoni;* the closest point of nesting for the race is in the San Bernardino Mountains. Heretofore vagrants of this subspecies have not been noted beyond the desert bases of these mountains. The bird was found on Eagle Mountain in mountain mahogany brush with a group of bushtits.

Gray Vireo. *Vireo vicinior*

Description. A moderate-sized (12–13 gram) vireo, gray above with only slightly olive tinge. Below white, slightly tinged with pale yellow on flanks; no conspicuous wing bars; eye ring white.
Range. Breeds from southern California, southern Nevada, southwestern Utah, central Arizona, and western Oklahoma south to northwestern Baja California, southern Arizona, and western Texas. Winters chiefly in southern Baja California and Sonora.
Occurrence in Monument. Scarce summer resident. Recorded from: Black Rock Spring, Sept. 3 *; Lower Covington Flat, May 19, Aug. 23, 27; Smithwater Canyon, May 19; Quail Spring, May 19, 20, 22, Sept. 9, 10.

This vireo has been found as a summer resident from May to early September, although undoubtedly it is present in April also. In its family it is a species that represents the extreme in tolerance of high temperature and low humidity while on its nesting ground, for its preferred habitats are the dry, hot, and open piñons and junipers of the desert ranges and the hot chaparral of the semidesert, inner coast district of southern California.

In Joshua Tree Monument, gray vireos have been found chiefly in the piñon and juniper tracts of the area from Quail Spring south to Covington

Flat. At Quail Spring and on the slopes of Smithwater Canyon in May, when gray vireos would be singing regularly, we were able to detect at most only two males at a time. In August after the nesting season at Lower Covington Flat, an occasional song was heard.

The two specimens taken on September 3 and 10 at Black Rock Spring and Quail Spring were adult males (testis 2 mm.) in late stages of the annual molt in which the new body plumage was nearly fully developed and the primary molt still involved only the outer three primaries. Evidently the species completes the molt before migration.

At Quail Spring in September we twice found gray vireos in the catclaw thickets of the bottom land, but this was only a few hundred yards below the scattered junipers in which they sing in May.

On May 19 while traversing Smithwater Canyon a gray vireo was flushed from its nest in a small piñon on a rocky hillside 50 yards above the canyon bottom. The nest, which held three eggs, was 5 feet up on the north side of the tree and was hung in the outer foliage. The nest was conspicuous, as the color of the nest material contrasted with the dark green foliage.

This vireo, although seen in the vicinity of Quail Spring and the water sources in Smithwater Canyon, is not known to come to water. Its attraction to this area seems clearly to be the favorable nesting habitat on the mountain slopes. Insect food, doubtless much of it soft larvae, must provide the necessary moisture throughout the hot summer period of residence.

Fig. 63. Warbling vireo (left), gray vireo (right), and solitary vireo (below).

Solitary Vireo. *Vireo solitarius*

Description. A large (14—16 gram) vireo with heavy bill. Gray head marked with conspicuous white eye ring connected with white line running to base of bill; two prominent white wing bars on secondary coverts; back and flanks greenish (in western race *V. s. cassinii*); cheeks gray, sharply delimited from white throat; breast and belly white; wing and tail feathers dusky edged with dull green and white.
Range. Breeds from southwestern British Columbia, southwestern Mackenzie, and central interior and southeastern Canadian provinces south to northern United States and in mountains to northern Georgia; in the west through coastal mountains and Rocky Mountains to Cape district of Baja California and to Guerrero and Veracruz; also mountains of Guatemala, British Honduras, and El Salvador. Winters from southern United States south to Nicaragua.
Occurrence in Monument. Sparse but regular spring migrant. Recorded from: Lower Covington Flat, May 19; Smithwater Canyon, Apr. 24[1], May 13[1]; Twentynine Palms, Mar. 24, 28, Apr. 2, 3, 4, 12—14, 17, May 7, 9, 12, 17; Eagle Mountain, May 16; Cottonwood Spring, Apr. 26, May 13, 15 *, Sept. 17[1].

The spring migration of solitary vireos extends from late March until mid-May. The specimens taken are of the race *Vireo solitarius cassinii*. The protracted period of passage through the desert, shown by our records in the spring, suggests the presence of populations moving both to southern and northern parts of the breeding range of the race. The species is to be expected in fall, but there is only one record, on September 17, for that period.

Migrants have been seen moving singly in the junipers and piñons and about cottonwoods and other vegetation around springs. Usually they are detected by means of their occasional songs, which they give while in spring migration, even at the earliest dates of record.

Warbling Vireo. *Vireo gilvus*

Description. A moderate-sized (10—12 gram), slim vireo with white superciliary line bordered by gray above; wing coverts without white bars. Gray or gray-green above; under parts whitish, the flanks pale yellow-green; wing and tail coverts dusky, narrowly edged with gray-green (fig. 63).
Range. Breeds from northern British Columbia, southern Mackenzie, and southern part of transcontinental timber belt of interior and eastern Canada south to southern United States and in mountains of México to Cape district of Baja California and Chiapas. Winters from northern México (southern Sonora) to El Salvador in Central America.
Occurrence in Monument. Fairly common migrant, spring and fall, passing through all sections. Recorded from: Little Morongo Canyon, Apr. 2 *; Black Rock Spring, Aug. 29, Sept. 1 *, 4 *; Lower Covington Flat, June 2[1], Aug. 23 *, 27 *, Sept. 23[1]; Smithwater Canyon, Apr. 24[1], May 19[1], July 10[1]; Quail Spring, May 14, 20 *, Sept. 9 *, 10 *; Stubby Spring, Sept. 5, 7 *, 8 *, 9 *, 11; Indian Cove, May 13, 14; Twentynine Palms, Apr. 4, 12, May 6, 8, 10, 12, 24; Eagle Mountain, Oct. 20 *; Cottonwood Spring, Apr. 26, 30, May 13, 14, 18 *, Sept. 7[1], 14 *, 17[1]; Lost Palm Canyon, Sept. 14 *.

Spring dates for this vireo range from April 2 to June 2, and fall dates from August 23 to October 20. Migrants of two different races and of several populations with different breeding seasons are involved, which probably accounts for the extended migratory season in the Monument.

Although this species is the vireo most frequently seen in the area, seldom does one encounter more than a single individual in a day of watching. Warbling vireos are most often noted in willows and cottonwoods about springs and in desert willows and catclaw, but they have been seen in mesquites and piñons. Occasionally the birds sing while en route in the spring. This has been noted at Twentynine Palms by Miss Carter on dates from April 12 to May 24.

In the early fall migration 2 out of 15 birds taken were in poor condition, one an immature, being conspicuously emaciated and weighing only 7.5 grams. Both these thin birds were encountered in the first week of September, 1950, at the close of the extremely hot weather of that year in the piñon belt. They were of the race *Vireo gilvus leucopolius* and would thus have already been forced to traverse large areas of arid terrain in the Great Basin and Mohave Desert in moving south from the breeding range. Almost certainly they would have succumbed before completing the migration into México. Most of the other individuals taken in spring and fall show abundant fat, even as late as October 20, and were well prepared for the remainder of the desert crossing.

The race *Vireo gilvus leucopolius* of the Great Basin Mountains is the only race collected in spring migration (see dates listed), but doubtless the coastal *V. g. swainsonii* passes through the area at least in April. *V. g. leucopolius* nests later than these populations of *swainsonii,* which breed at low elevations in coastal California, and hence is the form most to be expected in migration in late May. Of the fall migrants, the 6 taken on August 23, September 1, 4, 7, 9, and October 20 are *leucopolius.* Nine specimens ranging in dates from August 27 to September 14 are *swainsonii.* One straggler of *leucopolius* was obtained on July 10.

WARBLERS

This group consists of fourteen species, three of which are rare and have been detected only once or twice each—Virginia warbler, American redstart, and painted redstart. Two of the species may be resident for parts of the year—the Audubon warbler and the black-throated gray warbler. The other species are exclusively migrant and, apart from the rarities, pass through the Monument in numbers each spring and fall, moving between their breeding ranges on the Pacific coast and in mountains to the north and west of the deserts and their wintering ranges chiefly in México, from southern Sonora southward.

All the migrating warblers are almost exclusively insectivorous and they derive some of their necessary moisture from the insects they take while crossing the deserts. They may supplement this a little by visits to springs when they encounter them, but many individuals do not find the springs as they cross the desert mountains.

The critical aspects of desert crossing for migrant warblers would seem to be: (1) obtaining sufficient insect food, as much for its water as for its energy content, in vegetation to which they are not especially adapted and in which they are not accustomed to search, and from species of insects whose scarcity or mode of life does not make them readily available; and (2) encountering periods of severe heat and sun exposure, especially in the fall, which would augment the rate of water loss beyond that which the reduced intake of moist insects could offset.

The migrant warblers both in spring and fall deposit large amounts of fat before taking off on migration, and they usually appear in the Monument in extremely fat condition. From the energy standpoint, these birds probably can go several days without a normal daily food intake, utilizing the safety supply of fat and gaining metabolic water from it. Migratory movements across 700 miles of desert are usual for many of them, and in the fall these warblers still may have the larger part of the flight ahead of them (fig. 65). A little feeding, some resting or seeking shade during the worst of each daily cycle, and the fat reserve will ordinarily see them through.

The autumnal migration is the most taxing. It takes place chiefly in August and September, when shade temperatures may exceed 100° F. through much of the day and for many days in succession and when air humidity is down to the vanishing point. The birds cannot afford to use much water in lowering temperature by increased evaporation from lung and mouth surfaces. Moist ground or exposed water is available only in a very few localities, and subsurface retreats are manifestly unavailable in the way that they are for many desert animals. Remaining inactive in the coolest places they can find is their chief method of conserving water during midday.

In August and September we have seen clearly the effect of extreme conditions in causing the death of migrants. We found dead birds, dried and emaciated; we saw others, scarcely able to fly, so anxious for food, water, and shade that in the heat of the day they came into the shelter of our camps almost under our feet; and we took specimens that were in such poor condition that they almost certainly would have succumbed before they could travel the remaining 500 miles southeastward across the Colorado Desert basin. Immatures are conspicuous among casualties, but at this season they predominate so greatly over adults in the total population that we cannot demonstrate that they suffer disproportionately in migration, although we strongly suspect that this is true. More weakened individuals and casualties have been found among the warblers of the genera *Vermivora, Oporornis,* and *Wilsonia* than in those of other genera.

Thus, natural selection works before our eyes in these warblers, lopping off those that have ineffective physiologic mechanisms for fat storage, or those that were weaker nestlings a month or so earlier, or those that have inadequate behavioral responses to meet the emergencies of the desert crossing. These warbler species must maintain a fitness not only to live and breed on their winter and summer ranges but to get back and forth over a desert migratory route, a particular problem of existence which the more numerous warbler species of the eastern United States do not have to contend with. Is this one reason why there are fewer warblers in the West?

Our information about breeding and winter ranges of the common migrating warblers of the Monument indicates that most of them must follow, in the main, the northwest-southeast course of the flyway across the area. To a limited extent this can be supported by field observations. On August 28, 1950, orange-crowned and hermit warblers were moving across the crests of the mountains south of Quail Spring, following the general southeast trend of the axis of the Little San Bernardino Mountain Range, in long flights, sometimes as far as the eye could see, on other occasions in flights of 100 to 200 yards at a time. In other places the attraction of canyon bottoms or lines of piñon trees, or springs, seemed to cause detours from this general direction and even occasional reversals of it. The impression was gained that there might be much ranging about, with delays or reversals, perhaps sometimes caused by thunderstorms or winds, which would set the birds back several days in their travels.

Typically, migrating warblers in the desert are on the move, somewhat erratically as to direction, and with occasional foraging, yet often only stopping for rest without food searching. Seldom are they seen for several hours in the same place. Weakened birds or those loafing about oases are an exception, and some species are more prone to tarry than others.

The several kinds of warblers show some contrasts in predilections for different types of plant cover. Although all the desert plants differ from those on the breeding grounds of these warblers, some of the instincts of habitat selection of the nesting period are faintly or partly manifest during migratory passage. Thus hermit and Townsend warblers, both devoted to conifers in the north, are usually seen in piñons or junipers on the desert if any of these conifers are in the area, whereas Wilson warblers will stay more often in lower brushy cover, even though dry, and MacGillivray warblers favor the densest desert shrubs and remain close to the ground wherever possible, thus paralleling their adherence to low, dense, though moist, chaparral on the nesting grounds.

In spring many warblers may be seen in the canyon mouths and even at times on the open desert flats. They are likely to be absent on the higher ridges, at least in April. Contrarily, in the fall, migrants cross high over or inconspicuously through the low-lying areas and tend to follow ridge tops, windy crests, and shaded higher canyons. This general, and crude, altitudinal separation correlates broadly with the temperature and moisture differences of the spring and fall migration seasons and is a phenomenon sometimes seen also in more coastal areas. The deserts, however, are especially favorable for observing these as well as other aspects of migratory behavior, for there is no confusion of resident and migrant individuals. All members of the wholly migrant species here treated are exposed to view, so to speak, as migrants, their chief objective in this area being travel.

Concerning dates of warbler migration, only the broadest generalization can be offered, as each species has its own norm of seasonal movements, complicated in some by the presence of separate races and populations with different travel schedules. Little migration can be detected before the last week in March, and not until mid-April do numbers reach a peak in the relatively high desert areas comprising the Monument. Numbers continue to be large until May 20, with shifts in emphasis to later migrating species. Some warblers are in migration until the end of May, doubtless members of extreme northern breeding populations that do not normally reach their breeding grounds until early June. The conspicuous warbler movements, from mid-April to mid-May, are later than might be supposed in view of early arrival of several of these species on the Pacific coast. Either the early migrants do not pass in quantity through the elevated Monument area or they pass quickly and unnoticed, perhaps largely at night. Probably most of the spring migrants are those of the latitudes of northern California at least, to judge from their late dates of passage through the Joshua Tree country. In the autumn, migration is in progress from late August (probably earlier) through mid-October, with peak numbers from August 25 to September 20.

Concerning numbers of warblers in transit, the impression should not be gained that there is conspicuous mass movement. Typically one must listen and look closely to detect moving warblers even in the open desert ter-

TABLE 6

Span of Migratory Seasons of Ten Species of Migrant Warblers in Joshua Tree National Monument

Species of Warbler	*Spring*	*Fall*
Orange-crowned	Feb. 24–May 16	Aug. 24–Oct. 12
Nashville	Apr. 10–May 14	Aug. 27–Sept. 13
Yellow	Apr. 28–May 18	Aug. 28–Sept. 15
Black-throated gray	Apr. 3–May 19	Aug. 23–Oct. 23
Townsend	Apr. 24–May 20	(data inadequate)
Hermit	Apr. 22–May 13	Aug. 27–Sept. 3
MacGillivray	Apr. 8–June 2	Aug. 23–Oct. 16
Yellowthroat	Mar. 24–May 21	(data inadequate)
Yellow-breasted chat	Apr. 26–May 18	(data inadequate)
Wilson	Mar. 21–May 30	Aug. 24–Oct. 14

rain. Occasional flight notes may be heard overhead, day or night, but where individuals can actually be traced down, as along favorable canyon bottoms, only two to four warblers of all species may be seen in a couple of hours. At points of greatest concentration, as on the ridge crests above Quail Spring in late August, counts showed that warblers passed within sight range of one spot at the rate of one to three every two minutes for several hours during the morning. Also, occasionally at springs a true wave or concentration develops for a few days. For example, on May 12 and 13, 1951, as many as 60 warblers occupied the trees and bushes about Cottonwood Spring.

Key to the Warblers of the Monument

- A. Conspicuous white or yellow areas present in tail feathers (=rectrices, not the rump feathers necessarily).
  - B. No yellow in plumage (except for small spot in front of eye) .... *Black-throated gray warbler*
  - B′. Yellow or orange present in plumage.
    - C. Yellow areas on head.
      - D. Yellow on head restricted to patches on crown and throat; also yellow rump spot........ *Audubon warbler*
      - D′. Yellow stripes or patches above and below eye.
        - E. Yellow on sides of head in definite stripes; sides striped with black .... *Townsend warbler*
        - E′. Yellow on sides of head in large patch, or diffuse; no stripes on sides........ *Hermit warbler*
    - C′. Head black, gray, or white, without yellow.
      - D. Tail with white lateral feathers; body black and red...... *Painted redstart*
      - D′. Tail with yellowish or orange subterminal spots; body with black in male only........ *American redstart*
- A′. Tail without conspicuous light patches
  - B. Tail yellow-green, with longitudinal yellow areas on inner webs of tail feathers .... *Yellow warbler*
  - B′. Tail green, without yellow areas.
    - C. Throat and all of breast clear yellow.
      - D. Small (usual warbler size).
        - E. No black mask through eyes.
          - F. Crown either glossy black or dusky, contrasting with yellow of face .... *Wilson warbler*
          - F′. Face and sides of head gray........ *Nashville warbler*
        - E′. Black mask across eyes bordered by white above......... *Yellowthroat* (male only)
      - D′. Large (larger than house finch)........ *Yellow-breasted chat*
    - C′. Under parts dull yellow-green, or with restricted yellow spot, or gray or blackish throat.
      - D. Restricted yellow spot on lower throat and upper breast; rump yellow contrasting with gray back........ *Virginia warbler*
      - D′. Throat and breast dull yellow-green, gray, or blackish.
        - E. Throat and breast dull yellow-green without gray or blackish.
          - F. Color of under parts blending with that of sides and back .... *Orange-crowned warbler*
          - F′. Yellow-green of under parts contrasting with dull gray-green of sides and back and with whitish belly... *Yellowthroat* (female only)
        - E′. Throat, breast, and head black or gray contrasting with yellow belly .... *MacGillivray warbler*

Orange-crowned Warbler. *Vermivora celata*

Description. A moderate-sized (8–9 gram), yellow-green or gray-green warbler with little pattern visible, duller above than below. Poorly defined gray cheeks and ventral streaking present in some races but visible only at close range or in hand; concealed orange crown patch; ventral coloring varies with race and sex from bright greenish yellow to dull gray-green; wings and tail dusky, edged with greenish.
Range. Breeds from northern tree belt of Alaska and western Canada south to islands off northwestern Baja California, southeastern Arizona, and southern Manitoba. Winters from northern coastal and southeastern California, southern Nevada, central Arizona, and the Gulf states south through México to Guatemala.
Occurrence in Monument. Common migrant, noted passing through all sections. Recorded from: Little Morongo Canyon, 2500 feet, Mar. 31 *, Apr. 3; Black Rock Spring, Aug. 30 *, 31 *, Sept. 2 *, 3, 4; Upper Covington Flat, Apr. 9, Sept. 30 [1]; Lower Covington Flat, Apr. 29 [1], Aug. 24 *, 25, 28 *; Quail Spring, Apr. 22 *, 23, Sept. 9 *; Stubby Spring, Sept. 7, 11 *; Indian Cove, Apr. 8 *, May 13, 14; Twentynine Palms, Feb. 24, Apr. 22, 26; Pinyon Wells, Apr. 10, Oct. 12 *; Pinto Basin, Apr. 11; Virginia Dale Mine, Apr. 25 *; Eagle Mountain, May 16; Cottonwood Spring, Apr. 5, 26, May 13, 14–18, Sept. 13 *.

Because migrants of three geographic races of this warbler which breed from Alaska to southern California are involved, and because the Monument lies close to or perhaps even within the northern part of the wintering range of the species, the season of occurrence is extended. Extreme dates are February 24 and May 16 in the spring and August 24 and October 12 in the fall. Doubtless earlier fall dates occur.

Next to the Wilson warbler, this is the most common transient warbler. Usually the maximum numbers seen within an hour are four to six. The birds move individually, drifting through the piñons and junipers, the willows and cottonwoods of oases, and also low bushes, especially in the absence of trees. Many are seen, although often with some doubt as to identity, as they fly over or as they leave a distant tree after a brief rest.

The most extensive foraging noted has been in willows, as along Little Morongo Canyon, where these trees were coming into leaf in late March of 1951, and in dense, low green bushes as those of *Prunus* in canyon bottoms. The only time we saw these warblers on the ground was after a heavy rain, on September 3, 1950, when they and several other species turned to feeding on insects on and near the ground, largely termites which were swarming.

Sometimes we could see that the flight followed the northwest-southeast fly line, and doubtless this is the general line of travel of many. If individuals of the races other than the insular *Vermivora celata obscura* are to move toward the lower Colorado River valley or México in autumn, and most of them do, they are to be expected to fly south or southeastwardly through the Monument.

Four out of thirteen fall migrants collected in August, September, and October were in poor condition and light of weight. Two of these weakened birds were of the race *V. c. celata* from breeding grounds in Alaska and northern Canada, and one each was of the Great Basin race *V. c. orestera* and the

Fig. 64. A group of migrant warblers:

Wilson warbler

MacGillivray warbler

Yellowthroat

Orange-crowned warbler

Nashville warbler

coastal race *V. c. lutescens*. All were immatures. One of the examples of *V. c. celata* taken on September 2, 1950, weighed but 5.5 grams, was obviously weakened, and would not have survived long. This was at the close of the extremely hot, dry weather of August of that year.

The races involved in the migration can be determined only from preserved specimens. *V. c. lutescens* of the Pacific coast is represented by nine specimens, from August 28 to September 11 in the fall, and in the spring on March 31 and April 8. These last are dates when *lutescens* is well-established on its breeding grounds in California, and the specimens may represent populations of this race that move later in spring into coastal British Columbia or Alaska. Seven of the northern race *V. c. celata* have been taken once each on September 2, 13, and 30, twice on October 12, and once each in the spring on April 22 and 25. The late spring dates for the species may be chiefly of this race, as it is known to move north much later than *lutescens*. Contrarily, the earliest spring arrivals seen doubtless are *lutescens*, unless they possibly represent winter residents. Only three examples of the race *V. c. orestera* breeding in the Great Basin and Rocky Mountain areas have been taken, on April 29, August 24, and October 12.

Nashville Warbler. *Vermivora ruficapilla*

Other name. For long known in the West under the race name of Calaveras warbler.
Description. A small (8 gram), gray and yellow warbler. Head and sides of face gray, with concealed chestnut crown patch (in males and adult females) and white eye ring; throat and breast clear yellow; belly white, at least posteriorly; rump and under tail coverts yellow-green; back green, tipped with gray; wings and tail dusky, edged with green (fig. 64).
Range. Breeds from southern Canada south to Pennsylvania and Illinois and in the West to the southern Sierra Nevada of California. Winters from northern México to Guatemala, but apparently not in Baja California.
Occurrence in Monument. Migrant. Likely to occur throughout, but seen chiefly above the creosote bush belt. Recorded from: Black Rock Spring, Aug. 30 *, Sept. 1 *; Upper Covington Flat, Apr. 22 [1], May 14 [1], Sept. 9 [1]; Lower Covington Flat, Aug. 27 *, 28 *; Smithwater Canyon, Apr. 23 [1]; Quail Spring, Sept. 9 *; Stubby Spring, Sept. 6 *, 7 *; Barker's Dam, Apr. 28; Pinyon Wells, Apr. 10; Cottonwood Spring, Apr. 26, May 13, Sept. 13.

Spring dates of this migrant warbler range from April 10 to May 14, and fall dates from August 27 to September 13. The seasons of migration are, therefore, more narrowly limited than are those of the orange-crowned and Wilson warblers. Most of the movement each season occurs in a two-week interval centering, respectively, around the end of April and the first of September. Although more extreme dates than any thus far recorded may be expected on the basis of reports elsewhere in California, the main period of movement is probably accurately reflected by the records at hand. The Nashville warblers presumably are moving chiefly between the Sierra Nevada and the west coast of the mainland of México.

At the height of its fall movement this species for a time may equal the orange-crowns in numbers. About springs, such as Black Rock and Cottonwood, Nashville warblers were seen, one to three at a time, coming to water. Otherwise they were seen moving singly through scrub oaks, junipers, and piñons, or flying over, giving their almost bunting-like flight note as they followed their general route southeast or northwest. They show no distinct preferences for vegetation types, except that they usually move through tall tree cover, when available, at middle heights or through the tops.

During the hot weather of the fall of 1950 several specimens were taken which were moderately or heavily fat, weighing up to 10 grams in some instances. In contrast to these vigorous birds were 3 out of the 11 taken—one an immature, one an adult, and one of undetermined age—that were in poor condition with weights of 5.3, 6.8, and 6.3 grams, respectively. The immature was picked up dead on August 27 in an open wash at Lower Covington Flat, where it evidently had succumbed the day before to drought and starvation. Its breast muscles were much depleted and the lungs were pale pink rather than normal red. It is unlikely that the other two lightweight birds would have been able to complete their migration across the severe desert country ahead. Thus, as in the orange-crowned warbler, somewhat more than a fourth of the migrants we handled seemed doomed to failure.

The warblers of the species here reported belong to the western race *Vermivora ruficapilla ridgwayi.*

Virginia Warbler. *Vermivora virginiae*

Description. A small (7–8 gram), gray warbler showing only limited areas of yellow. Head, chin, back, and sides gray, the crown with a concealed chestnut patch (in males and some females) and eye ring white; yellow spot on lower throat and upper breast; rump and under tail coverts yellow; belly whitish; wings and tail dusky.
Range. Breeds from east-central California, northeastern Nevada, southern Idaho, and northern Colorado south to southern Nevada, southeastern Arizona, and central northern New Mexico. Winters in southern México south to Morelos and Guerrero.
Occurrence in Monument. A single record: Quail Spring, Sept. 8, 1950 *.

This inconspicuous warbler may occur more often as a migrant than the one record indicates. The Monument lies west of the main line of migratory movement of this species to and from the wintering ground, but it is directly south of the western border of the breeding range.

The bird taken was an adult female traveling for the moment with a group of bushtits in the catclaw thickets below Quail Spring. It weighed but 6.8 grams and may not have been in good condition.

Yellow Warbler. *Dendroica petechia*

Description. A moderately small (8–9 gram), pale to richly yellow warbler, without head markings, the back variously greenish. Wings dusky, broadly edged with yellow-green; tail dusky with inner vanes clear yellow, occasionally only dusky yellow; breast and flanks of males streaked with chestnut, the streaks usually lacking in immature males of the fall and in females.
Range. Breeds from northern tree belt in Alaska and Canada south to southern Baja

California, Galápagos Islands, central Perú, and coastal Venezuela. Winters chiefly from southern México to northern South America.

Occurrence in Monument. Seen occasionally in migration, chiefly about springs. Recorded from: Black Rock Spring, Sept. 3, 4 *; Upper Covington Flat, May 14 [1]; Lower Covington Flat, May 14 [1], Aug. 28 *; Stubby Spring, Sept. 3, 6 *, 10 *; Twentynine Palms, Apr. 28, May 2, 3, 4, 13; Cottonwood Spring, Apr. 30, May 14—18 *, Sept. 15 *.

Yellow warblers are sparse among the migrating warblers, and the few records made doubtless fall short of marking the limits of the migratory season for the species. However, they probably reflect the principal time of movement. The dates for the spring are April 28 to May 18 and for the fall are August 28 to September 15.

In most instances this species has been seen at springs, and approximately two-thirds of the occurrences were in cottonwoods or willows at such places. Thus the preference of yellow warblers for riparian growth on the nesting grounds is reflected in the stopping points of the species in the Monument. Rarely is more than one individual detected in a morning's watching. On September 3 a yellow warbler in a piñon area fed on insects on or near the ground after the heavy rain of the night before.

Specimens of yellow warbler from the Monument are all members of populations breeding in the coastal districts of California and Oregon and in the Great Basin. Therefore they are moving south or southeast in the fall toward mainland México and Central America.

All seven specimens have been fat and in good condition, reflecting no difficulties in the migratory passage, even though the species especially seeks and presumably needs well-watered places. A male taken on May 18 had testes 4 mm. long and thus was not yet in full breeding condition. It probably was a member of a northern population that would start nesting in late May or June.

The yellow warblers of the Great Basin in recent years have been classed as *Dendroica petechia morcomi* in distinction from *D. p. brewsteri* of the coastal breeding areas. These races have proved difficult to distinguish in practice in that only adult males in spring plumage can be partly differentiated and primarily on the basis of the breadth of the ventral stripes. In accord with the American Ornithologists' Union Check-list (1957), we therefore employ the name *morcomi* for populations of both the Coast and Great Basin, and all birds taken in the Monument are similarly classed.

Audubon Warbler. *Dendroica auduboni*

Description. A large (11—14 gram) warbler, most constantly characterized by large white subterminal spots in the tail and the presence of five yellow spots, on crown, chin, rump, and the two flanks. Body gray or gray and black according to sex and season. In general, females and birds in fall and winter plumage have mottling and streaking of dark gray on light gray above and below, and the yellow of the five spots is dull. Males in spring plumage show brilliant contrasts; the back is streaked with black and the throat, breast, and part of the flanks is solid black; wings with conspicuous white bars on secondary coverts.

Range. Breeds from central British Columbia, southern Alberta, and southwestern Saskatchewan south in the mountains to northern Baja California and southern Durango. Winters from southwestern British Columbia, southwestern Utah, central New Mexico, and southern Texas south to Costa Rica.

Occurrence in Monument. Common migrant; winter resident in lesser numbers. Recorded from: Little Morongo Canyon, Mar. 30, 31, Apr. 1, 2 *, 3; Black Rock Spring, Aug. 31 *, Sept. 3, 4 *; Lower Covington Flat, Apr. 9; Smithwater Canyon, Apr. 23 [1], Oct. 14 [1]; Quail Spring, Apr. 22, Oct. 15; Stubby Spring, Apr. 9, Sept. 11, 12, Oct. 15; Twentynine Palms, Oct. 22, Jan. 26, Feb.-Apr., Mar. 18, Apr. 12, 17, 22; Pinyon Wells, Oct. 9, 12, 13, 17; Pinto Basin, Apr. 11, 27; Eagle Mountain, Oct. 19; Cottonwood Spring, Feb. 7–9, Apr. 11, 22, 26, 30, May 13, 14, Sept. 15, Oct. 13, 23; Pinto Wash Well, Jan. 28 [1].

From the end of August through the middle of May this species may be expected in the Monument, but the numbers and concentrations vary distinctly with season and location. In the fall, migrants appear in numbers in September and October, and indeed in those months they are widespread in the piñon belt especially, many of them not actually moving but tarrying as though in fall residence. In midwinter the piñon belt is, so far as our records show, forsaken, but concentrations may be found at oases at lower elevations, as at Cottonwood Spring and Twentynine Palms. At this latter station Miss Carter rated them "very common" but indicates that this may mean two or three seen at once, contrasted with flocks when in migration, as on March 18. The winter pattern of occurrence consisting of a few scattered individuals at any one point accords with our experience at Cottonwood Spring and in October in piñon-covered canyon slopes. We think it significant that Sibley during his sojourn in cold weather at Quail Spring in February did not record this species once.

In the spring in late March, Audubon warblers most often are seen in the lower canyons, as in Little Morongo Canyon, along the watercourse. Here in the first three days of April, the numbers built up quickly, as many as half a dozen being seen in rapid succession. Birds flying long distances on northwesterly courses were noted on April 9 and 11. Mid-May seems to mark the last of these spring migrants.

On August 31 at Black Rock Spring, at the end of a period of several days of high temperature, two appeared at our camp in the piñons, flying into the shade of the camp in the afternoon, apparently in weakened condition and seeking relief. One of these taken was an immature male with no fat and dangerously low weight of 9.7 grams. An adult female taken here on September 4 was in somewhat better condition and weighed 10.2 grams.

Audubon warblers in winter show considerable diversity in food usage, obtaining insects by flycatching methods as well as by gleaning them in open or outer foliage of trees and bushes and indeed from the ground. Also they may turn to feeding on berries. This versatility and the resorting to several types of moist foods enables them to do well about desert oases in winter. The temperatures also are not severely low, nor, with the exception just noted, are they often dangerously high for days at a time.

Plate IV. Male warblers: Yellow warbler

Black-throated gray warbler

Audubon warbler

Hermit warbler

Townsend warbler

GMC

The senior author has regarded the racial subdivision of Audubon warblers north of México as unwarranted (Grinnell and Miller, 1944; Johnson, Bryant, and Miller, 1948). Although outvoted on this matter in check-list compilations (A.O.U. Check-list, 5th ed., 1957), he has yet to see evidence produced which shows more than a quite gradual cline of increasing wing length southward and eastward in the species' range as far as Arizona. Accordingly it is wholly impractical to subdivide the populations of the United States into two racial groups. The segregation of winter visitants and migrants into meaningful categories is even more hopeless. Therefore birds from the Monument are recorded as *Dendroica auduboni auduboni;* in any event the two male specimens available are rather short-winged like many of the individuals of the northwestern segment of the breeding range, whence the species was originally named.

Black-throated Gray Warbler. *Dendroica nigrescens*

Description. A small (7–9 gram) black, gray, and white warbler. Male with head and throat black, a white line above eye and below and around ear area; back between wings gray with black streaks; flanks streaked with black, the belly white; wings black and gray with two white bars on secondary coverts; tail black with longitudinal white areas on inner webs of four lateral pairs of feathers, the areas becoming larger laterally. Females similarly marked but black areas grayer or often overlaid with gray and white.
Range. Breeds from southwestern British Columbia, central Oregon, southern Idaho, southwestern Wyoming, and central Colorado south to northern Baja California, northern Sonora, and southern New Mexico. Winters from central California and southern Arizona south to Guerrero, Oaxaca, and Veracruz.
Occurrence in Monument. Common migrant. Recorded from: Little Morongo Canyon, Apr. 28; Black Rock Spring, Apr. 3, 4 *, Aug. 30 *, 31, Sept. 1, 2, 3 *, 4; Lower Covington Flat, Aug. 23 *, 24 *, 25, 28 *, Oct. 14[1]; Smithwater Canyon, May 19, Oct. 14[1]; Quail Spring, Apr. 22; Stubby Spring, Sept. 6, 7 *, 11; Twentynine Palms, Apr. 6, 24; Pinyon Wells, Apr. 10, Oct. 10 *, 15; Pinto Basin, Apr. 11; Eagle Mountain, Oct. 20; Cottonwood Spring, Apr. 22, Sept. 13, 17[1], Oct. 22, 23 *.

Black-throated gray warblers occur in migration for protracted periods, and there is a possibility that occasional individuals winter in the area, as the date of October 23, our last for the species in the fall, is late for migration. Also we have thought it possible that the species breeds in some of the heavier piñon and oak cover in the western end of the Monument, as in Smithwater Canyon, where it was observed on May 19. Regardless of these possibilities for winter and summer residence, the species in the main is a migrant, with moderate numbers recorded in the fall from August 23 to October 23; there is no conspicuous concentration period in this interval. In spring the occurrences, with one exception, are in April.

The birds are seen most often in the piñons and junipers, a habitat type in which the species may nest farther north and west. Here they characteristically stay within the foliage, at times creeping through the stiff needle tufts. At Black Rock Spring, on September 3, after a rain, they, less often than other warblers, came down to the open ground to forage on the abundant insects there. On other occasions black-throated grays have been found in catclaw and desert willows.

In the heat stress period of the early fall of 1950, eight specimens were taken. Two of these, on August 30 and September 3, an immature female and an adult male, weighed 5.7 and 6.0 grams, respectively. This is roughly 25 per cent underweight. The birds were in poor condition and probably would not have survived further migration. A normal bird at this time weighed over 8 grams, and one that was very fat weighed 10 grams. A moderately fat female spring migrant weighed 9.0 grams. On August 31, black-throated gray warblers were among those species that came into our camp, possibly attracted by the shade and the shiny black cars in seeking relief from the intense heat of the afternoon.

Townsend Warbler. *Dendroica townsendi*

Description. A moderate-sized (8—9 gram) yellow and black striped warbler with white longitudinal areas in outer tail feathers. Head with yellow stripes above and below a black or dusky green eye and auricular area; crown, throat, and chin black in males, yellow or green in females; breast yellow; sides striped with black or dusky; belly white; back green, with or without black streaks; wings black, the wing coverts with two white bars; tail black except for white areas (pl. IV).

Range. Breeds from southern Alaska and southern Yukon south to northeastern and central Oregon, northern Idaho, and northwestern Wyoming. Winters from central California and Nuevo León south to Nicaragua, but not in desert areas.

Occurrence in Monument. Sparse migrant. Recorded from: Black Rock Spring, Sept. 3 *; Lower Covington Flat, Apr. 30 [1], May 13 [1]; Smithwater Canyon, Apr. 24 [1]; Quail Spring, May 14, 20; 5 mi. N Key's View, May 14; Twentynine Palms, Apr. 30 to May 17; Pinto Basin near dunes, Apr. 27 *; Eagle Mountain, May 16; Cottonwood Spring, Apr. 30, May 13, 16.

Townsend warblers have a fairly restricted period of passage through the Monument in the spring, from April 24 to May 20. There is only one autumn record, September 3.

These warblers appear, one or two at a time, at oases like Cottonwood Spring and Twentynine Palms, and they also have been seen moving through the piñons. One was taken while it was traversing the creosote bush flats of Pinto Basin on April 27. It was a brilliantly plumaged male, with testes 2 mm. long and thus not ready to breed. One observed by Frances Carter at Twentynine Palms (1937:216) came down to feed on the lawn at the Inn, an unusual activity for this species, which prefers dense foliage for foraging, particularly that of conifers.

Probably the piñon belt of the Monument is too open and windswept for winter residence. Thus Townsend warblers use the Monument only in transit and occupy areas to the northwest and southeast in winter.

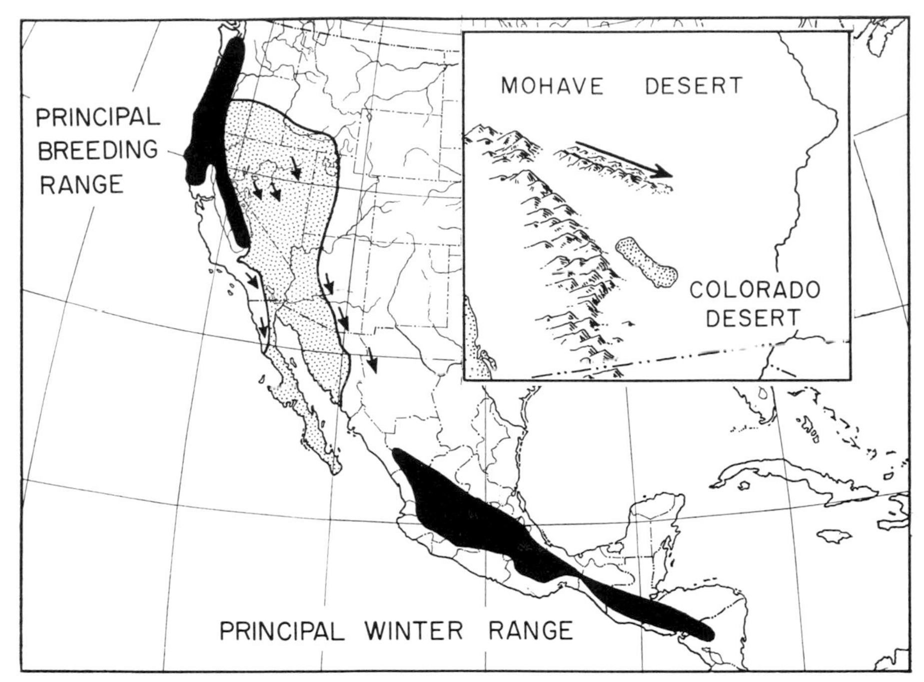

Fig. 65. Map showing principal breeding and winter ranges of the hermit warbler and some known routes (arrows) followed in fall migration. As in many species of warblers, the extensive desert and arid Great Basin areas (stippled) must be traversed during the heat and drought of late summer; insert shows one route of crossing along the Little San Bernardino Mountains.

Hermit Warbler. *Dendroica occidentalis*

Description. A moderate-sized (8–10 gram), gray-bodied and yellow-headed warbler, with longitudinal white areas in outer tail feathers. Sides of face and forehead yellow; crown also yellow in spring; auricular area with ill-defined dusky area in autumn in young birds; throat either black or yellowish; under parts otherwise whitish without stripes; back gray with variable amounts of black striping; wings black, the wing coverts with two white bars; tail black except for white areas (pl. IV, opp. p. 212).
Range. Breeds from southwestern Washington south through coast ranges and Cascade Mountains to southern Sierra Nevada of California. Winters chiefly from southern México to Nicaragua, rarely in coastal California.
Occurrence in Monument. Common migrant. Recorded from: Black Rock Spring, Sept. 3 *; Lower Covington Flat, Aug. 27 *, 28 *; Indian Cove, May 13 *; Twentynine Palms, Apr. 22, May 2, 3; Cottonwood Spring, May 1.

Hermit warblers more than any other migrant warblers appear in waves and have narrowly restricted periods of transit. On August 27, 1950, their appearance in southward migration (fig. 65) was sudden at Lower Covington Flat, none having been seen there in the several days preceding. For about a week they were then seen commonly, as many as ten crossing a given high ridge in an hour. On May 1, 1947, they arrived at Cottonwood Spring, the species not having been there earlier that spring.

Sometimes two individuals travel together. On August 27 two in company were taken, one an adult male, the other an immature male. The immature had the bill smeared with pitch from foraging in the piñons. No weakened birds were encountered, and all those taken in autumn were heavily supplied with fat. A male taken on May 13 had testes 3 mm. long and thus was not yet in breeding condition.

MacGillivray Warbler. *Oporornis tolmiei*

Other name. Also known as Tolmie warbler.
Description. A fairly large (9–12 gram), though short-winged, warbler with either a gray or black (adult male) hood forming a convex posterior border across breast. Back, rump, and tail green; lower breast, belly, and under tail coverts clear yellow; head dorsally dark gray (adult males) or green; eyelids white, forming conspicuous white upper and lower spots in adults (fig. 64, p. 208).
Range. Breeds from southeastern Alaska, southwestern Yukon, central Alberta, and southwestern Saskatchewan south to central California and the southern Sierra Nevada and to central Arizona and New Mexico; also mountains of Nuevo León. Winters chiefly south of United States in México and Central America, from southern Sonora and Baja California to Panamá.
Occurrence in Monument. Common migrant for protracted periods in spring and fall. Recorded from: Black Rock Spring, Aug. 30, 31 *, Sept. 4 *; Lower Covington Flat, Apr. 8, 29 [1], June 2, Aug. 23 *, 26 *, 29 *; Smithwater Canyon, Apr. 29 [1]; Quail Spring, Apr. 22 *, May 20 *, Sept. 10 *; Stubby Spring, Apr. 11, Sept. 10 *; Twentynine Palms, Apr. 23, May 11, 12, 14; Pinyon Wells, Oct. 16 *; Cottonwood Spring, May 14–18, Sept. 15 *; Lost Palm Canyon, Sept. 14 *.

The migration season extends at least from April 8 to June 2 (Rainey MS). In the autumn the movement probably starts by mid-August; our records extend from August 23 to October 16.

Seldom is more than one bird seen at a given locality in each day's observing, but occurrences are widespread in the Monument and an occasional bird can almost always be found by special search in the migration periods.

This warbler carries over into the migration period its instinctive behavior of staying close to the ground in dense cover. It usually frequents thick shrubs and moves between them rather than through trees even of such low stature as piñons. Many are seen in the low cover that is frequent about springs, but they are also noted in brush bordering dry washes. Chrysothamnus, low scrub oaks, willow clumps, and catclaw are shrub types in which we have specifically recorded MacGillivray warblers.

The direction of migratory flight is not usually suggested by the skulking movements of this species. Probably most individuals are moving along the southeast-northwest flight line en route to or from México.

The breeding population to which the migrants belong is that of the Pacific coast and the western Great Basin. Among the 18 specimens taken we find no individuals with long tails and the especially dull green color of the race *Oporornis tolmiei monticola* of the eastern Great Basin and southern Rocky Mountains. We do not agree with Phillips (1947:296–300) that the Sierran birds are racially separable from those of the northwestern Pacific coast, although the latter may be slightly more brilliantly colored on the average. Accordingly we use the race name *O. t. tolmiei* for birds of both areas and for the migrants thus far taken in the Monument.

While in passage through the desert, most MacGillivray warblers are in good condition, many of them being well supplied with fat. However, one immature female taken on August 29, 1950, had no fat and weighed but 7.8 grams, and one other immature female weighed but 8.0 grams on September 4, 1950. These two seemed in danger of dying in the period of intense heat and drought of that year.

Males in transit in the spring showed two stages in progress toward full gonad development. One taken on April 22 had testes 2½ mm. long, and one on May 20 had testes 5 mm. long.

Yellowthroat. *Geothlypis trichas*

Description. A moderate-sized (9–10 gram), short-winged warbler, with strikingly different pattern and color in male and female. Male with black facial mask bordered above by white or light gray; throat and breast clear vivid yellow, the flanks duller; belly yellow or white; under tail coverts yellow; back green or gray-green; wings and tail dull green. Female green or gray-green above and on face and sides of head; chin, throat, and breast pale yellow, forming fairly sharp border with darker dorsum; under tail coverts yellow; wings and tail dull green (fig. 64, p. 208).
Range. Breeds from southeastern Alaska, northern Alberta, and southern parts of eastern Canada south to Oaxaca and southern Florida. The races breeding north of about latitude 35° are wholly or partly migratory, wintering south to Panamá and the West Indies.
Occurrence in Monument. Sparse migrant; present for protracted periods in the spring. Recorded from: Little Morongo Canyon, Apr. 1 *, 2, 3 *; Smithwater Canyon, Apr. 10 *; Twentynine Palms, March 24, 25, 31, Apr. 3, 4, 10, 15, 24, 26, May 4, 5, 13, 21; Pinyon Wells, Oct. 16 *; Cottonwood Spring, May 1; Pinto Wash Well, Oct. 16[1].

Yellowthroats are infrequently seen in the deserts, and, in accord with their instinctive attraction to water and water-dependent vegetation, they appear chiefly at oases. In these places, and occasionally along dry washes, they seek the densest low vegetation available, staying near the ground and using to the maximum degree possible such shade and screening foliage as there is. The sedge beds at the oasis at Twentynine Palms and the sedge and arrowweed thickets bordering the meadows in Little Morongo Canyon have been the places where most of the yellowthroats in the Monument have been recorded. In the latter place two or three a morning could be found in early April. One was taken in thick bushes of *Baccharis sergiloides* in the dry canyon bottom at Pinyon Wells and another in low junipers in a dry section of Smithwater Canyon.

The migration season in the spring extends from March 24 to May 21, and involves at least two races, perhaps more, with different migration schedules. One taken on April 3 in Little Morongo Canyon was of the race *Geothlypis trichas scirpicola,* which breeds in southern California; this individual was probably en route to the coastal district. Two others taken, on

April 1 and 10, and the only two taken in the fall, October 16, are of the race *G. t. occidentalis,* which breeds from central California and central Nevada northward to Washington. We do not find it feasible to recognize *G. t. arizela* of the Pacific Northwest as distinct, and agree in this matter with Swarth (1912: 71–73) and Grinnell (1914: 202) rather than with Behle (1950: 206–207), who has lately reviewed this group of races.

The migrant yellowthroats taken have been in good physical condition. The males of April 3 and 10 had testes 2 and 2½ mm. long, respectively, and thus were far short of breeding dimensions.

Yellow-breasted Chat. *Icteria virens*

Description. A very large warbler (23–27 grams; 17 cm. long), green above and rich yellow on throat and breast. Sides of head blackish and gray; supraloral stripe, malar area, and eye ring white; belly white; lower flanks and under tail coverts buff; wings and tail dusky, with dull green feather edges.

Range. Breeds from southern British Columbia, southern Alberta, northernmost United States, and southern Ontario south through United States to Mexican Plateau as far as the Valley of México. Winters in México and Central America from Chihuahua to western Panamá.

Occurrence in Monument. Scarce migrant, noted chiefly in spring. Recorded from: Lower Covington Flat, Aug. 26 *; Twentynine Palms, May 1, 2–10, 12, 13, 17; Cottonwood Spring, Apr. 26, 30, May 14–18.

The spring migration of the chat seems to be fairly well defined by the dates of April 26 and May 18. There is only one autumn record. Typical of the migration of this species, individuals often tarry en route at favorable places. Frances Carter (1937:216) banded four individuals in a willow thicket at Twentynine Palms in early May and one repeated daily from May 2 to 10. Song outbursts, often given in migration, were heard on May 1 and 2.

The specimen taken in the fall was in fresh plumage. The bird was seen in chrysothamnus bushes and in the lower part of a desert willow.

The race represented is *Icteria virens auricollis.*

Wilson Warbler. *Wilsonia pusilla*

Other name. Pileolated warbler.

Description. A small (6–7 gram) warbler, bright, clear yellow on entire under surface and bright green or greenish yellow above. Crown black (adult males and some females), dusky, or dark green, always in contrast with yellow of face; wings and tail dusky with greenish edges (fig. 64, p. 208).

Range. Breeds from tree limit in Alaska and Canada south to coastal southern California, the mountains of New Mexico, northern Minnesota, and northern New England states. Winters chiefly from northern México to western Panamá.

Occurrence in Monument. Abundant spring and fall migrant, noted in all sections. Recorded from: Little Morongo Canyon, 1500-2500 feet, Apr. 2 *, 6, 27, 28; Black Rock Spring, Aug. 28, 30 *, 31 *, Sept. 1–4; Lower Covington Flat, Aug. 24 *, 25, 26 *, 27 *; Smithwater Canyon, May 13 [1]; Quail Spring, Apr. 23, May 19–21, Sept. 9; Stubby Spring, Sept. 6 *, 8 *, 9 *, 11 *; Indian Cove, May 13; 1 mi. NW Fortynine Palms, Sept. 24 [1]; Twentynine Palms, Mar. 21, 23, 28, Apr. 12, 16, 22–May 30; Pinyon Wells, Apr. 10, Oct. 13, 14; Eagle Mountain, May 16; Cottonwood Spring, Apr. 23, 25, 26, 30, May 12–14, 15 *, 17, Sept. 14, 17 [1].

This is the species of warbler most frequently seen as a migrant in the Monument. Its bright colors and inclination to move through the vegetation less rapidly than other transients, together with its jerky fly-catching actions, make it conspicuous. Although one or two individuals of this species may be seen at a time in a given locality, occasionally waves or concentrations are noted. For example, on May 12 and 13, 1951, at Cottonwood Spring as many as 60 Wilson warblers were counted, up to 10 in each tree. Also, "large numbers" occurred at Twentynine Palms on April 23, 1934, and the species was numerous there on April 16, 1935. At Little Morongo Canyon this warbler appeared suddenly, commonly but scattered, on April 2, 1951. Thus, although the migration periods are indicated as extending at least from March 21 to May 30 in the spring and from August 24 (probably a little earlier) to October 14 in the fall, there are times of concentration which suggest the height of movement of different races and populations.

A specimen of the race *Wilsonia pusilla chryseola* taken on April 2 suggests that the early spring migrants are of this race, which indeed reaches its breeding grounds along the Pacific coast often in late March and in the mountains in April. Two examples of *W. p. pileolata,* the race breeding in the Great Basin mountains and on north, interiorly into Alaska, were taken on May 13 and 15. Doubtless the waves of Wilson warblers in May are made up partly or largely of this late-migrating and late-nesting interior form. Most fall migrants taken in late August and early September have proved to be *chryseola,* suggesting that the principal group moving then is from the Pacific coastal areas en route southeastward. A female from Black Rock Spring on August 30, a male from Lower Covington Flat on August 26, and a male from Cottonwood Spring on September 17 are *pileolata.*

These warblers move through the country more slowly in migration than do orange-crowned and Nashville warblers. More often than the other species they are seen foraging, drinking, or even bathing. In the spring they occasionally sing, although the male taken on April 2, a day on which song was heard, had testes only $1\frac{1}{2}$ mm. in length.

One Wilson warbler, individually identifiable, adopted our camp at Black Rock Spring on September 1 and 2, catching flies and moths, particularly the former, about the garbage pit and some drying skeletons we had hung in a piñon. The bird moved about our feet at close range on the ground and even went into the tent. The temperature in the shade at camp was 101° F., but the heat was severe in the open. This warbler was caught on both days with a butterfly net, and we could tell, in hand, that it was of the race *chryseola.* On the second day we marked it for surer identification with a dark spot on the breast; it returned thereafter for the remainder of the afternoon.

A moist insect diet derived from considerable foraging in transit must serve this species in meeting the water problem of the desert passage. In temperatures of over 100° F. it has been seen definitely seeking shade. Some individuals were in a state of exhaustion during severe hot weather. Of twelve taken in the early fall period, five weighed less than 6 grams and were in poor condition. The adult female *pileolata* taken on August 30 in a severe heat period weighed but 4.5 grams and was considerably emaciated and desiccated. Both adults and immatures, to the extent of about 40 per cent of our migrant sample, seemed in danger of succumbing before completing their travels.

Wilson warblers use almost any available vegetation in the course of migration. But they seldom move to tree tops except in cottonwoods and willows about springs. Most of them forage near the ground. On September 3, 1950, they were seen feeding extensively on the wet ground following a nighttime rain. Swarming termites appeared to be part of the attraction on this occasion.

American Redstart. *Setophaga ruticilla*

Description. A small (7—9 gram), long-tailed warbler, with yellowish or pale orange patches in basal half of tail feathers and in bases of wing feathers (at least the secondaries). Adult male: head, back, and upper breast black, belly white, the anterior flanks with patch of orange; wing and tail black except for orange patches. Female and young males: dorsum olive-gray, head gray, throat light gray or white, belly white; flanks yellowish especially anteriorly; wings and tail dusky except for yellow patches.
Range. Breeds from southeastern Alaska, northern British Columbia, southern Mackenzie, central Manitoba, and central Quebec south to eastern Oregon, northern Utah, northern Colorado, Oklahoma, and northern parts of Gulf states. Winters chiefly from central México to Ecuador and British Guiana.
Occurrence in Monument. Two records: Smithwater Canyon, June 3, 1961 [1]; Twentynine Palms, May 28, 1935.

Frances Carter (1937:217) saw a female or young male American redstart characteristically spreading its tail to reveal the yellow tail patches as it moved about willow trunks at Twentynine Palms.

This rare migrant through California may be expected occasionally in transit to and from its breeding grounds in eastern Oregon, Washington, and British Columbia. Apparently most of the migrants of this species follow a more easterly route.

Painted Redstart. *Setophaga picta*

Description. A fairly small (8—10 gram), long-tailed warbler, with black body and large white patch on secondary wing coverts. Outer tail feathers (parts of 3), and eyelid white; lower breast and upper abdomen vermilion. Sexes alike.
Range. Breeds from mountains of northwestern and central Arizona and southern New Mexico south through the mountains of México, Guatemala, Honduras, and El Salvador to northern Nicaragua. Migratory in northern part of range, wintering chiefly south of United States but in small numbers in southern Arizona and coastal southern California.
Occurrence in Monument. A single record: near Stubby Spring, 4750 feet, Sept. 12, 1950 *.

The conspicuous painted redstart can hardly be regarded as a normal migrant in the Monument area, but only as a vagrant from migratory routes lying to the southeast. Its nearest nesting point is in the Hualpai Mountains east of Needles across the Colorado River valley. It has been recorded wintering occasionally to the westward in coastal southern California.

A painted redstart found near Stubby Spring on September 12 was heard giving its sharp, bleating note and was seen moving through the piñons at camp, coming to rest in a juniper. It was exceptionally lightweight for the species—7.2 grams.

The race of the painted redstart occurring in the United States is *Setophaga picta picta*.

ORIOLES AND ALLIES

Western Meadowlark. *Sturnella neglecta*

Description. Streaked, barred, and mottled with dark brown and grayish brown above and on wings and tail; lateral three pairs of tail feathers with longitudinal white stripes, chiefly on inner webs, but parts of feathers next the shafts always dark; head with central buff streak and yellow to buff supraorbital streaks; under parts conspicuously yellow with black "V" mark on throat and upper breast; flanks white streaked with dark brown.
Range. Central British Columbia east to western and southern Ontario and south to Baja California, northern Jalisco, southwestern Tamaulipas, and northwestern Louisiana. Partly migratory, extending south in winter to Michoacán.
Occurrence in Monument. Chiefly winter visitant, in small numbers. Occasionally present in spring and summer. Recorded from: Upper Covington Flat, July 24[1]; Quail Spring, Oct. 15; Queen Valley, Apr. 9 *; 2 mi. W Indian Cove, Oct. 13; Twentynine Palms, Oct. 17, Nov. 27, Jan. 5, 12, Feb. 21, Mar. 7, Apr. 18, May 14, July 13; Pinyon Wells, Oct. 14; Pleasant Valley, Apr. 10; Pinto Basin, Oct. 14.

Meadowlarks appear in both upper and lower levels of the Monument in the winter, principally from mid-October to mid-April, where they may be flushed from open ground. On April 9, 1951, we found a group of eight in Queen Valley, where in this year there had been sufficient rain to have induced a growth of filaree, sparse small primroses, and other flowers among the widely spaced Joshua trees. A female taken from this group had ova showing slight enlargement (1 mm.) but not enough to indicate local breeding, and moreover the birds were flocking rather than paired for nesting.

It is possible that in some years when there are good growths of low annual plants meadowlarks nest in the Monument. This may have been true in Pleasant Valley in April of 1960, where there was lush growth of *Erodium* and where a male was singing repeatedly. The species certainly nests in the cultivated areas and marsh borders of the Imperial Valley to the south. The occurrences we have noted on May 14 and July 13 at Twentynine Palms probably represent nonbreeding vagrants or laggards staying about the oasis. Except very locally, then, the desert in the Monument would not seem to provide sufficient ground cover of plants to induce this species to nest and to permit it to obtain for the raising of young a sufficient supply of insects by its ground-foraging methods.

The western meadowlark has been divided into two races (see A.O.U. Check-list, 5th ed., 1957), which would call for designation of the Pacific coastal birds as *Sturnella neglecta confluenta.* The Californian meadowlarks do not, however, conform well with typical members of this race, and we find the definition of the subspecies rather unsatisfactory at this time. Much more material taken on the breeding range needs to be analyzed before a useful racial taxonomy can be established.

Yellow-headed Blackbird. *Xanthocephalus xanthocephalus*

Description. A large (55–85 gram) blackbird, the males with yellow head and a white patch on the wing coverts. Females distinctly smaller than males, with yellow dull and confined to ventral and lateral parts of head; wing patch lacking.
Range. Breeds from central British Columbia, northern Alberta, central Saskatchewan, southern Manitoba, northern Minnesota, northern Wisconsin, and northern Ohio south to northeastern Baja California, central Arizona, southern New Mexico, northern Texas, and northwestern Arkansas. Winters from central California, central Arizona, southern New Mexico, central and southeastern Texas, and southern Louisiana south to southern Baja California, Guerrero, and Puebla.
Occurrence in Monument. Scarce vagrant. Recorded from: Quail Spring, July 2; Twentynine Palms, May 2, 4; Cottonwood Spring, Sept. 17[1].

The yellow-headed blackbird has been reported only at springs. Single males were recorded at Quail Spring and Cottonwood Spring. At Twentynine Palms this species was associated with red-winged blackbirds in the tules and mesquite there on May 2 and 4. The latter occurrences probably represent spring migrants, whereas an occurrence on July 2 doubtless should be classed as vagrancy, to which these and other blackbirds are much given.

Red-winged Blackbird. *Agelaius phoeniceus*

Description. A blackbird (39–62 gram) in which the males are conspicuously marked with red or red bordered by tawny at the bend of the wing. Females without red, the body more or less streaked ventrally with dusky and buff.
Range. Breeds from British Columbia, southeastern Yukon, central Mackenzie, north-central Manitoba, northern Ontario, southern Quebec, and central Nova Scotia south to northeastern Baja California and through mainland México and Central America to Costa Rica; also Bahamas and Cuba. Winters from northern United States southward.
Occurrence in Monument. Occasional spring and fall migrant. Recorded from: Twentynine Palms, Mar. 7, 23, Apr. 18, May 6, 17, 31, Oct. 25, Nov. 7; Cottonwood Spring, May 18, Oct. 13, 14.

Blackbirds range far in flights to foraging grounds, and apparently they cover great distances with ease. They seem readily to locate oases while flying high across desert country, and they may appear at such places to rest or feed. The occurrences in the Monument are chiefly at times of migration, but blackbirds may also be expected at other seasons, even though they are not regular summer or winter residents of the area. The dates of occurrence span a period in spring from March 7 to May 31, but the species was not continuously in evidence, although it apparently was fairly regularly present in late April and through May. The fall dates extend from October 13 to November 7.

Of the three species of blackbirds recorded in the Monument, the redwing has been seen on the largest number of dates and in greatest numbers. Most of the occurrences have been at Twentynine Palms, where in the wet places there Frances Carter saw as many as 12 on April 18 and mentioned that there were "numbers" on November 7. They were seen foraging about pastured horses in the grassy swamp that existed in the 1930's. Apparently the redwings were standing ready to snatch food uncovered by the horses as they pulled up tufts of grass. At other times single birds only would appear, as is true of the records of occurrence at Cottonwood Spring.

No specimens have been obtained, and consequently it is not known which of the several races of redwing have visited the Monument.

Hooded Oriole. *Icterus cucullatus*

Description. A small (24–28 gram) oriole with slender, moderately curved bill. Adult male: body yellow except for black chin and throat and area in front of eye and black patch in middle of back; wings black with large white patch on middle and lesser secondary coverts and white edges on greater secondary coverts and secondaries; tail black with small buff tips on rectrices, largest on lateral pair. Female: dull green above and on tail, the center of back somewhat dusky; below, dull yellow; wings dark brown with white bars on middle and greater secondary coverts. Males in first year may resemble females but show at least partial development of black throat.

Range. Breeds in southwestern United States from central California east to western and southern Texas and south through México to British Honduras. Winters chiefly from 30° N southward, occasionally north to southwestern California and southern Arizona.

Occurrence in Monument. Summer resident in small numbers, locally. Recorded from: Little Morongo Canyon, Mar. 31, Apr. 3; Quail Spring, Apr. 22, May 19, 20 *, 21, July 1; Indian Cove, May 13, 14; Twentynine Palms, Apr. 15, 26, May 2, 16, 21, 23, 27, July 12, 13 *; Cottonwood Spring, Apr. 11, 25, 26, 29, 30, May 12, 14, 15 *, 18 *, July 14, Sept. 13, 14, 15 *; Lost Palm Canyon.

Hooded orioles station themselves in summer chiefly in the fan palms (*Washingtonia filifera*) that are present at oases such as Twentynine Palms, Cottonwood Spring (fig. 12), and Lost Palm Canyon. However, they remain through the summer at Quail Spring, where no palms occur, and they apparently do likewise in the canyon at Indian Cove. The fan palms are especially attractive, as the leaves in the crowns of these trees afford elevated hiding places for these rather shy birds and, more critically, excellent nest locations and fiber material for the nests. The pouched nests are characteristically hung on the under side of a fan and sewed through it at the rims for secure attachment. Here the nests are shaded and unapproachable by nonflying predators, and they can withstand, because of the secure attachment, the frequent lashing of desert winds. Nests, either active or old, thus situated have been seen at the above-mentioned stations where palms are present. Old nests persist for a year or more, but nests of the year, no more than a few weeks or months old, are identifiable by the more yellow, less gray or weathered appearance of the palm fibers composing them.

On July 13 at Twentynine Palms, hooded orioles were scattered along the palms and cottonwoods at the oasis, and a pair was scolding as though concerned for young hidden in the palm crowns. An adult female taken here on this date had a postbreeding brood patch, but two males had testes at full size (9 mm.). Miss Carter noted the song of this species here from early May until the end of the month, when she left.

At Cottonwood Spring two females on May 15 and 18 had not yet laid, whereas two males on the 15th had testes 9 and 10 mm. long. It is judged that laying probably begins in May and possibly continues into July, as suggested by the breeding condition of the males, which is sustained until at least the middle of that month. Hooded orioles are present by the end of March and continue in the area until mid-September. On the 15th of the latter month an adult female was taken in worn breeding plumage which had not yet begun the annual molt. Evidently it would not have molted before migration.

At Quail Spring hooded orioles have been seen in May and through midsummer. A male in full breeding condition (testis 10 mm.) was taken here on May 20. We do not know where they nest at this station, as cottonwoods, willows, and palms are lacking. Possibly they use desert willows or Joshua trees, although this practice would be atypical for the species and hooded orioles are not in summer residence elsewhere in such habitat unless at Indian Cove.

Hooded orioles, by confining their summer residence to oases and by chiefly using shady, tall trees, avoid the extreme rigors of summer heat and drought. Their insect food, gathered in these trees or in forays out into the desert scrub of the canyon walls, is supplemented by nectar and fruit. At Twentynine Palms, Miss Carter captured two hooded orioles for banding by baiting a trap with sliced oranges and placing it on a platform in a mesquite bush. Probably these orioles do not often resort to water at the springs, although they could do so at every station where they are known to nest; in any event we have not seen them drinking.

The specimens of hooded orioles from the Monument as a group fall in the race *Icterus cucullatus californicus* (van Rossem, 1945), but some individuals are not in fact clearly separable from *I. c. nelsoni* of the Colorado River valley. This may reflect merely the incomplete differentiation of these races rather than an intergradational trend.

Scott Oriole. *Icterus parisorum*

Description. A large (32–40 gram) oriole (see frontispiece) with nearly straight bill. Adult male: black on head, back, and breast; rump and belly yellow; wings black, the bend of the wings yellow, bordered below with white and with white bar on tips of greater secondary

Plate V. A pair of hooded orioles.

GMC

coverts; tail black centrally and at tip, the basal two-thirds of lateral four pairs of feathers yellow. Female: dusky olive above, the rump dull green; below dull greenish yellow; wings dull black, with two white bars on wing coverts; tail dark brown centrally and at tip, grading to green laterally and basally. Young males and old females variously intermediate toward coloration of adult males.

Range. Breeds from southeastern California and southern Nevada east to north-central New Mexico and western Texas and south to Cape district of Baja California, Michoacán, and Oaxaca. Winters chiefly from northern Baja California and Sonora southward.

Occurrence in Monument. Fairly common summer resident. Recorded from: Morongo Pass, Apr. 28; Black Rock Spring, Apr. 3, July 6, Sept. 1 *, 3 *; Upper Covington Flat, Apr. 9, June 11 [1], 16 [1]; Lower Covington Flat, Mar. 27 [1], Apr. 7, 8, 15 [1], 23 [1], 30 [1], May 19, June 25 [1], Aug. 27; Quail Spring, Apr. 11, May 19–21, June 3, July 1, 3, 4; Queen Valley, Apr. 10; Indian Cove, Apr. 7 *, 8, May 13, 14, June 3; Fortynine Palms, Apr. 7; Twentynine Palms, May 7, 10; 1 mi. NE Split Rock, June 11 [1]; Pinyon Wells, Apr. 10; Cottonwood Spring, Apr. 30, May 12, 13, 17 *, July 7, 8, 14.

Scott orioles have been found in the Monument from March 27 to September 3, almost always in the piñon and Joshua tree habitats or in canyons where smaller arborescent yuccas, such as *Yucca mohavensis,* are to be found. Only occasionally do they appear in the open flats as at Twentynine Palms, where on May 7 and 10 Miss Carter recorded a pair and trapped a female, which she banded.

In the piñon and Joshua tree woodland, the loud rollicking song of this oriole may be heard principally in April and May. The conspicuous meadowlark-like song and the flashy appearance of the males may give a false impression of abundance. At Cottonwood Spring the species occurs in the several branch canyons and at times comes into the trees at the oasis. Here on May 12 four males were detected, which probably represents maximum concentration in the ¼ square mile area centering at the spring.

In early April, on the 8th and 10th, respectively, at Indian Cove and Queen Valley, males were seen making flights of 300 yards north and northeast which suggested migration. A male taken on the 7th at Indian Cove had a 7 mm. testis and a little subcutaneous fat, circumstances that suggest the concluding stages of migration. However, Scott orioles fly far in their established summer areas, and these males might have been settled at these places for breeding.

In foraging, Scott orioles visit clumps of tree yuccas (frontispiece), alighting on the spines and working up through them to the flowering or fruiting stalks. The flowering stalks are sources of insect food and possibly nectar, and the soft flower parts and fruiting pods are also levied for food and moisture. We have watched the birds fly to flowering clusters of ocotillos also for feeding purposes. On September 3 at Black Rock Spring, Scott orioles dropped to the ground to feed on swarming insects that were active after a rain in the night. Fruit, nectar, and insects keep the Scott orioles in water supply during their hot summer residency. They do seem attracted to springs, where available, although we have not actually seen them drinking.

The nests of this species are shallow hanging cups of fibers placed in the shade beneath a cluster of Joshua tree spines. On April 10 at Pinyon Wells a pair was in the early stages of nest construction. The site was 8 feet up in a small, five-branched Joshua tree near some scrub oak thickets. The nest was of yucca fibers slung under dead spines beneath a yucca branch. Splits in some of the dead spines afforded places where the strands of the nest could be inserted and would hold fast. The nest had one side and the bottom sketchily represented. The male came with fibers, about four in number, and spent four minutes at the nest, standing at its side and thrusting and pulling at loose ends. It pulled vigorously at times as though drawing and tightening an element in the fabric. This male sang on the wing going to and from the nest, and both male and female scolded at our approach.

Specimens taken on September 1 and 3 were independent juvenal females which had not yet started to molt and apparently would not have done so before migration.

Bullock Oriole. *Icterus bullockii*

Description. A large (32 to 40 gram) oriole, with short, straight bill. Adult male: crown and back black, rump yellow; throat with narrow black patch; under parts and side of face orange-yellow interrupted by black line through the eye; wings black with large white patch on middle and greater secondary coverts and white edges on secondaries; tail black centrally, the lateral four tail feathers mostly yellow, the tips dusky or gray. Female: dusky olive above, the tail dull greenish; below dull white with yellow area on breast and sides of head; wings dark brown with white edgings on secondary coverts and secondaries.

Range. Breeds in western North America from southern British Columbia, southern Alberta, and southwestern Saskatchewan south to northern Baja California and the southern part of the Mexican Plateau. Winters south of the United States, extending to northwestern Costa Rica.

Occurrence in Monument. Fairly common migrant. Small numbers summer resident locally. Recorded from: Little Morongo Canyon, Apr. 1 *, 2, 28; Lower Covington Flat, July 11 [1]; Black Rock Spring, Aug. 30 *, 31 *; Upper Covington Flat, Apr. 9; Quail Spring, July 1–5; Twentynine Palms, Mar. 14, 29, Apr. 1, 27, May 1 [1], 8, 31, June 29 [1], July 13; Cottonwood Spring, Apr. 26, July 7, 8, 14.

Bullock orioles are seen in spring and fall migration, being more conspicuous in the former. In Little Morongo Canyon on April 1 two males were taken that were apparently in migration. They still had some of their migratory fat left, and the testes were far below maximum length at 3½ and 4 mm. At Twentynine Palms Miss Carter in successive years found birds appearing on April 1 and on March 14; probably some of them were en route to the north. In fall on August 30 and 31 migrants of this species appeared in areas not suited for summer residence. One, an extremely thin juvenile in much weakened condition, flew into camp on the latter day seeking and taking proferred water. It obviously was exhausted from the heat or from migratory flight or the two in combination. It weighed only 25.8 grams, about 20 per cent under normal weight. Probably it would have succumbed in the course of further migratory effort.

At two oases at least, the species apparently nests in the cottonwoods. It is less certain that nesting occurs at Quail Spring and Lower Covington Flat, where this oriole has been seen in early July but where habitat is thought not to be suitable for breeding. Little Morongo Canyon, where Bullock orioles have been noted on April 28, may support nesting pairs. However, clear evidence of breeding has been found at Cottonwood Spring. The species has been heard singing there in late May and the characteristic old nests have been seen. The nests here in the cottonwoods are deep pouches made of fibers which persist a year or more after their use. At Twentynine Palms on July 13 a family group with the young still dependent and begging was watched.

Probably the Bullock oriole, which is less inclined to occupy desert areas than are related western species of orioles, finds the aridity and lack of surface water limiting in the Monument. The oases at Twentynine Palms and Cottonwood Spring, because of available places for drinking water, are the best places for them in this section of the desert.

The emaciated juvenal male of August 31 and the adult female of August 30 show no molt, the feathers of the latter being extremely worn. Therefore, molt would evidently occur after migration. The juvenile was surprisingly long-winged (98.5 mm.) and accords with van Rossem's figures (1945:238–239) for the large race *Icterus bullockii bullockii*. It presumably therefore had moved some distance from the northern interior breeding range of this race, the closest part of which is in central Nevada. The other specimens taken are of the more southern race *I. b. parvus*, which breeds in southern California and northern Baja California.

Brewer Blackbird. *Euphagus cyanocephalus*

Description. A large (50–75 gram) blackbird, the males glossy black throughout, the females uniform dark gray; iris of males white or yellowish.
Range. Breeds from central British Columbia east to southwestern Ontario and south to northwestern Baja California, southern Nevada, central Arizona, southern New Mexico, and northern Texas. Winters from southwestern British Columbia, central Alberta, central Oklahoma, and Arkansas south to southern Baja California, Oaxaca, central Veracruz, and the Gulf coast as far east as western Florida.
Occurrence in Monument. Occasional vagrant, especially in fall. Recorded from: Lower Covington Flat, Oct. 15 [1]; Quail Spring, July 3; Pinyon Wells, Oct. 13 *; Pinto Basin, Oct. 14; Cottonwood Spring, Sept. 15 *; Coxcomb Mountains, Sept. 29 [1].

No large flocks of this blackbird have been seen, but on several occasions single birds or two together have been found, chiefly at springs. The bird noted at Pinyon Wells was flying high overhead but descended to a boulder, calling; no water was present in the vicinity. Two in Pinto Basin were seeking the shade of a creosote bush at midday. At Quail Spring the birds came to the vicinity of the open water, and at Cottonwood Spring they came into the tops of the tall cottonwoods. Both specimens taken were females in fresh fall plumage.

The birds that occur in the Monument in summer and fall probably represent casual dispersal from breeding grounds in coastal southern California. Winter concentrations build up in the lower parts of the Colorado River basin, and fall migratory flight across the Monument may lead to these areas from the Great Basin to the north.

Brown-headed Cowbird. *Molothrus ater*

Description. A small (28–36 gram) blackbird, the male with a dark brown head and glossy black body, wings, and tail. Female nearly uniform light brown, very faintly streaked.
Range. Breeds from central and northeastern British Columbia southeast to central eastern Quebec and south to northern Baja California, Durango, northern Tamaulipas, and the Gulf coast of the United States. Winters from central California, southern Arizona, northeastern Texas, Missouri, the southern Great Lakes area, and Connecticut south to southern Baja California, Guerrero, Oaxaca, and southern Florida.
Occurrence in Monument. Scarce summer resident. Recorded from: Lower Covington Flat, May 19, 28 [1], 29 [1], June 25 [1]; Quail Spring, July 3; Lost Horse Valley, May 14; Queen Valley, Apr. 10; Twentynine Palms, July 13; Pinto Basin, Apr. 11; Cottonwood Spring, Apr. 26, June 22.

Brown-headed cowbirds range into the Monument in spring and summer, appearing singly or in pairs. This widely dispersed occurrence in spring is typical of the species as it scatters from winter flocking locations and begins patrolling suitable areas in search of the nests of other species of birds, which it parasitizes. Even so, the cowbirds in the Monument are especially widespaced and we have seen them chiefly near oases. The earliest date in spring was on April 10 in Queen Valley, where a single male was associated with a group of western meadowlarks and was giving its song performance. In Pinto Basin on April 11 and at Cottonwood Spring on April 26 males were similarly singing, at the latter station in the tops of the cottonwoods. On July 13, our latest date, at Twentynine Palms a pair flew by the oasis and the female several times gave its breeding chatter. We have found no parasitized nests in the Monument, but the patrolling of males and females and their nuptial calls leaves little doubt that the species breeds there.

Cowbirds fly long distances easily and could move to water during hot periods, although we have not actually seen them drinking. Probably they move out of the area entirely some time in July.

The brown-headed cowbirds occurring in the Monument are of the race *Molothrus ater obscurus*.

TANAGERS

Western Tanager. *Piranga ludoviciana*

Description. A predominantly yellow or yellow-green (28–32 gram) tanager, with black or dark brown wings and tail, the wings marked by two yellowish bars on the coverts. Adult male: anterior part of head red to orange; body yellow, except for black central section

Fig. 66. Western tanager (left) and black-headed grosbeak (right).

of back. Females and immatures: olive on back, green to dull yellow elsewhere on body. Bill dark brown to yellow.

Range. Breeds from southern Alaska, northern British Columbia, southern Mackenzie, and central Saskatchewan south to northern Baja California, northeastern Sonora, and western Texas. Winters from southern Baja California, Jalisco, and southern Tamaulipas south on Pacific side of continental divide in Central America through Guatemala and El Salvador to northwestern Costa Rica.

Occurrence in Monument. Common spring and fall migrant. Recorded from: Morongo Pass, Apr. 28; Black Rock Spring, Aug. 29, 30, 31, Sept. 2, 3 *; Upper Covington Flat, May 15 [1], 29 [1], July 24 [1], Aug. 28 *; Lower Covington Flat, May 15 [1], 19, Aug. 22, 23, 25, 26 *, 28; Smithwater Canyon, Sept. 9 [1]; Quail Spring, Apr. 22, May 14, 19–21; Stubby Spring, Sept. 5, 6, 7, 8 *, 9 *, 11; Indian Cove, May 13, 14; Twentynine Palms, Apr. 28, May 12, 14, July 13 *; Pinto Basin, Apr. 26, May 19 [1]; Eagle Mountain, May 16; Cottonwood Spring, Apr. 25, 26, 30, May 12, 13, 14, Sept. 13; Lost Palm Canyon, Sept. 14.

Western tanagers are readily detected migrants that move through the desert in late spring and early fall. Often small groups of three or four travel together. Migration in the spring has been recorded from April 13 to May 21, but it may be expected to continue to early June. However, the main concentration of migrants is in the last week of April and the first two weeks of May. In fall they have been recorded from August 22 to September 14, and this interval doubtless encompasses the main period of migration.

These tanagers use the larger trees and shrubs for resting and foraging whenever they are available along their route. Thus we have seen them most often in piñons, junipers, scrub oaks, cottonwoods, willows, and palms. They take long flights between rest points in the trees when they are actually moving, and the direction of these movements is conspicuously southeast and south in the fall, although at the time of a storm disturbance on September 2 northward flight was noted. No great attraction to springs has been noticed, although the trees at such places tend to concentrate them there.

The desert crossing in the stress period of late summer heat and drought is accomplished without noted recourse to water. But unlike the migrating finches and sparrows, which they resemble in many respects, they feed on insects while migrating, thus tapping an important water resource. We specifically noted one catching large insects in a cottonwood tree on August 26, and others came to the ground for insects, probably termites, following a nighttime rain on September 3. Juicy fruits and buds may be taken. Frances Carter found them attracted by grapefruit placed out for bait at her banding station in May.

Migrants normally are very fat. We recorded no tanagers in the fall that were in weakened condition. However, an adult male taken on July 13, 1946, had distinctly weak, flat breast muscles, although there was subcutaneous fat. Probably this bird had had a history of disease or prior starvation, although it was not in critical condition at the time. Its presence in midsummer tends to bear this out; possibly it had never completed its spring migration to the mountain breeding range to the west and north. On this midsummer date the testes were 5 mm. long and thus not in full breeding condition; there was a little scattered body molt in progress, but the plumage was still essentially that of the worn adult type.

In the fall western tanagers migrate south to southern México and Central America, rarely wintering any farther north than southern Baja California. Thus a long migration is undertaken through desert areas. Adults may leave the latitudes of California without having changed the spring plumage. The young of the year acquire a postjuvenal body plumage, beginning in the period while they are still attended by their parents, and they always seem to travel in this new first-winter body plumage. Of the fall migrants taken in the Monument, two adult females, on August 26 and 28, were in worn breeding plumage, with only a little body molt beginning, whereas one taken on September 3 had completely new plumage. Three immatures were, as was to be expected, in postjuvenal plumage.

Summer Tanager. *Piranga rubra*

Description. A large-billed tanager (33–38 grams) with the entire body plumage red in adult males and olive above and yellow below in females. Young males may be partly red. Flight feathers dusky, edged with red or yellow. Bill brown to yellow.

Range. Breeds from southeastern California, extreme southern Nevada, central Arizona, central New Mexico, southeastern Nebraska, central Illinois, central Ohio, Maryland, and Delaware south to Durango, Nuevo León, and southern Florida. Winters from southern Baja California and Veracruz south throughout México and Central America to Perú, Bolivia, and western Brazil.
Occurrence in Monument. Recorded once in spring: Cottonwood Spring, May 14.

This species normally does not range west of the Colorado River valley, where it nests. Occasional vagrants occur on the Pacific coast. The presence of this species on May 14 at Cottonwood Spring as reported to us by Dr. Wade Fox probably represents a spring migrant that had strayed westward from its normal course. Late April and early May is the normal period of arrival along the Colorado River.

FINCHES AND SPARROWS

There are some species of the large family Fringillidae that are known only as transients through the Monument and that migrate far southward, and, like many of the warblers, winter in western and southern México. These are principally the grosbeaks and the lazuli bunting. They must often cross 700 miles of desert country in their movements, some 500 miles of it after leaving the Monument on their southward journey.

These migrating fringillids are not insect feeders to any significant degree when they are away from their nesting grounds, but depend on seeds and fruits. Birds of this type have been shown (Bartholomew and Cade, 1956) to require drinking water when on a seed diet in order to avoid a water deficit, since they do not have sufficiently effective water-conserving mechanisms to permit prolonged dependence on metabolic water derived from food utilization, as do some rodents. Occasionally the taking of insects may aid these fringillids in obtaining water during their desert crossings but certainly not in the way that it aids warblers, vireos, and flycatchers. Succulent fruits are scarce on desert plants except on the cacti, and we have not often seen migrating fringillids eating them. But in the autumnal dry period this source of water may occasionally be important, particularly to grosbeaks. The chief ways in which the migrant species cope with the water deficit problem seem to be (1) rapid passage through unwatered areas, (2) frequent drinking at springs, and (3) in certain cases, delay of fall migration until the extremes of heat and drought have passed.

Although metabolic water is insufficient for the total water need, much water must be derived from this source through utilization of fat reserves. Like the warblers, migrant fringillids are often extemely fat, and food and water intake can be lessened in transit as a consequence of burning this reserve. However, when the early fall migration is going on, although heat and drought are severe, seeds and fruits are in comparatively good supply in the desert so that opportunities for feeding are present. Still, as in the warblers, some of these species are forced to forage in the desert on or about vegetation to which they are not accustomed or especially adapted; they must find types of food that are strange to them and that may require adjustments of habits of searching and manipulating with the bill. Occasional examples of autumnal migrant fringillids succumbing or about to succumb to desiccation or starvation have been seen, but the instances are not so frequent as among the insectivorous flycatchers and warblers.

The direction of migratory movement has been observed only in those relatively common species that fly well above ground in the open. Thus the black-headed grosbeaks and lazuli buntings have been seen in the fall making prolonged flights that have a southeastward direction, such that would probably take them across the lower Colorado River basin to southern Sonora. In spring little has been recorded of the direction of flight.

Ways of utilizing cover while in transit reflect the more rigorous habitat selection of the birds' breeding season. Thus black-headed grosbeaks tend to move through trees at middle and upper levels, and, when they are loafing, they favor broad-leaf, water-dependent trees like willows and cottonwoods, which they favor on the nesting grounds. Contrarily, green-tailed towhees stay low in the plant cover in dense bushes that are as nearly like the mountain chaparral and sagebrush of their breeding grounds as anything available in the Monument.

There are seldom any great concentrations of purely transient fringillids; conspicuous aggregations are those of the winter resident species of this family. Of the migrants usually only two or three individuals are seen together. Occasionally as many as a dozen of one species may be found about an oasis but not associated as a flock.

House Sparrow. *Passer domesticus*

Other name. English sparrow.
Description. Male: crown gray, bordered by chestnut; back streaked black and rusty; chin, throat, breast, and orbital area black; under parts otherwise gray, becoming white on cheeks and belly. Female: head brownish gray with lighter line behind eye; back streaked as in male; under parts entirely gray-brown, lighter posteriorly. Bill yellowish horn color, turning black in spring in male.
Range. Native in Europe, Asia, and northern Africa. Introduced in North America and many other parts of world. Now established from central Canada south to Chiapas on the North American continent.
Occurrence in Monument. Resident in the towns, vagrant elsewhere. Recorded from: Twentynine Palms; Pinto Basin, Apr. 13 *; Pinto Wash Well, Oct. 16 [1].

This species usually does not invade the desert except locally about towns where it is resident, as in Twentynine Palms. At times birds do cross through the Monument, as did a male at Pinto Basin on April 13; this sparrow was in breeding condition but obviously was not established locally for nesting. It flew about our parked truck for several minutes.

Miss Carter reported that in the spring of 1934 house sparrows made themselves unpopular by appropriating or destroying nests of the Say phoebe under the eaves of the Inn porch at Twentynine Palms. Flycatching habits were observed, as a sparrow would sometimes rise from the ground to pursue a moth for some distance through the air and finally capture it.

Black-headed Grosbeak. *Pheucticus melanocephalus*

Description. A large (35—40 gram) finch, of stocky build and large head and bill. Male: head, including chin, black, a cinnamon stripe back of eye and in middorsal line present in young males; throat, breast, and flanks cinnamon; belly yellow anteriorly, white posteriorly; back black with streaks of white and cinnamon; wings and tail black with middle and greater secondary coverts and outer tail feathers broadly tipped with white; primaries with large white patch basally; upper mandible black in breeding season, the lower mandible brown and whitish. Female: head streaked with dark brown and cinnamon or white in same pattern as in young males; back and rump streaked; under parts as in males but paler and with streaks on flanks; white markings on wings and tail smaller and less contrasting than in males (fig. 66).

Range. Breeds from southern British Columbia, southern Alberta, southern Saskatchewan, and North Dakota south through Pacific coast districts, Rocky Mountains, and western Great Plains to northern Baja California and to the uplands of mainland México as far as Oaxaca. Winters from southern Baja California, southern Sonora, and Nuevo León south to Chiapas.

Occurrence in Monument. Common spring and fall migrant; some stragglers present throughout summer. Recorded from: Black Rock Spring, Aug. 29, 31, Sept. 1 *, 2 *, 3 *, 4 *; Upper Covington Flat, July 8 [1]; Lower Covington Flat, Apr. 30 [1], Aug. 22—29, 24 *, 26 *, 27, 28 *, 29; Smithwater Canyon, May 13 [1]; Quail Spring, May 14, July 1—5, Sept. 9 *; Stubby Spring, Sept. 6, 8, 9 *; Indian Cove, Apr. 27, May 13, 14; Twentynine Palms, Apr. 24, May 6, 8, 12, 14, 17, 30; Pinyon Wells, Apr. 10; Pinto Basin, Apr. 11; Eagle Mountain, May 16; Cottonwood Spring, Apr. 14, 26, 30, May 13, 14, 18 *, June 3, July 8, 11 [1], 14, Oct. 13, 14.

Black-headed grosbeaks are conspicuous migrants through the Monument and are present as a species for extended periods. Since representatives of both early-breeding lowland birds and later-breeding mountain birds probably pass through, the period of spring migration is fairly long. These grosbeaks begin to leave their breeding grounds soon after young are raised and either move about locally or migrate. Thus grosbeaks detected in midsummer may be fall migrants; but apparently some individuals never reach the breeding grounds, and such stragglers may also account for the summer occurrences in the Monument. The dates that relate to normal spring movements range from April 10 to May 18; those for the fall range from August 24 to October 14.

Grosbeaks are seen usually in the piñons and junipers and in the willows and cottonwoods about springs. This manner of occurrence reflects their preference for trees for concealment and foraging. They show greatest affinity for willows and springside vegetation, probably finding here a type of foliage instinctively attractive, as are also the associated water sources. They loaf or forage much more at oases than in the piñon woods in which they are more likely to be on the move, drifting along through the tree tops and giving occasional faint whistles or less often the sharp *click* note. Frequently the birds associate in groups of two or three as they move in migration. In mid-May as many as thirteen have been seen in the course of one morning at Cottonwood Spring.

At the Hulsey Mine close to Cottonwood Spring they at one time foraged in the grape arbor, probably taking buds and fruit, and at Twentynine Palms, Frances Carter found that chicken feed, citrus fruit, and water were effective baits for them at the banding station. Probably the use of springs for drinking and such succulent vegetable material as they can find about oases aid in solving their water problem during desert crossings. The summer, nonbreeding records for June 3 and July 1–5, 8, and 14 are all at water sources, to which the birds were probably closely limited at this season.

In migrating to and from the breeding grounds of the Pacific coast and Great Basin, black-headed grosbeaks traverse at least 500 miles of desert or semidesert terrain each way; some must travel 700 miles of such country. The wintering grounds in southern Sonora are probably reached by a southeast flight through the Monument and beyond, from coastal California. Along the crests east of Lower Covington Flat on August 28, flight lines over and through the piñons were generally in this direction, and the same was true at Stubby Spring on September 6. However, some grosbeaks move more directly north and south as shown by the significant banding record obtained by Miss Carter: a "young," apparently first-year, male banded at Twentynine Palms on May 12, 1934, was found dead the subsequent spring about 500 miles due north at Battle Mountain, Nevada, on May 18, 1935. It was probably a member of the rather ill-defined Rocky Mountain race, *Pheucticus melanocephalus melanocephalus,* rather than the smaller-billed coastal race, *P. m. maculatus,* which seems to contribute most of the transient grosbeaks in the Monument.

Of the 13 specimens taken in the fall migration, all but one were immatures. The single adult male was in worn breeding plumage on September 2. None of the transients showed any sign of molt in progress. The birds of the year had apparently all undergone the postjuvenal molt, involving the body plumage, before departure; this is in accord with the findings of Michener and Michener (1951:94–96) at Pasadena on the breeding grounds. Of those birds for which there is adequate record bearing on fat condition, 5 appeared to be well supplied with fat reserves, whereas 3 were thin and of light weight, including the adult male. One of the immatures weighed only 27.0 grams on September 4 and was very thin; probably it would not have survived long.

The specimens taken thus far are all referable to the Pacific coast race, *Pheucticus melanocephalus maculatus,* although, as noted in connection with the banding record, *P. m. melanocephalus* which breeds in the Great Basin and Rocky Mountains probably also occurs.

Blue Grosbeak. *Guiraca caerulea*

Description. A fairly large (25–30 gram) finch, the adult males predominantly blue, the females brown. Male: head, upper surface, and under surface dark blue, variously tipped with brown and white in fresh plumage; chestnut bar on middle wing coverts and narrow cinnamon bar on greater secondary coverts; flight feathers and tail black, bordered with blue; upper mandible black in breeding season, lower mandible brown. First-year males with variable amounts of blue. Females and juveniles: brown above and below, but paler ventrally; wing markings as in male but both bars dull cinnamon.
Range. Breeds from central California, southern Great Basin, northern Colorado, central South Dakota, southern Illinois, and southeastern Pennsylvania south to northwestern Baja California, the Gulf states, and central Florida and through mainland of México and Central America to Costa Rica. Winters from southern Baja California, southern Sonora, and Veracruz south to northwestern Panamá.
Occurrence in Monument. Reported only in spring: Twentynine Palms, May 7–17.

This grosbeak was recorded almost daily in 1934 from May 7 to 17 at Twentynine Palms by Frances Carter. She found the birds exceedingly shy but did observe them feeding on scratch feed scattered on the ground. This species is a late spring migrant and on the dates observed probably was en route to breeding grounds in the Owens Valley or the Great Valley of California. The birds probably belonged to the race *Guiraca caerulea salicaria* which breeds west of the Colorado River valley and migrates south to southern Baja California and southern Sonora in winter. However, no specimens from the Monument are available to verify this supposition.

Lazuli Bunting. *Passerina amoena*

Description. A small (13–15 gram) finch, either brightly marked with blue or of various tones of brown throughout. Adult male: head, neck, throat, and rump, bright turquois blue; center of back, lesser wing coverts, and margins of flight feathers and tail dull blue; wing coverts with two bars of buff or white; breast cinnamon, fading to dull white on belly and lower flanks; bill black in breeding season. Female: upper surface and wings dark brown except for light wing bars; under parts pale brown, somewhat darker on breast and nearly white on belly; bill dark brown. Immature males and some females with variable amounts of blue.
Range. Breeds from southern British Columbia across the southern border of Canada to central North Dakota and south to northwestern Baja California, central Arizona, and central Texas. Winters from southern Baja California and southern Arizona (sparingly) south to Guerrero and central Veracruz.
Occurrence in Monument. Common spring and fall migrant. Recorded from: Black Rock Spring, Aug. 29, 31 *, Sept. 1 *, 3 *; Lower Covington Flat, Apr. 8, Aug. 23, 25, 26; Smithwater Canyon, Apr. 24[1]; Quail Spring, May 15, 19–21, Sept. 8; Stubby Spring, Sept. 5; Indian Cove, May 13, 14; Twentynine Palms, Apr. 17, May 2, 6, 17; Pinyon Wells, Apr. 10; Eagle Mountain, May 16; Cottonwood Spring, Apr. 26, 30, May 15 *, Sept. 13.

Lazuli buntings and black-headed grosbeaks are the most common finches that are purely transients through the desert area, and the buntings by reason of shyness and the dull plumage of most of the individuals are less conspicuous than the grosbeaks. The dates for the spring migration of the lazuli bunting in the Monument are April 8 to May 21, and for the fall are August 23 to September 13.

The buntings are most often detected by identifying their nasal buzzing note given as they fly past. At times they may be seen feeding on or near the ground or resting briefly at middle heights in the vegetation between the long direct flights that they take through the open scrub or woods. They frequently come in about springs, although we have not seen them at the water, probably because they are too shy to approach it when a person is in view. The species may be expected at all levels of the Monument, and probably it occurs somewhat earlier in the fall migration period than our dates of record indicate.

Foraging in bunch grass in the course of the autumnal passage has been noted, and probably such food as is taken in migration consists of seeds of grasses and annual vegetation. We have no evidence that the birds augment their water intake by feeding on fruits or insects in the critical stress period of late summer and early fall. Drinking at springs probably is important to them. Migrants we have taken have been fat and in good condition. However, on August 23, at Lower Covington Flat, in a period of very dry, hot weather in which daytime temperatures exceeded 110° F., a mummified lazuli bunting was found in a dry wash where the bird had recently crawled in under some campground debris. This bird had apparently been in bad condition either from lack of water or food during the intense heat period.

The migration of lazuli buntings takes most of them in winter to central and southern México, short of the Isthmus of Tehuantepec. The northern points for regular wintering are southern Sonora and southern Baja California. Probably the buntings that pass through the Monument move chiefly to the west coast of México, at least as far as 500 miles away in southern Sonora; their breeding areas lie rather nearby in coastal California and the Great Basin. The direction of fall flights was often noted to be southeastwardly, as along the mountain axis in the vicinities of Black Rock Spring and Stubby Spring in early September.

The early fall migration of the lazuli buntings is undertaken by the adults before much if any of the annual molt has been carried out. One adult taken on September 1 was in worn spring plumage with no molt in progress, whereas an adult male taken on August 31 had partly replaced the body plumage and was beginning the molt of the primaries. Birds hatched the preceding spring quickly enter a postjuvenal molt and when in transit have at least a partly new body plumage; such was true of one taken on September 3, in which no further molt was in progress and in which the conspicuous juvenal streaking of the breast had been lost. Young of the year, as also probably many adults, do not, then, undertake the taxing process of molt while migrating across the deserts in the critical fall period.

Fig. 67. Cassin finch (upper left), purple finch (right), and pine siskin (below).

Purple Finch. *Carpodacus purpureus*

Description. A finch of the same size as the common house finch (p. 238) but bill less stubby, tail distinctly forked, and head marked in females and many males by light supra- and postorbital streak. Adult males dull red or rosy-red on head, breast, and rump. Females and first-year males without red, the body streaked, but streaks less sharply defined than in house finch.
Range. Breeds from northern British Columbia east in timber belt to Newfoundland and south on Pacific coast, west of Cascade–Sierran axis, to northern Baja California; in east extends south to Great Lakes area and western Maryland. Winters from southern parts of Canadian provinces south to southern Arizona, south-central and southeastern Texas, the Gulf coast, and central Florida.
Occurrence in Monument. Winter visitant, usually in small numbers. Recorded from: Twentynine Palms, Nov. 4–18, Apr. 2, 5, 9, 13; Cottonwood Spring, Oct. 24 *, Apr. 26, 30, May 13.

The specific records of occurrence of purple finches extend from November 4 to May 13 and probably represent winter residence in the main rather than transients in migration, even though we have no midwinter observations. The frequency of occurrence of the species may be somewhat obscured by possible confusion of some records with those of the similar-appearing Cassin finch. However, most of the occurrences cited above are attested by identifying calls and songs peculiar to this species and known well to the observers.

Miss Carter recorded considerable numbers associated with large flocks of goldfinches in November. At Cottonwood Spring one to five individuals have been reported at a time. Singing before departure in the spring was noted on April 2, 5, 9, 13, and 30, and on May 13.

This species and the Cassin finch occur in the Monument during cool weather when water problems are not severe. Probably the birds eat fruits as well as seeds.

The specimen taken on October 24 was an adult female, fat and in good condition. It is of the race *Carpodacus purpureus californicus.*

Cassin Finch. *Carpodacus cassinii*

Description. Similar to purple finch, but about 20 per cent larger, the bill more slender and longer, and reds of males paler, especially on breast. Females more boldly striped ventrally, the ground color whiter.

Range. Breeds in mountains of interior western North America from southern British Columbia and southwestern Alberta south to northern Baja California, southern Nevada, northern Arizona, and northern New Mexico. Winters at lower elevations and south to southeastern California and to Zacatecas, rarely farther.

Occurrence in Monument. Rare winter visitant. Recorded from: Quail Spring, Dec. 20 *; Eagle Mountain, Oct. 19, 20.

Twice this species was seen and heard plainly in the piñon belt on Eagle Mountain, and one was taken at Quail Spring. The numbers may be somewhat greater than our records show; see account of purple finch. In similar desert mountains to the northeast, the Cassin finch also frequents the upper vegetation belts, foraging in piñons and junipers in winter (Johnson, Bryant, and Miller, 1948).

House Finch. *Carpodacus mexicanus*

Description.—A finch $5\frac{1}{2}$ inches long and weighing 17 to 21 grams. Male marked with red in variable amounts on head, throat, breast, and rump, the color varying individually and with season from dull rose to crimson and rarely to orange or yellowish; otherwise body dusky, obscurely streaked above and prominently streaked on under parts. Tail essentially square-ended. Female similar but lacking red.

Range. Resident from southwestern and southern interior British Columbia, central Idaho, and northern Wyoming south to southern Baja California and over Central Plateau of México to Guerrero and Oaxaca.

Occurrence in Monument. Common resident. Recorded from: Morongo Pass, Apr. 28; Little Morongo Canyon, Mar. 31—Apr. 3; Black Rock Spring, Apr. 3, July 6, Aug. 29, 31 *, Sept. 3; Upper Covington Flat, Nov. 4 [1]; Lower Covington Flat, Apr. 8, 9, 15 [1], 24 [1], May 15 [1], 19, Aug. 24, 29 *; Smithwater Canyon, Apr. 8, May 13 [1]; Quail Spring, Jan. 25, Feb. 11, Apr. 10, 11, 22, 25, May 19—21, July 1, 3, 4, Sept. 8 *, 9 *, 10 *; Stubby Spring, Apr. 9, Sept. 5, 9 *, 10 *, 11 *; Indian Cove, Apr. 7, 27, May 13—14; Fortynine Palms, Apr. 7; Twentynine Palms, Jan. 16 [1], Mar. 17 *, Apr. 29, May 13, 14, July 13, Oct. 7, 17, Dec. 30; Pinyon Wells, Apr. 10, Sept. 13, Oct. 9, 10, 12; Live Oak Tank, Apr. 6—8; Pinto Basin, Apr. 5, 6; Eagle Mountain, Oct. 18—21; Cottonwood Spring, Feb. 7—11, Apr. 11, 25, 26, 29, 30, May 12, 13, 14—19, June 3, 22, July 7, 8, 14, Sept. 13, 14, Oct. 14, 17 [1], 21—24; Lost Palm Canyon, Oct. 23; Pinto Wash Well, Jan. 27 [1], Feb. 17 [1].

House finches are common in the Monument but are not evenly distributed over it. They concentrate conspicuously in the vicinity of water sources, and in the open desert two or more miles from drinking water they only rarely may be found. If they occur in such places other than as vagrants, they apparently have moisture sources in the form of cactus fruits or other such succulent material. In the floor of Pinto Basin, for example, on April 5 and 6 the greatest number seen was 3 in 1½ hours, and these were flying overhead, possibly not even settling in this arid basin. At Key's View, where there is no water but abundant Joshua Tree and juniper cover available, we have failed to find house finches, nor have we found them in large tracts of Joshua trees on Upper Covington Flat, which is far from a water source.

Concomitant with this obvious concentration within cruising distance of water is our repeated observation of house finches engaged in drinking at all seasons of the year. At most watering spots, and especially in hot weather, they come and go throughout the day, and we have the impression that the birds must take water daily and probably many times a day. Four to twelve are often present at once, sitting or fluttering about the edge of a water tank or flow. Bartholomew and Cade (1956) have shown in experimentation with house finches that in order to maintain body weight and good health they took 16 per cent of their body weight in drinking water a day at air temperature of 43° F. At 68° F. they took 22 per cent, and at 102°, 44 per cent. This was when they were on a diet of dry seeds, a normal diet for the species at some periods of the year. Conversely, if kept on this diet without water the finches at 68° F. lost 15 per cent of body weight in 48 hours, but they could regain the loss in 24 hours when given water. When on succulent food, there was no significant loss of weight in the absence of drinking water.

Salt (1952), as a result of investigations of the metabolic efficiency of house finches at high air temperatures (up to 104° F.) in dry air (relative humidity 12 per cent) reports: "It has a body-cooling system which represents a compromise between water conservation and maximum cooling power. As a result it is able to withstand high temperatures so long as the vapor pressure of the air remains fairly low but it requires drinking water to replace moisture lost in evaporative cooling. . . . Compared with the other two species [the purple finch and Cassin finch] it can operate" more efficiently in hotter and drier climates.

Bartholomew and Cade (1958) have found in testing the salinity of water used by house finches that they are less able than mammals occupying desert regions to use saline waters. They cannot survive on sea water. However, in the Joshua tree area we are not aware of any highly saline water sources that might complicate the drinking problem for house finches.

As already suggested, house finches are basically ground-foraging seedeaters. Seeds form the bulk of the diet at all seasons. When less dry vegetable food is available, they do, however, turn to this in part. On April 3 in Little

Morongo Canyon, house finches were plucking at small popcorn flowers (*Plagiobothrys*), obtaining either flower parts or leaf buds. On April 10 at Quail Spring, they were feeding on green filaree and peppergrass (*Lepidium*). On July 14 they were feeding on grapes that were beginning to ripen in an arbor at a house near Cottonwood Spring. In nearby desert areas we have seen them eating cactus fruit.

At the oases, palms are particularly favored for roosting and nest sites. On September 14 we found that many were settled for the night under palm fronds within 10 feet of the ground in a low clump of these trees at Cottonwood Spring. One was captured with the aid of a butterfly net.

Full song is much in evidence at the oases in March, April, and May, when nesting is underway. On April 3 in Little Morongo Canyon, a bird was tugging at dry fibers it had found among the green leaves of *Yucca mohavensis,* busily gathering nest material. At Indian Cove on April 7, house finches had a nest with 5 young in a tall cholla cactus. The young varied conspicuously in size and all had distinctly whitish down. The young had been fed before sunrise, as masses of seeds could be seen through the skin of their necks when we flushed the female at 6:30 A.M. The three larger young had their eyes open; the others did not and were only about half as large. The nest was made chiefly of peppergrass but with some other dried plant stalks with cottony seed covers. The nest was 5 feet up, well supported in a crotch in the cholla. Its dimensions were 5 x 3 inches deep outside, and the cup was 2½ x 1½ inches deep. Already the nest rim had some accumulated droppings, a situation normal to the nests of this species. Another nest at this station the same day held 3 newly hatched young and one egg without an embryo. This nest, also in a cholla, was 3 feet up and well shaded in the bush. The dimensions were closely similar to those of the other nest. A nest at Pinyon Wells on April 10, likewise in a cholla, held 4 young about four days old and one unhatched egg.

In Smithwater Canyon on April 8, a pair of house finches was vigorously scolding a scrub jay that evidently was threatening their eggs or young. A coachwhip snake (*Masticophis flagellum*) had captured and eaten 4 nestling house finches on April 30, 3½ miles north-northeast of Cottonwood Spring.

In the period from September 8 to 11, 9 adult house finches were taken which showed stages of the annual molt. Six of these had essentially finished the molt and showed only traces of sheaths on the outermost primaries and a small scattering of feathers with sheaths on the body. Two others were in late, but less advanced, stages of body molt, one with the outer 3 primaries still growing and one with old primaries 8 and 9 not yet dropped. The ninth bird had only begun the body molt, the plumage being predominantly old and worn, but the primary molt had progressed to the point of dropping number 8. Two birds of the year on August 29 and 30 were in early and

Fig. 68. A pair of house finches at their nest in a Bigelow cholla at the edge of Pinto Basin.

middle stages of postjuvenal body molt; no molt of the remiges or rectrices was taking place.

The house finches of the Californian deserts are of the race *Carpodacus mexicanus frontalis.*

Pine Siskin. *Spinus pinus*

Description. A small (10—12 gram) finch, streaked, dusky on gray, above and below, and with pale yellow "windows" at the bases of the inner primaries and the secondaries. Tail short and slightly forked (fig. 67)
Range. Breeds from Alaska to Labrador in the transcontinental timber belt and south in the mountains to northern Baja California, Chiapas, and Guatemala; in eastern North America south to Kansas, the Great Lakes area, and Pennsylvania. Winters north to southeastern Alaska and southern Canada and extends south, outside of breeding range, to the Gulf coast and southern Florida.
Occurrence in Monument. Winter visitant in small numbers. Recorded from: Little Morongo Canyon, Mar. 31 *; Lower Covington Flat, May 14 [1], Aug. 23 *; Twentynine Palms, Oct. 30, Nov. 21, Dec. 19, May 7, 12; Pinyon Wells, Oct. 14; Cottonwood Spring, May 14—19.

Winter visitant siskins have been detected as early as August 23 and as late as May 19. The species is notoriously irregular in time of breeding, and vagrants or laggards might be expected at almost any time in the desert, except that in the heat of midsummer these wide-ranging birds would probably quickly leave. As a species primarily adapted to the cool, higher or northern conifer belts, it is not well suited to the desert area of the Monument and numbers are small.

The occurrence on August 23 was of a single vagrant only. This bird was a fully independent juvenile, and it showed no beginning of postjuvenal molt. It was found in the piñons at Lower Covington Flat. At Twentynine Palms, Miss Carter saw single individuals or recorded "a few" feeding with goldfinches "on the beds of yellow composites." In the spring in Little Morongo Canyon, siskins were feeding on willow catkins—a very common feeding practice of this species. At Cottonwood Spring from May 14 to 19 siskins were associated with cedar waxwings in the cottonwoods, and at times they visited the flowing water to drink.

The siskins occurring in the Monument are of the race *Spinus pinus pinus.*

American Goldfinch. *Spinus tristis*

Description. A goldfinch with two prominent white or light buff bars on secondary coverts and white of tail in form of longitudinal areas on inner webs of feathers. Male in spring plumage with black crown and bright yellow body above and below; wings black except for white patterning. Males in winter plumage and females dull olive-green above; below dull yellow on chin and throat, the breast and belly gray or buffy (pl. III, opp. p. 182).
Range. Breeds across southern Canada and south to northern Baja California, central Utah, southern Colorado, northeastern Texas, central parts of Gulf states, and South Carolina. Winters from southern British Columbia, northern interior United States, and southeastern Canada south to northern Baja California, northern Sonora, Coahuila, central Veracruz, the Gulf coast, and southern Florida.

Occurrence in Monument. Occasional winter visitant. Recorded at: Lower Covington Flat, Aug. 23; Quail Spring, Sept. 8; Twentynine Palms, Nov. 23–Dec. 3, 12—16.

Miss Carter observed a flock of about 12 American goldfinches feeding on the ripe seeds of jackass clover for a period of about 10 days in the fall of 1934, and a smaller flock, with as few as three individuals, up to December 13. Otherwise, occurrences of single individuals have been noted at the stations and dates listed. This goldfinch is much less inclined than its congeneric relatives to range into deserts. Usually it stays in river bottoms and near water sources.

Lesser Goldfinch. *Spinus psaltria*

Other names. Green-backed goldfinch, Arkansas goldfinch.
Description. A small (9—11 gram) goldfinch with one distinct bar of white or light buff on wing coverts below a second, poorly defined bar. Male: green above with black crown, the back with variable amounts of dusky and black intermixed; sides of face green; under parts clear yellow; wings black with white window at base of primaries and white edgings on inner secondaries; tail black with white areas on inner webs of outer three pairs of feathers, the areas bordered terminally by black tips. Female: dull olive-green above on head and back, brighter on the rump; under parts greenish yellow; wings and tail dark brown, the white patches reduced in area and contrast compared with male (pl. III, opp. p. 182).
Range. Southwestern Washington, northern Nevada, northern Utah, northern Colorado, and western Oklahoma south through México, Central America, and South America to northwestern Perú and northern Venezuela.
Occurrence in Monument. Common resident. Recorded from: Little Morongo Canyon, Mar. 31; Black Rock Spring, Aug. 30, 31, Sept. 4; Upper Covington Flat, Aug. 24; Lower Covington Flat, Apr. 8, Aug. 23, 25 *, Sept. 9 [1]; Quail Spring, Jan. 19, 20 *, 23, 25, 26, 27, Feb. 13, Sept. 10 *, Dec. 4; Stubby Spring, Sept. 5, 6 *, 7 *, 11 *, Oct. 15; Key's View, July 16; Indian Cove, Apr. 7, 27, May 14; Twentynine Palms, Feb. 19, Mar. 23, 30, Apr. 27, May 17, Oct. 17–Nov. 15, Nov. 1, Dec. 30; Pinyon Wells, Apr. 10, Oct. 12; Split Rock Tank, Jan. 24; Live Oak Tank, Jan. 24, Apr. 6–8; Eagle Mountain, Oct. 19; Cottonwood Spring, Feb. 7—11, Apr. 11, 25, 26, 29, 30, May 1 *, 12, 13, 14, 19, Sept. 13, 15 *, Oct. 13, 14, 21—24; Pinto Wash Well, Sept. 24 [1].

The lesser goldfinch is a species which is strongly attached to the water sources in the Monument. It is seen at the water perhaps even more commonly than the house finch and seems to be more inclined to cruise great distances over dry terrain in going to and from the water. Typical of its visits to drink is the following notation made by Sibley at Quail Spring on January 20: "Each time we have gone to the tank at the spring a flock of greenbacks has either been perched in the adjacent catclaw or has appeared to drink before we departed. Their routine seems to be . . . to fly in and perch in a catclaw where they sit for a few minutes, then one by one they drop down to the puddles of the overflow and drink. Gradually they return to the catclaw and apparently wait until the entire flock has drunk; then off they go once more." This observation also points up the highly social urge of this species. When these goldfinches are flying overhead, even if far from springs, an imitation of their clear call notes will usually bring the birds down, especially solitary ones, in obvious search for a companion.

Censuses at the Quail Spring water tank on January 25, 26, 27, and February 13 showed many small roaming flocks coming in; these flocks numbered from 5 to 16. During four one-hour counts, 11 A.M. to 12 M., the total individuals per hour were 25, 25, 29, and 40. At Cottonwood Spring in May and October the counts of birds at one time have on several occasions been 5 and 6. Here they were bathing as well as drinking.

This species has even heavier dependence on seed for food than does the house finch. Only once have we seen them feeding on buds, when a single individual joined some pine siskins feeding on willow catkins in Little Morongo Canyon on March 31. Under these circumstances, and in view of the high relative water loss associated with their small size, this species is very sharply limited by the availability of water. Its capacity and inclination to fly far to and from it is a correlated and doubtless vital attribute.

Characteristic foraging was watched at Twentynine Palms, when on April 27 two or three dozen of these goldfinches frequented beds of a small, yellow composite in the swamp to feed on the seeds. And from October 17 to November 15 they were feeding daily on the ripe seeds of jackass clover. When disturbed they would fly up and arrange themselves in rows along the strands of a barbed wire fence, only to drop down again as soon as the observer was quiet.

Lesser goldfinches will on occasion sing while in flocks. We have record of this on March 30, and occasional song may be expected at any season. Song is most prevalent in April and May, when the species may be expected to breed and when flocks are less often seen or are smaller. Our only specific evidence of nesting was the appearance on April 26 at Cottonwood Spring of 3 stubby-tailed young, still weakly flying, that were crowding about a male and begging for food. Although this would indicate egg-laying in early April, it is possible that nestings occur through the summer and even in the fall. This is suggested by the collecting on September 7 and 15 of four adult males in fresh plumage with testes 3, 3, 4, and 4 mm. long, since functional size may be no greater than 5 or 6 mm. in this species. Moreover, autumnal (October) nesting in this goldfinch is known in coastal southern California (see Willett, 1933:163). The fall breeding activity is probably a case of resumption of breeding state following a quiescent period in summer. This is suggested by the fact that an adult male on August 25 in the late stages of annual molt (9th primaries in sheaths) had a testis of only 1½ mm.; of course it is not known whether breeding activity in both spring and fall occurs in the same individuals.

Although the adults collected in the first half of September were, as noted, essentially or entirely finished with the annual molt, two immatures taken on September 6 and 10 were in postjuvenal body molt that was only about half-completed; no molt of remiges or rectrices was in progress.

A male in full breeding condition (testis 7 mm.) taken at Cottonwood Spring on May 1 is unusual in that the back is almost completely black, as in the race *Spinus psaltria psaltria* of Texas, eastern New Mexico, and central and southern México. The bird was evidently stationed for breeding and as such was part of the local population, whereas all others which we have at hand are of the green-backed race *S. p. hesperophilus.* The black-backed bird is thus, on a population basis, best regarded as a variant individual of *hesperophilus,* and yet it might represent a vagrant from the range of *S. p. psaltria* that had settled in the range of *hesperophilus.* Actually black versus green backs of adult males may represent a polymorphism in which the proportion of the two types varies geographically, with black backs being almost but not completely absent in the northwestern segment of the species' range. We therefore favor the variant or polymorph interpretation and use the formal race designation of *hesperophilus* for all the lesser goldfinches of the Monument.

Lawrence Goldfinch. *Spinus lawrencei*

Description. A small (9–12 gram) finch, differentiated from the other goldfinches by gray or gray and black body and by bright greenish yellow wing bars and margins. Male: crown and chin black; neck, cheeks, and flanks gray; back green, the feathers tipped with gray; rump bright greenish yellow; breast yellow; belly white; wing coverts and margins of outer black flight feathers greenish yellow; tail black with subterminal white patches on outer three feathers; bill flesh-colored. Female: gray on head and body, except for dull yellow breast patch; wings and tail as in male, although markings less distinct and contrasting (pl. III, opp. p. 182).

Range. Breeds in California and northern Baja California, chiefly west of the Sierra Nevada and the southern deserts, and occasionally in western Arizona. Winters sporadically through most of breeding range and in southern Arizona, northern Sonora, southern New Mexico, and western Texas.

Occurrence in Monument. Summer resident and migrant. Common in some years. Recorded from: Little Morongo Canyon, Mar. 31, Apr. 1, 3; Black Rock Spring, Sept. 4; Lower Covington Flat, Apr. 9, Aug. 23; Smithwater Canyon, Apr. 8 *, 10; Quail Spring, Apr. 11, 25, Sept. 8, 10 *; Stubby Spring, Sept. 6, 8, 10 *, 11 *; Indian Cove, Apr. 7, Oct. 16; Twentynine Palms, Oct. 17–Nov. 14, Mar. 23–May 26 (intermittently); Pinyon Wells, Apr. 10, Oct. 12, 13, 15; Live Oak Tank, Apr. 6–8; Eagle Mountain, Oct. 19, 20; Cottonwood Spring, Sept. 14, 15 *, Apr. 25, 26, 29.

Lawrence goldfinches are usually seen as scattered individuals, singles or pairs, but once up to 32 (three-fourths of them males) were at a watering trough at Pinyon Wells on April 10. Most of the occurrences are in the migratory periods when this species is moving to or from its principal breeding area on the coastal slopes and the part of its wintering range that lies in southern Arizona and beyond to the east and south. In the fall period our dates range from August 23 to November 14 and in spring from March 23 to May 26. But the late April and May occurrences represent nesting individuals in part. Miss Carter in her continued watching of goldfinches through the winter at Twentynine Palms recorded none from mid-November to March 23, and Sibley in daily censuses at Quail Spring in winter found none. It would not be surprising if occasional Lawrence goldfinches did occur in midwinter as also in midsummer, even though records are now lacking.

The observations in fall often suggest migratory movement, although this is difficult to prove in such a freely moving species as a goldfinch. For example, on September 15 one was flying southeast and when collected it was found to be heavily loaded with migratory fat. On September 11 again, long southeasterly flights were noted and specimens taken on the 10th and 11th were fat.

Not infrequently this species associates with lesser goldfinches and house finches both in coming to water and in foraging. This was true at Twentynine Palms when goldfinches were feeding on jackass clover, and at water at Pinyon Wells on October 12 and 13. But the association is not close, and this species will separate out and leave a mixed flock. On April 3 in Little Morongo Canyon a pair of Lawrence goldfinches was feeding in company with house finches on popcorn flowers on a steep, sunny slope.

In Smithwater Canyon on April 8, 1960, pairs were sorted out and one of these was found building in the tip of a piñon 30 feet above ground. The nest looked like a round ball, 3 inches in diameter, in the dense needle cluster. The female was gathering lining material; she obtained a feather on a nearby slope and flew to the nest, the male singing vigorously below her in the tree while she put it in place. Members of another pair collected here were at the egg-laying stage, the female with the first ovum 5 mm. in diameter and the male with testes at 5 mm. At a waterhole on Lower Covington Flat on April 9 (fig. 69) a male of a pair drove off other males from the vicinity of the female.

At Cottonwood Spring in late April, 1953, males were singing conspicuously, and on April 29, 1952, a nest was found a quarter of a mile southeast of the camping area in a juniper tree. The nest was 5 feet above ground and held four white eggs on this date. The pair was seen in courtship feeding 75 feet from the nest. Possibly the species nests at Twentynine Palms, since the birds stay there into May and sing, and we suspect breeding in Little Morongo Canyon. In each of these possible nesting areas water is available on which this species, like other goldfinches, is dependent.

Two specimens, an adult and an immature on September 10 and 15, had slight traces of molt but their plumage was predominantly new and in good condition, as it was in two other specimens taken in this period.

Fig. 69. Lawrence goldfinch in a waterhole dug by coyotes in the stream bed on Lower Covington Flat, April 9, 1960.

Green-tailed Towhee. *Chlorura chlorura*

Description. A large (25–30 gram), unstreaked sparrow. Crown chestnut; sides of head gray, the lores whitish; back dull greenish gray; throat white, sharply set off from gray of neck, breast, and flanks; belly whitish; wings and tail dull green to bright green; margins of wing at wrist yellow.
Range. Breeds in interior mountains from southern Oregon, southeastern Washington, southern Idaho, and western and southern Wyoming south to southern California, central Arizona, and southern New Mexico. Winters from southern California, central western and southern Arizona, and western and southern Texas south to southern Baja California and Morelos.
Occurrence in Monument. Spring and fall migrant. Occurs chiefly in piñon belt. Recorded from: Black Rock Spring, Aug. 29, Sept. 3 *; Quail Spring, Apr. 22; Pinyon Wells, Apr. 10, Oct. 13, 14 *, 15 *; Eagle Mountain, May 16; Cottonwood Spring, Apr. 25, 26 *, May 13.

Green-tailed towhees are seen fairly commonly in migration. The dates for the spring are April 10 to May 16 and for the fall are August 29 to October 15. The birds are seen singly as a rule, but sometimes they are associated in pairs in the spring; as many as 8 were seen in the vicinity of Cottonwood Spring on May 13.

These towhees usually are detected in and about the cover afforded by the larger, denser plants, such as scrub oaks, junipers, *Prunus fasciculata,* mesquite, and catclaw. Thus they show the tendency, so conspicuously manifest on their breeding grounds, to seek concealment among plants close to the ground. In migration, by day at least, they move from one bit of cover to the next, at times of course necessarily crossing open desert areas. The shade of the cover which they seek as well as the springs which they frequent may help in the solution of their problem of water loss during the heat of the early part of the autumnal migration.

The three birds that have been taken in the fall have all been immatures which had finished the postjuvenal molt involving the body plumage; they were to varying degrees fat. This species winters in small numbers in the Colorado Desert but also farther south through northern and central México. Accordingly, migrants through the Monument may not be far from their wintering grounds. Since as a species they are able to cope readily with desert conditions in the winter season, they may not face severe problems in regard to food and climate as they pass through the Joshua Tree area. Indeed, they might occur in the Monument in the winter, although they normally seek at this time the lower desert levels that are warmer. Some of the breeding areas to the north and northwest are nearby, as in the higher mountains of southern California.

The migrants we see, then, may either be individuals engaged in long-

range movements, passing from the northern Great Basin to western México, or those making shorter movements within California. At present we have no means of distinguishing between such individuals or of tracing the extent of their journeys. One green-tailed towhee taken on April 26 at Cottonwood Spring had fully developed testes, 7 mm. long; it probably would soon have reached its breeding grounds.

Rufous-sided Towhee. *Pipilo erythrophthalmus*

Other names. Spotted towhee; Red-eyed towhee.
Description. A large (35–40 gram) sparrow, the head, back, tail black, the scapular areas and secondary coverts streaked and tipped with white, and the lateral three pairs of rectrices white-tipped on the inner webs. Flanks rufous; lower breast and belly white. Bill black. Eyes red.
Range. Breeds from southern British Columbia, central parts of prairie provinces of Canada, northern Michigan, southern Ontario, central New York, and northern Vermont south to southern Baja California and through parts of México to Guatemala; in the eastern United States south to northeastern Oklahoma, northern Arkansas, the Gulf coast from southeastern Louisiana eastward, and southern Florida. Winters from southern British Columbia, northern Idaho, Utah, Colorado, Nebraska, the southern Great Lakes region, southern New York, and Massachusetts southward.
Occurrence in Monument. Sparse resident. Numbers are somewhat augmented by winter visitants. Recorded from: Little Morongo Canyon, Apr. 2; Black Rock Spring, Apr. 3 *, 4, Aug. 29, Sept. 1, 3 *, 4; Upper Covington Flat, June 16 [1], Oct. 8 [1], 14 [1], 15 [1], Nov. 18 [1]; Lower Covington Flat, May 19 *, July 10 [1]; Smithwater Canyon, Oct. 14 [1]; Quail Spring, Feb. 13 *, Apr. 22, May 21 *, Sept. 19 *, Dec. 4 *; Stubby Spring, Feb. 10, Sept. 5, 8 *, 11; Lost Horse Valley, Dec. 4 *; Indian Cove, Apr. 8 *, May 13, 14 *, Sept. 7 *, 8 *, 12 *; Pinyon Wells, Oct. 9, 10, 12 *, 13, 15 *, 16 *, 17; Eagle Mountain, May 16 *, Oct. 19 *; Cottonwood Spring, Sept. 15.

Rufous-sided towhees find limited areas of suitable brush cover in the Monument for their permanent residence and breeding. Occasional nesting pairs are encountered along the upper parts of the Little San Bernardino Mountains and on Eagle Mountain. At Quail Spring there is a scattering of these towhees in the brush patches, and at Pinyon Wells members of the resident race, *Pipilo erythrophthalmus megalonyx,* were encountered regularly. As many as ten towhees might be seen here on a single morning in October, representing both resident and winter-visitant races. They were kicking in the litter of oak leaves in the manner highly characteristic of this species, whereby they forage under protection of the chaparral and gather food in the loose subsurface layer of vegetable material on the ground (see especially Davis, 1957). Also at Stubby Spring and at Black Rock Spring occasional birds were found in the scrub oak. Other cover in which we have occasionally found them are manzanita, juniper, catclaw, indigo bush, canyon-bottom baccharis, mountain mahogany, purshia, and desert willows. The lowest point zonally at which they nest and are permanently present is at Indian Cove. Here in the canyon mouth they extend down into dense desert willow thickets and on May 14 were evidently nesting there.

Fig. 70. Green-tailed towhee (left) and rufous-sided towhee (right).

The breeding season is in April and May. On April 3 a male with testes 7 mm. long was taken, and another was singing regularly in scrub oak and manzanita. On May 19 at Lower Covington Flat and on May 21 at Quail Spring males with 10 mm. testes were obtained. At Indian Cove on May 14 a male with 9 mm. testis and a female with a brood patch were collected. On May 16 on Eagle Mountain at 4000 feet a female had a brood patch and an enlarged oviduct, signifying laying within a period of a few days preceding.

Rufous-sided towhees seem to be most often seen in areas where there are springs, although they are not completely dependent on them and probably do not go to them daily even when near them. We have seen towhees drinking at Black Rock Spring on September 1 on two occasions in a period of hot weather.

Although this species of towhee is much more tightly bound to protective cover than is the brown towhee, which is by no means a close relative, it has spread to the eastern limits of the Monument in scattered brushlands, whereas the brown towhee has not. We think that rufous-sided towhees are more given to pioneering and redispersal than are the brown towhees, and this may be a factor in the present differences in their distribution in the Monument. Possibly the rufous-sided towhee, if as subject to heat and to water loss as the brown has been shown to be, compensates just enough by its day-long adherence to shading cover to be able to tolerate the hot period in the desert mountains. Also the proportion of animal food taken by it may be greater and the vegetable food may on the average be less thoroughly desiccated in the subsurface leaf litter than are the dry, exposed seeds used in large measure by brown towhees.

By the first half of September rufous-sided towhees have completed their molt. Only one of 4 specimens in this period showed traces of molt, this being a bird taken on September 3. Of the 8 specimens whose age was determined from skull examination or wing feather color in males, only 2 were birds of the year.

The resident towhees of the Monument were collected whenever feasible to determine possible deviations in characters from the race of the brushlands of the San Bernardino Mountains and the coastal slopes. No significant differences were found. In matters of tail spotting, size of hind toe and claw, and color of sides and crissum the examples of the Monument match *Pipilo erythrophthalmus megalonyx* typical of areas to the west.

In the winter, migrants from the Great Basin reach the Monument. Among the 23 specimens of the species from the Monument that we have examined, there are four which represent the race *P. e. montanus,* the nearest breeding areas of which are to the north in the Providence Mountains and the mountains of the Inyo district. The specimens are one male taken on December 4 at Quail Spring and one adult female and two immature females taken on

October 19, 12, and 13, respectively, at Eagle Mountain and Pinyon Wells. These birds have the long white spots of the outer tail feathers and somewhat paler rufous areas of the interior races and the distinctly long tails (♂ 100.2; ♀♀ 96.3, 100.8, 101.9 mm.) typical of *montanus* in contrast to *curtatus* of the northern Great Basin.

Brown Towhee. *Pipilo fuscus*

Description. A large (40–48 gram), brown sparrow without stripes or conspicuous markings. Above uniform brown, the tail somewhat darker; throat warm buff, dotted and bordered with black spots; breast and belly gray-brown, the flanks darker; crissum rusty.
Range. Resident from southwestern Oregon south to southern Baja California and from western Arizona, northern New Mexico, southeastern Colorado, northwestern Oklahoma, and central Texas south to Colima and Oaxaca.
Occurrence in Monument. Resident at western border of area. Recorded from: Little Morongo Canyon, Mar. 31 *, Apr. 1 *, 2, 3 *.

Brown towhees occur only in the western end of the Joshua Tree study area, extending from the borders of the San Bernardino Mountain region east as far as Little Morongo Canyon. In endeavoring to find the cause of their eastward limitation in range, we watched closely the habitat used in Little Morongo Canyon. Here on the steep canyon walls brown towhees worked among open bushes as far as 200 yards upslope from the stream course, but on other occasions we saw them retreat into the mesquite, arrowweed, and chrysothamnus thickets of the canyon bottom. They were watched foraging, by weak scratching methods, in the sand in openings among baccharis shrubs, and others were found ranging out into creosote bushes and chollas. The impression was gained that the dense patches of cover, whether of desert species like mesquite or of higher zonal plant types, was important to them. To the east in the Monument, chaparral is more open, although at some distance beyond, as near Pinyon Wells, there are patches of density comparable to that occupied by brown towhees in the eastern end of the Santa Rosa Mountains (3 mi. NW Rabbit Peak) across the Coachella Valley to the south. It seems probable that an isolation factor has influence here, the towhees failing to cross the unfavorable open chaparral belt in sections of the Little San Bernardino Mountains, whereas in the Santa Rosa Mountains they have a continuity of fairly dense brush extending east from the coastal areas. It is difficult to believe that brown towhees have never been in the isolated parts of the chaparral belt at Pinyon Wells, but probably repopulation of such small insular areas is occasionally necessary following local die-out in periods of severe desert conditions, and repopulation has not occurred across the present intervening barriers.

The towhee population under consideration is of course in an area where water is permanently available, although we did not see them going to it in the cool spring period of our observations in Little Morongo Canyon in late March and early April. Dawson (1954) has especially studied this towhee and its relative, the Abert towhee of the desert river valleys and bottomlands, to determine their tolerance of heat and water loss. He found in experimental testing that neither species can withstand exposure for more than two hours to midday sun in summer. Although the Abert towhee shows somewhat better metabolic efficiency at high temperatures and less elevation of its body temperature during sustained stress than does the brown towhee, both species nevertheless dissipated much water in evaporative cooling when held at environmental temperatures of 102° F. Both species are of course dependent on dry seeds for food through considerable parts of each year. When they are on such dry food and subjected to a temperature of 102° F., they may take as much as 50 per cent of their weight in water each day and use much of this to maintain good condition and keep their temperature within normal limits. It is clear, then, that brown towhees in the Monument must have good shade during the maximum heat period of each summer day and that even with this protection they may need to take on water if sustained high temperature prevails for many days. Possibly, then, this species can tolerate neither dense or open brush in the Monument without water sources. There is, then, a climatic factor, namely high summer heat, which in a sense may bound the species in the desert area and which can be overcome only by use of water. Dense brush may help but alone may not provide enough relief. Thus the brush areas about Pinyon Wells may not be reached because water sources and dense shade patches in the interval between are too few or widely spaced.

In Little Morongo Canyon in spring brown towhees were chiefly in pairs, as is the usual situation in this species throughout the year. Some early morning singing was heard on March 31 between 6 and 7 A.M., which may have indicated an unpaired male, as singing is not indulged in by mated males in the western races of this species. Four males taken at this time were in full breeding condition with testes 8 to 10 mm. in length, but the two females collected had not yet laid and the largest ova were only 1 mm. in diameter. Probably nesting would not have taken place until mid-April.

The brown towhees taken in Little Morongo Canyon belong to the race *Pipilo fuscus senicula* of coastal southern California.

Savannah Sparrow. *Passerculus sandwichensis*

Description. A small (16–17 gram) sparrow, streaked above and below with dusky; superciliary area light buff to yellow. Tail short, slightly forked, and feathers pointed.

Range. Breeds from northern Alaska east to northern Quebec and south, on Pacific coast, to southern Baja California and central Sinaloa; interiorly to mountains of Oaxaca and Guatemala; in east extends to Missouri, West Virginia, and western Maryland. Winters from southern British Columbia, southern Nevada, southern Utah, central New Mexico, Oklahoma, northern parts of Gulf states, and Massachusetts south to El Salvador, Cuba, and the northern Bahamas.

Occurrence in Monument. Sparse winter visitant and migrant. Recorded from: Morongo Valley, Apr. 15; Lower Covington Flat, Oct. 14 [1]; Quail Spring, Mar. 9 *, 21, Apr. 4; Twentynine Palms, Feb. 9, Mar. 24, Apr. 5, 18, May 7; Pleasant Valley, Apr. 10.

Savannah sparrows are never common in the Monument, as even in winter they seem to prefer grassy or moist terrain rather than dry, barren ground surfaces on which to forage. Occurrences in February and March appear to represent winter residence, whereas those in October, April, and early May could be individuals moving through in migration.

Miss Carter recorded the species bathing in tiny puddles of water in the swamp at Twentynine Palms on April 5. The birds would then fly up into the mesquites to dry. The fact that this sparrow was most often seen at this station, where there is damp ground, is probably significant. Occurrences in the open, dry desert are likely to be rare and to represent only brief periods of winter residence.

The specimen taken on March 9 is an example of *Passerculus sandwichensis anthinus,* the form that most commonly winters in this part of California and which breeds in the interior of northern Canada and northwestern Alaska.

Grasshopper Sparrow. *Ammodramus savannarum*

Description. A small (15–17 gram) sparrow, mottled above with gray, yellowish black, and brown on back; head brown with buff central crown area; below yellowish buff without stripes; belly white. Tail short, the feathers pointed.
Range. Southeastern British Columbia, southern parts of prairie provinces of Canada, southwestern Quebec, and Maine south to southwestern California, southern Nevada, central Colorado, central parts of Gulf states, and Florida; also in Central America, West Indies, and northwestern South America. Winters from central California, southern Arizona, Oklahoma, Tennessee, and North Carolina southward.
Occurrence in Monument. One record: Lower Covington Flat, Sept. 30[1].

In the collection of Long Beach State College is a female of this species taken on September 30, 1961. It evidently represents a migrant and belongs to the far western race *Ammodramus savannarum perpallidus.*

Vesper Sparrow. *Pooecetes gramineus*

Description. A fairly large (20–24 gram), gray sparrow, streaked above and below, with dusky outer tail feathers showing a stripe of white as the bird flies; bend of wing chestnut.
Range. Breeds from western Washington, central British Columbia, southwestern Mackenzie, and the prairies south of the transcontinental timber belt of interior and eastern Canada south to central California east of the Sierra Nevada, central Nevada, southeastern Utah, the southern Rocky Mountains, central Missouri, and North Carolina. Winters from central California, southern Nevada, central Arizona, central Texas, southern Illinois, and Connecticut south to southern Baja California, south-central México (Guerrero), the Gulf coast of the United States, and central Florida.
Occurrence in Monument. One record: Stubby Spring, Sept. 12 *.

A vesper sparrow was noted on open, level ground along the road near the trail leading down to Stubby Spring on September 12, 1950. This area had had heavy rain the week before, and the ground on which the bird was foraging was soft. No other sparrows were seen with it as it flitted short distances and worked closely along the ground. The bird proved to be an adult female in fresh plumage, with no subcutaneous deposits of fat.

This species winters in the lower levels of the Inyo district and the Colorado Desert, and it is possible that some birds may remain in winter within the Monument. The bird obtained had evidently just arrived in fall migration.

The specimen belongs to the race *Pooecetes gramineus confinis* which breeds over extensive areas in the Great Basin to the north and also in the Rocky Mountain region.

Lark Sparrow. *Chondestes grammacus*

Description. A large (25–30 gram), conspicuously marked sparrow; crown with broad chestnut stripes alternating with a dull white central stripe and white superciliary stripes; cheeks chestnut; black submalar stripe bordering throat; back streaked; under parts white except for small black spot on center of breast; tail feathers, except central pair, broadly tipped with white.

Range. Breeds from southwestern Oregon, central southern British Columbia, the southern plains of Canada, and the Great Lakes region south to southern California, Arizona, Zacatecas, eastern Texas, Louisiana, and western North Carolina. Winters from central California, southern Arizona, central Texas, the Gulf states, and Virginia south to southern Baja California, El Salvador, and southern Florida.

Occurrence in Monument. Seen occasionally in periods of spring and fall migration. Recorded from: Black Rock Spring, Aug. 29, Sept. 2 *; Upper Covington Flat, Aug. 28 [1]; Lower Covington Flat, July 9 [1]; White Tanks, Apr. 6–8.

Lark sparrows are likely to appear on areas of open ground near springs or open tanks. It was in such a place that a specimen was taken on September 2 at Black Rock Spring. This bird, a female, was an immature that was almost finished with the postjuvenal body molt.

Migratory movements in this species are not conspicuous in the Southwest. There is, however, a good deal of local movement and postbreeding wandering. Lark sparrows breed in the Coachella Valley, at least at Palm Springs, and in western parts of the Mohave Desert. They move in winter in small numbers to the Imperial and Colorado River valleys and may yet be found in winter in the lower levels of the Monument area.

The lark sparrows of the western United States belong to the race *Chondestes grammacus strigatus.*

Black-throated Sparrow. *Amphispiza bilineata*

Other name. Desert sparrow.

Description. A small (12–14 gram) sparrow, with large black area occupying face, chin, and throat, and two white lines, one above eye, the other bordering throat below the eye and ear. Back and wings brownish gray, unstriped; tail black with white lateral border and tip on outermost feathers; under parts posterior to throat white, blending to gray on flanks. Juveniles lack black on head and throat and have streaked breast.

Range. Breeds from southeastern Oregon, northern Utah, southwestern Wyoming, western and southern Colorado, northwestern Oklahoma, and central northern Texas south to southern Baja California, northern Sinaloa, interior Jalisco, Hidalgo, and southern Tamaulipas. Winters from deserts of southern United States southward through breeding range.

Occurrence in Monument. Abundant resident. Recorded from: Morongo Pass, Apr. 28; Little Morongo Canyon, Mar. 30–Apr. 3, Apr. 28; Black Rock Spring, July 6, Aug. 29, 31; Upper Covington Flat, June 11 [1], Aug. 24 *, 28; Lower Covington Flat, Apr. 8, 9, 15 [1], May 15 [1], 19, Aug. 22, 24, 27 *, 28 *, Oct. 24 [1], Nov. 6 [1], 11 [1], Dec. 16 [1]; Smithwater Canyon, June 17 [1], Aug. 23; Quail Spring, Jan. 18, 21, 23, 26, 27, Feb. 10, 11, 13, Apr. 10, 22, 25, May 14, 19–21, June 3, July 1, 2, 4, Sept. 9 *; Stubby Spring, Apr. 9, Sept. 5, 6, 7, 8 *, 15 *; 5 mi. N Key's View, May 14; Key's View, July 3–4, 15; Indian Cove, Apr. 7, May 13, 14; Fortynine Palms, Apr. 7; Twentynine Palms, Mar. 7 [1], 12–24, Apr. 1, 4, May 7, 13 [1], 23, July 13; Pinyon Wells, Apr. 10, Oct. 12, 13 *, 16; Split Rock Tank area, Apr. 26; Live Oak Tank, Apr. 6–8; Pinto Basin, Apr. 10, 11, 27, July 7, 10, 14, 15; Virginia Dale Mine, Apr. 24, July 12, 14; Eagle Mountain, May 16, Oct. 18–21; Cottonwood Spring, Apr. 5, 25, 30, May 13, 14–19, July 8, Sept. 13 *, 15 *, Oct. 22 *, 23 *; Lost Palm Canyon, Oct. 22.

Black-throated sparrows among the birds of the area represent an extreme in desert adaptation, for they are in large part seed-eaters, and yet they do not require drinking water and occupy the most exposed and hot, open brushy areas. In the intense heat of July we have found them in the creosote bush flats and about the sand dunes in Pinto Basin, or along the scattered smoke trees of this basin. On heavily insolated rocky slopes, on which the scattered cover consists of low encilia bushes, pairs of these sparrows will be stationed. Likewise in the scattered bushes and cholla cactuses of alluvial fans black-throated sparrows occur regularly. Moreover, their activity in the hot summer days is sustained extraordinarily through midday. Thus in the Quail Spring area on July 1 these sparrows were specifically noted as active throughout the day, and in Pinto Basin on July 15 they were out in the open on bush tops as late as 10:30 in the morning when sun temperatures were so high as almost to be unbearable to humans.

In most of the places where black-throated sparrows are stationed in summer, no water is available, even if they were to fly long distances in search for it. Thus they are not dependent on it. However, they have been seen to come to springs, as at Cottonwood Spring on April 30 and at Quail Spring on June 3, July 4, and February 13, and at a water hole in Smithwater Canyon on August 23. On February 13 one was actually seen drinking from a seep in the water tank, and drinking can be inferred, although it was not actually witnessed, on the other dates just cited.

We suspect that in midsummer in most of the areas where these sparrows are stationed, some use of insects aids the water problem for them. Insects are fed to the young at least. Possibly cactus fruits are used, but we have no proof of this, and in many localities these are not available. Foraging is carried on by running over the open ground and about the bases of bushes where seeds fall. Even on desert gravel pavement, they seem to be able to glean such food, which must be thoroughly dried out. Thus, all evidence considered, it is still apparent that this species has great capacity to withstand heat,

sun exposure, and limited moisture intake. Experimental determination of its tolerances in these respects is much to be desired, as is closer attention to the proportion of different elements that make up its diet.

We have not seen evidence of conspicuous seasonal changes in numbers in the Monument. Winter or nonbreeding flocks are small and loosely organized, if they occur at all. At Quail Spring in January and February, when other sparrows, as winter visitants, were locally abundant, black-throated sparrows were present in two's and three's in the course of hour-long counts—much the same situation as would prevail in spring and summer. However, at Stubby Spring on September 6 and 8, loose flocks of 12 and 25 individuals were noted, and on October 23 a flock of 6 was seen near Cottonwood Spring. Flocks may at times join those of sage sparrows and white-crowned sparrows. But even in the fall we often saw birds apparently associated as pairs rather than in flocks. This was true on October 16 at Pinyon Wells at 3800 to 4100 feet and on October 23 at Cottonwood Spring. Probably there is some local movement in fall and winter, but midwinter records are scarce for the higher stations along the crests of the Little San Bernardino Mountains.

Black-throated sparrows are extremely widespread in the Monument, since open desert scrub of any type is acceptable to them and only the denser patches of piñon and the limited tracts of chaparral are avoided. However, in summer, when the greatest dispersal is evident, the numbers seem to be larger below the 4000-foot level.

The tinkling song of this species, rather lacking in vigor and persistence, is heard chiefly in April and May and occasionally later, as on June 3. Nesting must at times begin early, for Williams (1938) observed streaked fledglings in the Split Rock area on April 26, in company of adults that were carrying food. These young must have hatched from eggs laid in late March.

At the top of the large sand dune in Pinto Basin on April 11 a nest with 4 pale blue, unmarked eggs was found (fig. 72). The nest was 4 x 3½ inches deep outside and had a cup about 1½ inches deep lined with soft plant material and some feathers. The outside was compacted of grasses and small dried plants. The nest was beside a hummock of sand held by a creosote bush but was built in a burrobush (*Franseria dumosa*), partly shaded by the creosote. The burrobush formed a complete canopy for the nest, which was about 10 inches off the ground. On this same date another pair was carrying nest material to a site 8 inches off the ground in a creosote bush and partly supported by a small forb beneath. The nest consisted of only a few strands of dry grass at the time.

On May 16 at 3600 feet on the north side of Eagle Mountain an adult was flushed from a nest placed 1 foot up in a small bush. The nest held 3 young that had conspicuous white natal down; probably they were no more than 3 days old. The bush was a *Coleogyne ramosissima,* the blackbush, with short, rigid twigs and reduced leaves, a species providing low, open cover on

Fig. 71. Black-throated sparrows in smoke tree.

Fig. 72. Nest and eggs of black-throated sparrow in a burrobush on the sand dune in Pinto Basin, April 11, 1960.

the lower slopes of the mountain. The nest was imperfectly shaded by this plant but firmly supported below its crown, which was only knee high.

Young black-throated sparrows are conspicuously slow in beginning the postjuvenal molt. We took three juveniles with still no traces of this molt on August 24 and 28 and on September 8, and two others on August 27 and September 8 showed only slight beginnings of new feather growth on the throat and breast. Some of these young could have been from late nests, although we doubt that any were hatched in this area later than June. Adults on September 9, 13, and 15 were well along in the annual molt, with the outer three primaries in various stages of replacement and the rectrices partly replaced. As late as October 22 some body molt could still be detected.

The black-throated sparrows of the Monument belong to the northwestern race of the species, *Amphispiza bilineata deserticola.*

Sage Sparrow. *Amphispiza belli*

Other name. Bell sparrow.
Description. A medium-sized (13—18 gram), long-tailed sparrow, the tail black in contrast with gray back. Back light to dark gray, according to race, with either faint dark stripes or none at all; white frontal spots and eye ring; black mustache marks and small black spot in center of breast; under parts otherwise white but with brown streaks on flanks; outer tail feathers white-edged (fig. 74, p. 265).
Range. Breeds from central interior Washington, southern Idaho, southwestern Wyoming,

and northwestern Colorado south to central Baja California, southern Nevada, northern Arizona, and northwestern New Mexico. Winters from central California, southern Great Basin, and southern Rocky Mountain regions south to central Baja California, northern Sonora, northeastern Chihuahua, and western Texas.

Occurrence in Monument. Fairly common winter visitant. Recorded from: Black Rock Spring, Aug. 30, Sept. 1 *, 2 *, 3 *, 4 *; 4 mi. N Warren's Well, Jan. 1; Lower Covington Flat, Aug. 24 *, 27 *, Dec. 2 [1]; Stubby Spring, Sept. 5, 6 *, 7, 8 *, 11 *; Lost Horse Valley, Oct. 31–Nov. 1 *; Twentynine Palms, Jan. 18, Mar. 21, Apr. 6; Cottonwood Spring, Sept. 14, Oct. 13, 14, 22 *; Lost Palm Canyon, Oct. 22.

Sage sparrows are probably somewhat more numerous than the listed records suggest, for they are inconspicuous in their behavior, especially in winter and in periods of migration, the only seasons when they occur in Joshua Tree Monument. Fall arrivals have been detected as early as August 24 and the latest spring date is April 6, although the latter probably does not represent the end of their presence.

Sage sparrows are ground feeders that run swiftly with uptilted tail in the openings between the bushes of the brushlands they frequent. Although they may fly when alarmed or closely approached, they often retreat on foot out of sight behind the bases of bushes. Loose flocks of 3 to 6 may be found, but at other times only single individuals or two in casual association are seen. On August 24 one was watched moving about over sand and among dead twig debris at the base of chrysothamnus bushes.

All the rather diverse races of this species occupy semiarid brush in the nesting season, but they avoid the more extreme southern deserts, such as those in the Monument. Although they migrate to the Monument early in the fall, they largely escape the intense heat periods of this area. Even though they favor arid regions, they commonly go to water when it is available on their wintering grounds. At Black Rock Spring, we saw them come to the water on September 1 and 2. The predominantly granivorous diet of this sparrow would seem to make drinking desirable even if this is not by any means always possible for them.

The two pale inland races of the sage sparrow are the forms of the species to be expected chiefly in winter. The larger of these, *Amphispiza belli nevadensis,* has more tendency to distinctly streaked back and is a summer resident of Nevada and northeastern California and other parts of the Great Basin. It has been taken in the Monument on November 1 in Lost Horse Valley and on October 22 at Cottonwood Spring. The race *canescens,* which breeds in the Inyo district and on the north side of the San Bernardino Mountains, is more commonly represented in our collections: Black Rock Spring, September 1, 2 (2 specimens), Stubby Spring, September 6, 11. Also there are specimens somewhat intermediate in size toward *nevadensis* from Black Rock Spring on September 1, 2, and 3; Lower Covington Flat on August 27; Stubby Spring on September 8; and Cottonwood Spring on October 22.

More surprising is the appearance in the Monument in early fall of the dark, small coastal race *A. b. belli,* which has generally been regarded as resident and not ranging outside its breeding area (see Grinnell and Miller, 1944:501). To be sure, the breeding grounds on the south side of the San Bernardino Mountains are not far distant. Three typical individuals of this distinct race were taken on August 24 at Lower Covington Flat and on September 3 and 4 at Black Rock Spring. *A. b. canescens* and *A. b. belli* appear to arrive in migration earlier than does *nevadensis* from its more distant breeding range.

The sage sparrow is a species that molts early and apparently before migration. All specimens taken, even those as early as August 24, were in fresh new plumage and without traces of annual or postjuvenal molt. Of the 16 specimens, 12 were females, one was a male, and three were unsexed. Of the 12 in which age was determined from the skull, 10 were adults.

Slate-colored Junco. *Junco hyemalis*

Description. A moderate-sized (16—18 gram) sparrow, with white lateral tail feathers common to the junco group. Head, breast, and flanks slate gray, duller and more mixed with buff and brown in females than in males; belly white; back slate gray to dull brown, not sharply contrasting with head and neck. Bill white (fig. 73).

Range. Breeds in transcontinental boreal forests of North America south to northern British Columbia, the northern Great Lakes region, the mountains of northern Georgia, and Connecticut. In winter from southern Canada south to northern states of México and the Gulf coast.

Occurrence in Monument. Sparse winter visitant. Recorded from: Little Morongo Canyon, Mar. 31 *; Upper Covington Flat, Nov. 18 [1]; Indian Cove, Apr. 8 *; Eagle Mountain, Oct. 21 *; Cottonwood Spring, Oct. 22 *.

This northern species may occasionally be seen among the abundant Oregon juncos that winter in the Monument. However, we have been sure of its presence only by singling out dark-sided birds and collecting them. On April 8, one was found foraging alone in the edge of a wash, and the bird taken on March 31 apparently also was alone. Possibly the spring migration schedule of this form is sufficiently later than that of certain of the flocks of Oregon juncos, with which it associates, for the slate-colors to be left behind for a short time at this period of the year. The bird of March 31 was fat and thus in readiness for migration, whereas the one of April 8 was not yet in this condition and the testes were only 1½ mm. in length.

The response of slate-colored juncos to the desert environment in winter is not noticeably different from that of Oregon juncos.

The specimens taken on October 21 and 22 and on November 18 are of the race *Junco hyemalis cismontanus* of northern British Columbia, whereas those of the spring period happen to belong to *J. h. hyemalis* and probably came from breeding grounds farther north, either in Alaska or western Canada.

Oregon Junco. *Junco oreganus*

Description. A moderate-sized (16–18 gram) sparrow, with the white lateral tail feathers of the junco group. Head black or dark gray, above and below, sharply contrasting with flanks, which are buff to tawny brown; back brown to reddish brown; belly white. Bill white. Females with heads usually less intensely black than in males.

Range. Breeds from southeastern Alaska, central British Columbia, west-central Alberta, and southwestern Saskatchewan south to central coastal California, the mountains of northern Baja California, western Nevada, northeastern Oregon, southern Idaho, and northwestern Wyoming. In winter from southeastern Alaska, southern British Columbia, western Montana, Wyoming, and South Dakota south to the northern Mexican states and central Texas.

Occurrence in Monument. Common winter visitant. Recorded from: Black Rock Spring, Apr. 3, 4 *, Sept. 4 *, 5 *, Oct. 16; Upper Covington Flat, Apr. 9, Nov. 4[1], 5[1], 18[1]; Lower Covington Flat, Apr. 8, Oct. 23[1], Nov. 26[1], Dec. 2[1]; Smithwater Canyon, Feb. 16[1], Apr. 10, Nov. 19[1]; Quail Spring, Jan. 19, 20 *, 22 *, 23, 25, 26, 27, Feb. 10, 11, 13, 17, 18, Apr. 10, Oct. 15; Stubby Spring, Oct. 15, Dec. 4–6; Lost Horse Valley, Nov. 1–2; Barker's Dam, Feb. 18; Indian Cove, Oct. 12; Twentynine Palms, Mar. 2; Pinyon Wells, Oct. 9, 12, 14 *, 15, 16, 17 *; Split Rock Tank, Jan. 23, Oct. 29–31; Eagle Mountain, Oct. 19 *, 20, 21 *; Cottonwood Spring, Feb. 7–9, Apr. 5, 30, Sept. 13, 15 *, Oct. 23; Lost Palm Canyon, Oct. 22.

Oregon juncos occur from September 4 to April 30, but the numbers are small in September and late April. The conspicuous concentrations are in midwinter, especially at Quail Spring, where juncos are abundant in January and February. Records are scarce or nonexistent below 3000 feet in flat terrain. Most favored is the piñon belt. At Pinyon Wells in mid-October Oregon juncos were common among the piñons and *Prunus* bushes, eriogonum, and scrub oak. After 10 A.M. in the morning there on October 16 they were seen adhering to shaded sections of the ground. Flocks of 6 to 10 were typical of this season there and on Eagle Mountain among the scrub oaks and piñons. At Cottonwood Spring flocks of no more than 6 each were seen on October 23 among palo verdes and mesquites of the wash.

The junco activity at Quail Spring in midwinter was especially watched by Sibley. A large flock of well over 100 was in the area. At times the estimated number visiting the water area ran up to 130, 150, and 200 an hour, some birds drinking and others preening. The birds were much attached to the water and evidently regularly used it. Their desire for water seemed to override their fear, and they would come within 10 feet of the observer. This large flock of Oregon juncos occasionally had chipping sparrows and white-crowned sparrows intermingled, but here and elsewhere the association was temporary or casual, the juncos moving off at other times in their own flock.

Foraging of juncos is entirely on the ground in the area of winter residence in the Monument. Here seeds are sought, although in early spring green, forming seeds or leafy parts of annuals may be taken. Such was true in a flock feeding among filaree and peppergrass at Quail Spring on April 10.

Juncos seem, then, to have several procedures which permit them, as a boreal-adapted species, to cope with the desert heat and drought: (1) they

Fig. 73. Juncos and white-crowned sparrow: Gray-headed junco Oregon junco
White-crowned sparrow Slate-colored junco

occupy the desert principally in the cooler months; (2) in the warm hours of the day they forage in the shade; (3) they form greatest concentrations near springs where they can drink regularly; (4) and they apparently supplement their dry food when possible with moist seeds, buds, or leafy parts.

The two races of Oregon junco that winter regularly in the Monument are *Junco oreganus thurberi,* which breeds in the Sierra Nevada and the nearby San Bernardino Mountains, and *J. o. montanus* from the mountains of eastern Oregon and Washington, northern Idaho, and southern interior British Columbia. *J. o. thurberi* seems to be the more common and is the only race taken in early September. Six examples of *montanus* have been taken as follows: Upper Covington Flat, November 4 and 18; Quail Spring, January 20; and Pinyon Wells, October 15 and 17. Other dates for which our specimens are listed are records of *thurberi.* Early fall arrivals, as on September 4 and 5, were birds that had completed the annual molt. One specimen taken by a party from Long Beach State College on November 4 on Upper Covington Flat was an example of the race *J. o. mearnsi,* which breeds principally in eastern Idaho and Wyoming.

Gray-headed Junco. *Junco caniceps*

Description. A sparrow with the white lateral tail feathers of the junco group. Head and flanks light, uniform gray, the flanks somewhat paler. Back gray, except for well defined rust-red patch in center; belly white; orbital area blackish. Bill white (in local wintering race; fig. 73).
Range. Breeds in coniferous and aspen forests of Great Basin and Rocky Mountain region from northern Nevada, northern Utah, and southern Wyoming south to central Nevada, central Arizona, southern New Mexico, and western Texas. Winters at lower elevations in or near breeding range and south to southern California (rarely), northern Sinaloa, and Durango.
Occurrence in Monument. Rare winter visitant. Recorded from: Upper Covington Flat, Nov. 4[1]; Pinyon Wells, Oct. 15 *, 16.

This distinctively marked junco occasionally migrates southwest into southern California, although chiefly it is found east of the lower Colorado River valley in winter. One was taken as it foraged on a gravelly surface with a flock of Oregon juncos on October 15 at Pinyon Wells. The next day another was seen in a junco flock in this area.

Chipping Sparrow. *Spizella passerina*

Description. A small (10–13 gram), slim sparrow, striped on the back with dusky on buff background. Under parts entirely clear gray or buffy (streaked in juveniles); crown chestnut, bordered laterally by white superciliary stripe; frontal area and line through eye black. Immatures with crown striped, but concealed chestnut present below feather tips.
Range. Breeds from central Yukon southeast to Quebec and south over United States and through mountains of México and Central America to northeastern Nicaragua. Winters from about latitude 36° southward, the migratory races moving as far as Oaxaca.
Occurrence in Monument. Common winter resident. Recorded from: Little Morongo Canyon, Mar. 31; Black Rock Spring, Aug. 30 *, 31 *, Sept. 2 *, 3 *, 4 *, Apr. 4; Upper Cov-

ington Flat, Aug. 27[1], 28 *; Lower Covington Flat, Apr. 15[1], 30[1], Aug. 28 *, Sept. 17[1], Nov. 26[1]; Quail Spring, Sept. 8, 10, Jan. 19, 20 *, 21, 23, 25, 26, 27, Feb. 10, 13, 17, Apr. 10 *, 22; Stubby Spring, Sept. 6 *, 10 *; Barker's Dam, Feb. 18; Twentynine Palms, Mar. 24, May 2–4, 17; Cottonwood Spring, Apr. 30.

Chipping sparrows have been recorded from August 27 to May 17, and through the winter they are common, at least near watering places such as Quail Spring and Barker's Dam. The occurrences in April and May that we list are clearly those of migrants, as Miss Carter's observations revealed an influx of spring migrants in late March at Twentynine Palms, where these sparrows were present in large numbers and continued to be until May 17; the species was apparently absent in winter at that locality.

In the fall the records probably reflect migrants in considerable part, from August 27 to September 10. However, some individuals taken then were adults in worn plumage which had not started or had barely begun the annual molt and others were juveniles, fully grown but with little progress made in the postjuvenal molt. It is possible therefore that at Black Rock Spring in the piñon belt, where most of these specimens were taken, there had been a summer, breeding population. If this is not so, the fall migration must begin before molt and be carried on during it. Migration could entail only short distances traveled from nearby breeding areas in the San Bernardino Mountains.

In the winter aggregations of sparrows at Quail Spring, chipping sparrows were less abundant than Oregon juncos. But as many as 50 were counted in an hour on one occasion. At other times counts were of 7, 25, and 30.

Drinking water clearly is sought by this species. At Black Rock Spring in late August and early September, chipping sparrows usually were loafing about the spring and obviously were attracted to it. Near Quail Spring on January 26 birds were moving to a water hole in a canyon and also going to drink at the water tank. Also at Barker's Dam on February 18 chipping sparrows were drinking.

Food is gathered on the ground and consists of weed seeds almost entirely in winter and in amounts up to 50 per cent in spring. On the desert wintering range, dry seeds probably are the staple food supply. Insects may occasionally be taken and aid in satisfying moisture requirements, but they are probably of little consequence until spring, when some green annual plants might be available anyway. On April 10 at Quail Spring, chipping sparrows were feeding among green filaree and peppergrass and could have obtained green seeds or moist green leaf fragments then. On September 3 after a heavy rain, these sparrows were among other species working busily on the ground among termites and other insects stimulated to emerge by the rainfall. On March 31 in Little Morongo Canyon, a flock of about 20 chipping sparrows was gleaning food on an almost bare sand surface.

Fig. 74. Species of sparrows: Lincoln, Chipping, Black-chinned, Sage, Brewer, Fox

Chipping sparrows, as seed-eaters, must at times require drinking water on their wintering grounds, and they tend to be seen chiefly near springs accordingly. In the heat stress periods of late summer also, springs are focal points for them.

Among the fall-taken specimens, the following conditions of plumage and molt were seen: 4 adult females, 3 in worn breeding plumage (August 28–30), and one in early stages of annual molt (primaries 5–9 old; September 4); 10 birds of the year, 3 in full juvenal plumage (August 28–September 6), 4 early in or halfway through postjuvenal body molt (August 31–September 4), 3 in complete new postjuvenal plumage (September 2–4). Two spring-taken males, March 31 and April 10, were in fat premigratory state and had testes 2½ and 2 mm. long.

The chipping sparrows from the Monument are all of the race *Spizella passerina arizonae,* the form occupying all the western section of the continent, according to our view.

Brewer Sparrow. *Spizella breweri*

Description. A small (9–12 gram) sparrow, finely streaked above on head and back with dusky on a gray background. Under parts entirely or essentially clear gray; wing coverts edged with pale buff. No sharply defined head pattern or rufous areas (fig. 74).

Range. Breeds in interior mountains and higher basins of western North America from southwestern Yukon to southern California, central Arizona, and northwestern New Mexico. Winters from southern California, southern Nevada, central Arizona, southern New Mexico, and central Texas south to southern Baja California, Jalisco, and Guanajuato.

Occurrence in Monument. Common winter resident. Recorded from: Morongo Pass, Apr. 28 *; Black Rock Spring, Aug. 30 *, Sept. 1 *; Upper Covington Flat, Apr. 9; Lower Covington Flat, Mar. 17 [1], Apr. 1 [1], Oct. 1 [1], 14 [1], Dec. 16 [1]; Quail Spring, Jan. 21 *, 23, 25, 26, 27, Feb. 13, Apr. 22 *; Stubby Spring, Sept. 9 *, 10 *; 5 mi. N Key's View, May 14; Barker's Dam, Feb. 18; Indian Cove, Apr. 8; 6 mi. S Twentynine Palms, Apr. 28; Pinyon Wells, Apr. 10, Oct. 9 *; Pinto Basin, Apr. 10, 11; 1 mi. N Sunrise Well, Sept. 28 [1].

This species winters in the Monument in substantial numbers, especially in the western section and the upper levels. We have records extending from August 30 to May 14, but the winter population is present chiefly from mid-September through April. In winter at Quail Spring flocks consisting usually of 6 to 12 individuals were seen. On other occasions in larger aggregations in common with Oregon juncos and white-crowned sparrows, as many as 25 to 75 were seen at one point in the course of one hour of observation.

In spring before departure flocks of Brewer sparrows engage in group singing, many individuals at once giving their buzzing canary-like song. This was noted on April 8, 9, 10, 11, and 28. The species nests nearby in the San Bernardino Mountains as well as to the north, and some individuals may be close to breeding condition before they move to nearby breeding areas in the higher artemisia brushlands. A male taken from a flock at Quail Spring had a testis 4 mm. long on April 22, and one taken in Morongo Pass on

April 28 had a 6 mm. testis, which is essentially full breeding size. There is some possibility that the species breeds in the latter area at the east end of the San Bernardino Mountains.

Brewer sparrows in winter are largely seed-eaters, working on open ground to glean their food from among the sparse remnants of annual vegetation. They also work through a variety of small bushes in the openings and flats. The food taken must often be dry and water supplements from dew or frost in winter may well be important to them, although, as a species, they are generally tolerant of rather dry, hot conditions. On February 18 Brewer sparrows actually were seen drinking at Barker's Dam, and on January 26 they were going to a water hole in a canyon near Quail Spring. Indeed, the large flocks at Quail Spring seemed to focus activity especially near the water tank.

An adult female taken at Black Rock Spring on August 30 at the end of a severe heat period and just after migration weighed only 7.4 grams, 2 grams below normal; it must have been in seriously depleted condition; also it had just completed the annual molt.

Brewer sparrows taken in the Monument are of the widespread southern race of the western United States, *Spizella breweri breweri.*

Black-chinned Sparrow. *Spizella atrogularis*

Description. A small (10–12 gram), slender, and long-tailed sparrow, with head, rump, breast, and flanks lead gray. Chin black in adult males and partly or entirely so in females; back rusty red striped with black. Bill dull red (fig. 74).

Range. Breeds from central California, southern Nevada, central Arizona, and southern New Mexico south to northern Baja California, Guerrero, and Oaxaca. Winters from southern California, southern Arizona, and central Texas southward.

Occurrence in Monument. Fairly common summer resident. Recorded from: Little Morongo Canyon, Apr. 3; Black Rock Spring, Aug. 31 *, Sept. 4 *; Upper Covington Flat, Apr. 9; Smithwater Canyon, Apr. 8 *, 10 *, 23 [1], 24 [1], May 19; Quail Spring, Apr. 6, July 3 *; Pinyon Wells, Apr. 10; Eagle Mountain, May 16 *.

These sparrows breed in moderate numbers in the higher parts of the Little San Bernardino Mountains and on Eagle Mountain. The environmental features of the areas occupied are open chaparral interspersed with junipers and piñons on steep canyon walls or mountain sides. In general, then, the species occurs in the piñon belt but not in all sections of it.

On the north-facing, brushy, and rock-strewn slopes of Eagle Mountain on May 16, black-chinned sparrows were found stationed for breeding. At 4500 feet a male sang steadily for 15 minutes, and at 4200 feet a male was taken. This latter bird was probably in its first year, as the chin was not fully black and the testes, at 3 mm. length, were not in full breeding condition. It was found perched in a *Yucca mohavensis.*

At Pinyon Wells on April 10 at least four males located on the canyon walls were singing regularly.

In Smithwater Canyon on April 8 and 10 birds were taken within half a mile of the springs. Some had reached full breeding condition (testis 7 mm.) while others had not, but on May 19 a male was singing steadily on the steep brush slopes above this area, and on July 3 at nearby Quail Spring a fully grown juvenile was taken. This individual had not yet begun the postjuvenal molt. It probably was not far from the place it had hatched. The nesting season thus seems to be May and June; eggs probably are laid chiefly in May as they are in the Providence Mountains to the northeast (Johnson, Bryant, and Miller, 1948).

In late August and early September in very warm weather occasional black-chinned sparrows were found at Black Rock Spring. They may have been migrants, but they could have represented birds raised locally. On August 31 one visited camp, attracted, we believe, by the shiny black surfaces and windshields of the cars, which may have suggested water. On September 4 one was taken near the spring down canyon from camp. These sparrows may at this time have felt water shortage, as the temperature ranged up to 115° F. in midday. In general this sparrow must withstand well the high temperatures characteristic of its summer range, and it is not then dependent on water, which is generally absent in the chaparral belt. Like other members of the emberizine subfamily, it feeds its young on insects and doubtless the adults use such food at the same time.

The four specimens collected at Black Rock Spring are all immatures as judged by condition of the skull, remnants of juvenal body plumage, or failure to have freshly replaced primaries. Fresh body plumage predominates in them, and the group thus reflects the occurrence of the late or concluding period of the postjuvenal molt at the end of August.

The sample of specimens is not entirely adequate to show the racial affinities of the breeding population of the Monument. The three spring-taken males in relatively unworn plumage lack brownish tones in the breast normal for *Spizella atrogularis cana* of the San Bernardino and San Jacinto mountains and probably because of this and their generally long tails are best placed in the race *Spizella atrogularis evura* of eastern California and Arizona.

White-crowned Sparrow. *Zonotrichia leucophrys*

Other name. Gambel sparrow.

Description. A fairly large (24–30 gram) sparrow, clearly marked in adult and spring plumages by two black crown stripes, one above the eye, the other through the eye, which alternate with white or light gray. Center of back streaked with dark brown on gray; breast clear gray, unstreaked, the flanks light brown; wings dark brown with two white wing bars on secondary coverts; tail dark brown. Bill pink (in races here involved). Immatures with crown pattern of dark brown on light brownish gray (fig. 73, p. 262).

Range. Breeds from northern Alaska east along tree line to northern Labrador and south to southern coastal California, southern Sierra Nevada, and northern New Mexico; in east, south to central Manitoba and southern Quebec. Winters from southern

British Columbia, southeastern Washington, southern Idaho, Wyoming, Oklahoma, the Ohio Valley, and North Carolina south to southern Baja California, Michoacán, Querétaro, the Gulf coast of the United States, and Cuba.

Occurrence in Monument. Common winter visitant; locally abundant. Recorded from: Little Morongo Canyon, Mar. 30–Apr. 1; Black Rock Spring, Apr. 3, 4, Oct. 16; Upper Covington Flat, Feb. 18 [1], Mar. 17 [1], 18 [1], Oct. 14 [1]; Lower Covington Flat, Apr. 8, Oct. 1 [1], Nov. 27 [1]; Smithwater Canyon, Feb. 16 [1], Sept. 30 [1], Nov. 19 [1]; Quail Spring, Jan. 19, 20, 21 *, 23, 25, 26, 27, Feb. 10, 13, Apr. 10, 11, 22, Oct. 15; Stubby Spring, Apr. 9 *, 10, Oct. 15; Lost Horse Valley, Nov. 1; 5 mi. N Key's View, May 14; Indian Cove, Apr. 7, 8; Fortynine Palms, Apr. 7; Twentynine Palms, Feb. 2–Mar. 27, Apr. 1–30, May 1 [1], 6, 14, Oct. 18, 23, 30, Nov. 17; Pinyon Wells, Apr. 10, Oct. 9, 10, 12, 13 *, 14 *, 16, 17; Pleasant Valley, Mar. 19 [1]; Live Oak Tank, Apr. 6–8; Pinto Basin, Apr. 6, 10, 11, Oct. 13; Virginia Dale Mine, Apr. 25 *; Eagle Mountain, Oct. 15 [1], 18, 19; Cottonwood Spring, Feb. 7–9, Apr. 25, 26, 29, 30, May 7 *, 13, Sept. 14, 15 *, Oct. 13, 14, 21–24.

White-crowned sparrows are the most conspicuous winter visitant sparrows to the Monument because of their numbers and tendency to settle in the heavier vegetation at the oases. Also, their inclination to form cohesive flocks and to sing while in winter flocks, especially at roosting time, make them evident even to the casual observer. Headquarters for flocks consist often of mesquite and catclaw thickets, and in the washes they seem to select the larger and denser shrubs such as daleas and chrysothamnus. From these retreats the flocks work out over the ground surface in foraging, to rush back to the cover in contagious alarm when danger threatens.

The white-crowns have been detected as early as September 14 and as late as May 14, but the numbers are not great until mid-October and most of the winter visitants leave in the last week of April. At Twentynine Palms, Miss Carter found these sparrows present in large flocks in February and March and was able to trap and band 54 individuals. At Quail Spring censuses at the water tank showed 50 in one hour at midday on February 13. They were often in association, temporarily, with Oregon juncos in the vicinity of the water, but on January 21 flocks of 10 to 15 were seen separate from the other flocking fringillids. The white-crowns were coming to the water to drink as well as to forage in its vicinity. On March 31 in Little Morongo Canyon, along the small stream, flocks of 10 to 20 were encountered approximately every 150 yards along the mesquite and willow thickets of a mile of canyon bottom. On April 10 near Quail Spring white-crowns were feeding in the low green filaree.

These sparrows by their late arrival avoid the heat-stress periods of the summer and autumn, and, although they are seed-eaters in the main, they seem able to supplement their diet with moist green material or possibly with some insects sufficiently to permit winter residence even where there are no springs. However, they will use drinking water when they can, and they tend to build up in numbers at oases either because of the better protective cover there or the available water, or both.

The two races of white-crowned sparrow that reach the Monument are *Zonotrichia leucophrys gambelii,* which breeds in western Canada and Alaska, and *Z. l. leucophrys,* the breeding bird present in the high mountains of the western United States (Banks, 1964), whose nearest breeding areas are in the southern Sierra Nevada and central Nevada (and at least sporadically in the San Bernardino Mountains). *Z. l. gambelii* is the predominant wintering race, *Z. l. leucophrys* occurring chiefly in late spring migration. This latter form, marked by a black loral area that reaches to the eye, can be distinguished in hand readily and at times even by close observation in the field.

Among the predominant *gambelii,* individuals of *Z. l. leucophrys* have been seen on March 3 and 16 (banded at Twentynine Palms), May 7 (collected at Cottonwood Spring), May 13 (judged from distinctive, non-*gambelii* song type at Cottonwood Spring), and May 14 (observed at Twentynine Palms). Miss Carter noted that her record on May 14 was long after the migration of the Gambel sparrows at Twentynine Palms. From this and other evidence (see Grinnell and Miller, 1944:520–521), it seems fairly certain that spring migration of *Z. l. leucophrys* through southern California is later in the main than that of *gambelii.*

Examples of *gambelii* taken on April 9 and April 25 show, respectively, early and late stages of the prenuptial molt of the head and body.

Golden-crowned Sparrow. *Zonotrichia atricapilla*

Description. A large (27–35 gram) sparrow, lacking stripes beneath. Head gray, without stripes through the eye, the head either mottled dull gold and dark brown or sharply marked with broad lateral black stripes and central gold area bordered by gray posteriorly; back striped with dark brown on light brown; breast gray or gray-brown, the flanks brown; wings and tail dusky, the former with two white bars on tips of secondary coverts.
Range. Breeds from northwestern Alaska southeast in mountains to extreme northern Washington, southeastern British Columbia, and southwestern Alberta. Winters chiefly in lowlands from southern British Columbia south along Pacific coast to northern Baja California; rarely interiorly to northern Sonora, Colorado, and New Mexico.
Occurrence in Monument. Scarce winter resident. Recorded from: Upper Covington Flat, Nov. 18[1]; Lower Covington Flat, Mar. 17[1]; Stubby Spring, Apr. 11; Pinyon Wells, Oct. 13 *, 17; Cottonwood Spring, May 9 *.

Several golden-crowned sparrows were seen in wintering flocks of white-crowned sparrows on October 17 at Pinyon Wells. The specimen taken there on October 13 was a female showing immature skull condition and mottled head patterning. In the shallow canyons above Stubby Spring on April 11 these sparrows were again noted, and van Rossem took one at Cottonwood Spring, a female in prenuptial molt of the head, on the late date of May 9.

This sparrow, much more than the related white-crowned sparrow, favors humid areas and those with heavy brush. Thus its winter range is chiefly on the coast, and only a few move into the desert, where even in winter the dryness of the region may be difficult for them. Like other, related sparrows, it must depend on dry seeds for food in fall and early winter, but it conspicuously turns to green plant material in late winter and spring in its coastal range. Such food is in poor supply in the desert, although in the higher parts of the desert mountains, where we have seen the birds most often, low green annuals and some dew and frost may critically aid those that winter there.

Fox Sparrow. *Passerella iliaca*

Description. A large (25–33 gram) sparrow with unpatterned brown or gray-brown back, tail, and wings; under parts white, heavily streaked and spotted with brown (fig. 74, p. 265).
Range. Breeds north to arctic tree line from northwestern Alaska to northern Labrador and south on Pacific coast to northwestern Washington; in mountains south to southern California, central Utah, and central Colorado; in eastern part of continent south to central Ontario, southern Quebec, and Newfoundland. In winter from southern British Columbia, southern Utah, Colorado, the lower Missouri Valley, the southern Great Lakes Basin, and southern New Brunswick south to northern Baja California, southern Arizona, southern Texas, the Gulf coast, and central Florida.
Occurrence in Monument. Scarce fall migrant; rare winter resident and spring migrant. Recorded from: Little Morongo Canyon, Apr. 1; Upper Covington Flat, Nov. 4[1]; Quail Spring, Oct. 15; Twentynine Palms, Nov. 4; Pinyon Wells, Oct. 9 *, 12 *, 13 *, 14 *, 15, 17 *; Eagle Mountain, Oct. 19, 20, 21; Cottonwood Spring, Oct. 13, 22.

Fox sparrows have been found chiefly in October in the period of fall migration. Apparently a few remain through the winter, as the November records suggest. There is only one spring record, that of two seen in streamside tangles and leaf litter in Little Morongo Canyon, on April 1 (see p. 273, song sparrow).

At Pinyon Wells in scrub oak thickets, fox sparrows were regularly seen from October 9 to 17. Beneath the protecting canopy in the fallen oak leaves, they were scratching heavily in the foraging manner typical of the species. As many as 5 to 20 solitary individuals might be seen in a morning's observation. Similarly on Eagle Mountain in sparse oak scrub and in shady draws beneath mountain mahogany, fox sparrows were found scratching.

These forage areas, frequented in October and possibly later also, are at these times cool and well shaded, and layers of duff that are slightly moist are often found. We may presume that the vegetable food taken there is somewhat softened and not completely dry. Insects are also taken from the litter, although the proportion of them in the diet is not known. Insects could aid materially as a moisture source for this sparrow.

At Twentynine Palms on November 4, a fox sparrow was watched feeding in the drifts of fine dust and leaves about a house and at the bases of mesquites. A few minutes later it was captured by an accipitrine hawk, presumably a Cooper hawk.

Of the eight birds collected in October at Pinyon Wells, six were known to be immatures and one an adult. Two of them, on October 14 and 17, including the adult, were fat, which suggests that migration had just been concluded or was still in progress. Two immature females taken on October 17 were very lightweight at 21.9 and 22.0 grams and had flat breast muscles and pale pink lungs as in birds that are sick or undernourished. The latter two possibly had neared the danger level in energy reserve in the recent migration across the Mohave Desert basin.

All specimens are of the races that breed in the interior to the northward. One of October 14 is of the race *Passerella iliaca olivacea* of central and interior Washington, one of October 12 is *P. i. canescens,* which breeds in central Nevada and the White Mountains of eastern California and southwestern Nevada; one of November 4 is of the race *monoensis* of the Mono Lake district; and the other six are *P. i. schistacea* from interior southern Canada, northern Idaho, and eastern Oregon.

Lincoln Sparrow. *Melospiza lincolnii*

Description. A fairly small (14–17 gram), short-winged and short-tailed sparrow, streaked above and below and with a diffuse band of buff across the breast. Above gray with fine black stripes on head and back, somewhat coarser on the latter; below white, except breast, the stripes confined to throat, breast, and flanks. Wings and tail dusky, edged with buff and brown (fig. 74, p. 265).
Range. Breeds in boreal zone of North America from northwestern Alaska to central Labrador and south in mountains to southern California, central Arizona, and northern New Mexico. Winters from northern California, central Arizona, Oklahoma, central Missouri, and northern Georgia south to Guatemala, the Gulf coast, and central Florida.
Occurrence in Monument. Winter resident; locally common at oases. Recorded from: Little Morongo Canyon, Mar. 31 *, Apr. 1, 2 *; Lower Covington Flat, Apr. 8; Fortynine Palms, Apr. 7 *, 8 *; Twentynine Palms, Feb. 25, Mar. 6–Apr. 26, May 6; Cottonwood Spring, Sept. 13 *, Apr. 30 *.

This sparrow winters regularly in the desert but establishes residence only about moisture sources, where it frequents low brush tangles and sedges near or at the water's edge. By confining its activities to such cover it avoids exposure and has drinking water available in the period of its sojourn, which extends from September 13 to May 6.

In Little Morongo Canyon in early April, Lincoln sparrows were common in the swampy meadow in the canyon bottom. Often two or three were together, and as many as 20 were seen in a quarter of a mile of stream course; probably twice as many were present, as they often hide undetected in the cover. An insect-like location note, *zeet,* was frequently heard. At the small oasis at Fortynine Palms two were seen on April 7, and one was watched foraging on open ground 8 feet from the pool of water. At Cottonwood Spring specimens also were taken at the water's edge in April and September. Miss Carter caught 18 and had many repeats in a trap placed, between March 6 and April 26, on a tongue of dry ground beside a sedge-filled pool at Twentynine Palms.

We noted that five specimens taken on and about the first of April had no fat or very little and thus had not developed their premigratory fat condition. Moreover, the testes of three males were still small, $1\frac{1}{2}$ to 2 mm. However, on April 30 a male had a 3 mm. testis, showing definite recrudescence (no record of its fat condition was made); probably it was in migration or was about to leave winter quarters. Miss Carter's records indicate that winter visitants departed by April 26, with one straggler noted on May 6.

The specimens from the Monument belong to two races. Seven of them are of the widespread race breeding in northern and interior North America, *Melospiza lincolnii lincolnii;* two are *M. l. alticola,* the long-winged race nesting in the mountains of the western United States. The examples of the latter were taken on April 2 and 30 at Little Morongo Canyon and Cottonwood Spring, respectively, and it is the latter which showed some vernal enlargement of the gonads.

Song Sparrow. *Melospiza melodia*

Description. A moderate-sized (18—22 gram), short-winged sparrow. Breast streaked, the streaks tending to concentrate in a central spot; head and back streaked dark brown on light brown or gray-brown; cheeks, superciliary area, and center of crown lighter-colored, forming a weakly defined system of head stripes.

Range. Breeds from Aleutian Islands, southeastern Alaska, southern Mackenzie, northern Ontario, and central Quebec south to south-central Baja California, Michoacán, Puebla, northern Arkansas, and northern Georgia. In winter leaves interior Canada and extends south to southern Texas, the Gulf coast, and southern Florida.

Occurrence in Monument. Scarce local resident, in Little Morongo Canyon; rare winter visitant elsewhere. Recorded from: Little Morongo Canyon, Apr. 1 *, 2; Barker's Dam, Feb. 18 *; Fortynine Palms, Oct. 12; Pinyon Wells, Oct. 14 *.

At the western edge of the Monument in Little Morongo Canyon, along the permanent stream, there is a limited amount of habitat for resident song sparrows of the race *Melospiza melodia cooperi* of southern coastal California, here at the eastern edge of its range. On April 1 a breeding male (testis 8 mm.) was singing with full power every 2 to 5 minutes in a swamp of three-foot-tall three-square rushes mixed with willow and baccharis brush. The bird was aggressive toward a fox sparrow and drove it back when it approached along the edge of the running water. The song sparrow was watched hopping slowly along the moist edge under the heavy shade of the willows. The next day, a mile down canyon, by the water, one was heard singing.

Resident song sparrows at the southern and desert borders of their range are strictly dependent on streams or marshes where they have access to damp ground, water, and good shade. In the Monument only Little Morongo Canyon seems adequate for them as permanent residents.

Occasional migrants of the races breeding farther north in western North America reach the Monument in winter. One was observed at Fortynine Palms on October 12, but it was not obtained and its race and area of origin thus remain unknown. The specimen taken at Barker's Dam on February 18 was foraging along the rivulet below the pond. It is a male of the race *Melospiza melodia fisherella,* which breeds in the western part of the Great Basin. The bird of October 14 at Pinyon Wells was in a dry, shaded canyon, where it retreated into some bushes of desert almond (*Prunus fasciculata*). It is an immature of the race *M. m. morphna,* which breeds in the Puget Sound area.

SIX

MAMMALS

BATS

California Leaf-nosed Bat. *Macrotus californicus.*

Description. A fairly large bat (13–16 grams, wingspread 12 inches) with large ears, about 35 mm. long, and small fleshy tubercle or leaf on nose. Buffy gray to brown above and pale drab below.
Range. Southern California, the Colorado River valley of southern Nevada, and southern Arizona south to Cape district of Baja California and southern Sonora.
Occurrence in Monument. Recorded from: Pinto Wash Well, Aug. 29 [1], Sept. 29 [1].

Parties from Long Beach State College have taken two specimens of this bat from Pinto Wash Well in the eastern end of the Monument.

Fringed Myotis. *Myotis thysanodes*

Description. A myotis or little brown bat with relatively long ears (18–20 mm.; when pressed forward, extending 3 to 5 mm. beyond nose) combined with conspicuous fringe of stiff hairs on edge of tail membrane (uropatagium). Foot relatively large (7.6–9.0 mm.). Coloration cinnamon-buff above, light gray below.
Range. Oregon, Okanagan Valley of British Columbia, southeastern Washington, northern Wyoming, and southwestern South Dakota south to southern California and western Texas, and on Mexican Plateau to Oaxaca.
Occurrence in Monument. Apparently scarce. One record: Lower Covington Flat, Aug. 27 *.

Most of the occurrences of this species in California and the Great Basin are from localities at middle elevations, above the lower desert basins. The one station known in the Monument, at Lower Covington Flat, is in the piñon belt at 5000 feet. Here on August 27 a bat of this species was hanging from the roof of a horizontal spring tunnel, 26 feet from the entrance where the light was sufficient so that the animal could be seen readily. This was at 4:45 P.M., and the air temperature at this point in the tunnel was 75° F. The bat was a male and weighed only 5.2 grams, lower than normal for the species; it was moving and restless and seemed ready to take flight.

The specimen of fringed myotis is representative of the northern race of the species, *Myotis thysanodes thysanodes.*

Hairy-winged Myotis. *Myotis volans*

Other name. Long-legged myotis.
Description. A myotis with under surface of wings furred as far out as the elbow; ears short (10–14 mm. when pressed forward, barely reaching nose); foot relatively large (6.5–9.6 mm.). Coloration tawny to dark brown above, brown to dull yellowish white below.
Range. Southeastern Alaska, northwestern British Columbia, central Alberta, and southwestern South Dakota south to southern Baja California, northeastern Sonora, central Chihuahua, and northwestern Coahuila; also Jalisco and Veracruz.
Occurrence in Monument. Probably fairly common. Recorded from: Lower Covington Flat, Aug. 25 *; Quail Spring, Sept. 9 *, 10 *.

Hairy-winged bats occur in western North America primarily in the Upper Sonoran and Transition zones. In the Monument the two record stations are in the piñon belt at 3900 and 5000 feet. At 8:45 P.M. on August 25 a bat of this species was flying back and forth in a horizontal spring tunnel. It had apparently been disturbed from a night roost as we entered. It was caught in a butterfly net. On September 9 at about 6:30 P.M. a mist net was set horizontally 3 to 9 inches over the water surface of the artificial tank at Quail Spring. One of these bats was taken in the net at dusk as it came to water, and the following morning another individual was found dead in the net. All three specimens were females without embryos and weighed 5.5, 8.6, and 5.6 grams, respectively.

The examples of the hairy-winged myotis are of the race *Myotis volans interior,* the subspecies that ranges over the Great Basin and interior and southern California.

California Myotis. *Myotis californicus*

Description. A small myotis with relatively small feet (5.5–7.0 mm.); ears fairly short (11.2–14.6 mm. when pressed forward, exceeding nose by no more than 1 to 3 mm.). Above light ochraceous buff; under parts whitish; hairs of back dull-tipped, not shiny.
Range. Extreme southern Alaska and western British Columbia south to southern Baja California, and from Utah and Colorado south to central Michoacán, northern Oaxaca, and Veracruz.
Occurrence in Monument. Fairly common. Recorded from Lower Covington Flat, Apr. 8; Quail Spring, July 1 *, Indian Cove, Apr. 8 *; Queen Valley, July 2 [1]; Pinyon Wells, Oct. 16 *; Cottonwood Spring, July 11 [1], 12 [1], Aug. 2 [1].

This myotis is to be expected in the Sonoran zones of the desert. We obtained one of these small, pale bats on October 16 at Pinyon Wells as it flew erratically at dusk along the wash at the mouth of a canyon. On April 8 one was taken under similar conditions at Indian Cove, at 3200 feet. On July 1 two were caught as they hit wires stretched across the water surface of the artificial tank at Quail Spring. On April 8 at Lower Covington Flat one was caught in the horizontal mine tunnel where occasionally bats of other species have also been found. At 9 P.M. it flew about as we entered but was finally located hanging from the roof, 60 feet back near the end of the tunnel, in a 3-inch deep crypt in the ceiling. The bat of April 8 was a male, but the other four were females without embryos. Weights were: ♂, 2.8; ♀♀, 2.6, 2.7, 3.1, 3.2 gm.

The specimens are examples of the pale desert race of the species, *Myotis californicus stephensi.*

Fig. 75. Pallid bat (upper left), California myotis (upper center and right), and big brown bat (below).

Western Pipistrelle. *Pipistrellus hesperus*

Description. A very small bat (2.5–4.1 grams); ear small and with short, rounded tragus. Above drab gray to smoke gray, paler below, the fur contrasting with black ears and wing membranes. The only bat of the area regularly seen flying by daylight.
Range. Sacramento Valley of California, southeastern Washington, northern Utah, and southern Colorado south through Baja California and through western and central mainland of México to Jalisco and Hidalgo.
Occurrence in Monument. Abundant. Recorded from: Black Rock Spring, Aug. 29, 30 *, 31, Sept. 1 *, 3 *; Upper Covington Flat, May 13 [1], June 29 [1], July 10 [1], 11 [1], Aug. 27; Lower Covington Flat, Aug. 22 *, 23 *, 25 *, 26 *, 27 *, 29; Smithwater Canyon, June 28 [1]; Quail Spring, Apr. 2, July 1 *, 4 *, Sept. 8, 9; Stubby Spring, Sept. 5; Key's View, July 15 *; Indian Cove, Sept. 5; Fortynine Palms, Apr. 7; 4 mi. N Twentynine Palms, July 24 *; Twentynine Palms, July 13; 3 mi. S Queen Valley, July 2 [1], 3 [1]; Cottonwood Spring, Apr. 29 *, July 7 *, 8, 11 [1], 12 [1]; Lost Palm Canyon, Oct. 22.

The pipistrelle is the most abundant bat in the Monument, and it is also the species most frequently seen by reason of its early emergence to forage in late afternoon and its frequent flight in the early morning after sunrise. In our usual twilight watches for bats, this small, erratic-flying species could be counted on to come into view first. The activity of many of these bats before sundown is attested by specific observation of their flight as follows: July 1, 6:00, 6:30; July 7, 6:10; July 8, 5:13, 5:47, 6:12; August 22, 6:30; August 29, 6:30; September 9, 6:00. Although at these hours there were often shadows in the canyons, most of these occurrences were noted with the specific comment that the bats were in full sun part of the time, and for periods of 10 minutes to half an hour.

In the morning we have recorded flying pipistrelles as follows: April 7, 11:00; August 26, 9:00, 7:00–9:45, 9:30; August 27, 7:40; August 29, 6:50; August 30, 8:30; September 8, 8:00. The observations on August 29 indicated that none was flying at 6:30 and that they appeared subsequently; whether this represents reactivation after dawn or is a consequence of local movements we cannot be sure. On August 27, the shade temperature at 7:40 during a flight was 84° F.

The numbers of this noncolonial bat seen range typically from one to 10 over a foraging area visible from one observation point, but about a watering tank as many as 40 may congregate at one time. This was true at Quail Spring on July 1. Perhaps water is sought more frequently in the heat period of summer, and the greater concentration at tanks may then be expected. Pipistrelles will take water by skimming the surface of a pond even by daylight, as noted at 8 A.M. at Quail Spring.

The season of activity or emergence extends from April 2 to October 22, according to our records. Probably occasional flight takes place in winter also, since this has been noted in early February in the Colorado River valley (Hall, 1946:150). Foraging is seen often along canyons, where possibly insect concentration is particularly good, but we have also seen pipistrelles

Fig. 76. Western pipistrelles in flight over Pinto Basin.

on flights over Joshua tree and piñon woods, as well as among the open cottonwoods and palms at oases.

On the morning of August 26 Hendrickson watched this species about a small water hole near Lower Covington Flat. Here from 7 to 9:45 in full sunlight at close range he saw them catch insects and fly toward him chewing on the prey with exaggerated jaw movements. On one occasion a fly was seen first before the bat closed on it and captured it. One bat was watched disappearing, without pause in flight, into a crack in rocks of the canyon wall (fig. 77), where later it was collected. Near here on the same day, a pipistrelle at 9 A.M. circled over another small water hole that was partly overhung by a rock wall. The bat settled on the rock in the shade, head up, then turned with head down. This temporary roost was only 2 feet above the water. Soon the bat flew away without drinking.

These small bats are subject to rapid drying out, as any small type of vertebrate would be. This problem is met in the way it is in other small bats, by

Fig. 77. Crevices in a canyon wall at Lower Covington Flat where a pipistrelle entered for daytime roosting on the morning of August 26, 1950.

nocturnal or largely nocturnal and crepuscular schedules, moist insect diet, resort by flight to drinking water, and by roosting in tight crevices where the humidity is not so low as in the open and where there is protection from drying winds and high temperatures.

Twice pipistrelles have been found dead near or in the water. On September 5 at Indian Cove in the narrow, rocky canyon gorge one was dead on the rock surfaces two feet from water in a pothole. On October 22 in Lost Palm Canyon, one was found floating in a small pool three feet across. Cause of death could not be ascertained in either instance.

In our efforts to capture bats by wires or nets strung across water surfaces, we have most often been involved with pipistrelles and have been impressed with their ability to detect these obstructions, evidently chiefly by their sonar system. In the heavy concentration of pipistrelles at Quail Spring in July in their sweeps toward the water, they often hit the taut wires strung there, but so lightly that only a few fell into the water. When they did fall they had to be struck down further to prevent take-off, as they were not seriously disoriented or stunned. At Black Rock Spring the water is at the base of a small cliff, and they flutter in close with especial care. Here a mist net strung vertically in front of the spring resulted in the pipistrelles brushing up to it only lightly, apparently detecting the fine taut mesh. Occasionally one could be caught by hand as it clung or fluttered on the net surface. The efforts at Quail Spring led to some experimentation with flashlights played on the wires. A light shining in the face of the oncoming bats served to reduce the number of strikes on the wires. The pipistrelles were rather consistently approaching from one direction, moving upwind. The light when at their backs resulted in relatively more strikes. The combined results of the experimentation suggested that facing a light deterred a bat from dipping down to drink but that light coming from behind the bat did not deter it from coming in, nor help it to avoid striking wires. Results were equivocal on the question whether the light from behind tended to concentrate numbers in the beam area.

Auditory sense, as in all bats, is of high order, and response to sound and navigation by it largely through echolocation is well authenticated through the work of Griffin (1941). We were interested to note that a wounded bat kept in a sack in the evening attracted three others, which flew down close over it. We held the bat to our ears and could barely hear a humming or buzzing sound, apparently close to the upper limit of human hearing. It seemed likely that the free-flying bats had responded to this. On another occasion a rather loudly ticking watch laid on a rock by a pool attracted pipistrelles, which hovered close over it.

Among the specimens taken were 13 males and 19 females. In general the sexes were about equally represented at our principal collecting stations, but at Quail Spring from July 1 to 4 females outnumbered males 14 to 3. No pregnant females were collected, probably because none happened to be taken in May or June. The 14 females from Quail Spring in early July were checked specifically for embryos in 12 instances and found to contain none, but three of these females were recorded as showing signs of nursing young.

Weights of males (10) range from 2.5 to 3.7 grams and average 3.03, and females (18 nonpregnant) ranged from 2.4 to 4.1 grams and averaged 3.33.

The pipistrelles of the Monument belong to the nominate race *Pipistrellus hesperus hesperus.*

Big Brown Bat. *Eptesicus fuscus*

Description. A fairly large (9–20 gram) bat, with black ears, the ears and flight membranes naked. Body brown, somewhat lighter below than above (fig. 75, p. 277).

Range. Central British Columbia and northern Alberta east to southern Quebec and western and southern New Brunswick and south through the United States, the Greater Antilles, México, and Central America to western Panamá.

Occurrence in Monument. Fairly common. Recorded from: Upper Covington Flat, July 23 [1]; Lower Covington Flat, Apr. 9 [1], May 29 [1]; Quail Spring, July 1 *, 4 *, Sept. 10 *; Stubby Spring, Sept. 6 *, 10; Cottonwood Spring, July 7; Pinto Wash Well, Aug. 18 [1], Sept. 29 [1].

These bats tend to move in relatively straight flight patterns. This fact coupled with their size distinguishes them to the experienced observer from the smaller bats of the genera *Myotis* and *Pipistrellus*. In numbers this species is far overshadowed by the abundant pipistrelles. At Quail Spring, where much attention was given to bats on July 1 and 4, only occasional big brown bats were noted among the 30 or 40 pipistrelles. Two of the former were caught as they hit wires strung over the surface of the water on July 1, and one was taken on September 10 in a net left set overnight. This species appeared as early as 5:45 P.M. on July 1 but otherwise was caught as late as 7:30 P.M. and during the night.

At Stubby Spring on September 6 a mummified bat of this species was found among the rocks at the base of the large water tank.

The ten specimens of big brown bats saved were all males, those in July with testes measuring up to 9 mm. in length. The skins are light brown, characteristic of the desert race of the species, *Eptesicus fuscus pallidus*.

Hoary Bat. *Lasiurus cinereus*

Description. A large (22—28 gram) bat, with tail membranes furred over entire dorsal surface; under surface of wing furred along bones of forearm; inside of ear pinna densely furred. Above brownish black, tipped with white; below brown tipped with white; yellowish on throat; fur of under wing area yellowish or tawny; ear black-rimmed; small tufts of whitish hair on dorsal wing surface at elbow and wrist.
Range. Canada, from southern British Columbia, southeastern Mackenzie, and the Hudson Bay area southward, the United States, and the Mexican uplands south to the valley of México. Also recorded in South America.
Occurrence in Monument. Recorded once: near Lower Covington Flat, Apr. 8 *.

This bat is a migratory species and probably visits the Monument in winter or passes through in migratory flights. The summering areas seem to be chiefly in the cooler conifer belts to the north or in the higher mountains. The hoary bat found near Lower Covington Flat was hanging upside down and solitarily, as normal for this species, in a clump of twigs and needles of a piñon, 10 feet aboveground. This roosting spot at midday, when it was found, was only about half-shaded. The bat looked like an inconspicuous gray fluff of nest material in the needle cluster where we were searching for goldfinch nests at the time. The bat was a female, very fat, and without embryos.

Spotted Bat. *Euderma maculatum*

Description. A large bat (wingspread about 11 inches), black above with three conspicuous white spots, one on each shoulder and one at base of tail; below white with a blackish collar. Ears wide and long, measuring about 40 mm.

Range. Southern Idaho and south-central Montana south to southern California, southern Arizona, and southern New Mexico.
Occurrence in Monument. One record: Twentynine Palms, May, 1939 *.

The natural history of this rare bat is almost unknown. The occurrences of it are chiefly in the arid or semiarid sections of the western United States. Benson (1954) reported on the specimen from the Joshua Tree area as follows, on the basis of information supplied by James E. Cole, former superintendent of the Monument. Cole stated "that Miss June Garvin found [in May, 1939] the bat in torpid condition hanging on the side of her porch in Twentynine Palms. . . . [This was] near the boundary of Joshua Tree Monument. . . . Miss Garvin took the bat to the local high school where it subsequently died. . . ." The specimen is now preserved as a complete skeleton in the Museum of Vertebrate Zoology. It was a fully adult male.

Fig. 78. Spotted bat in flight.

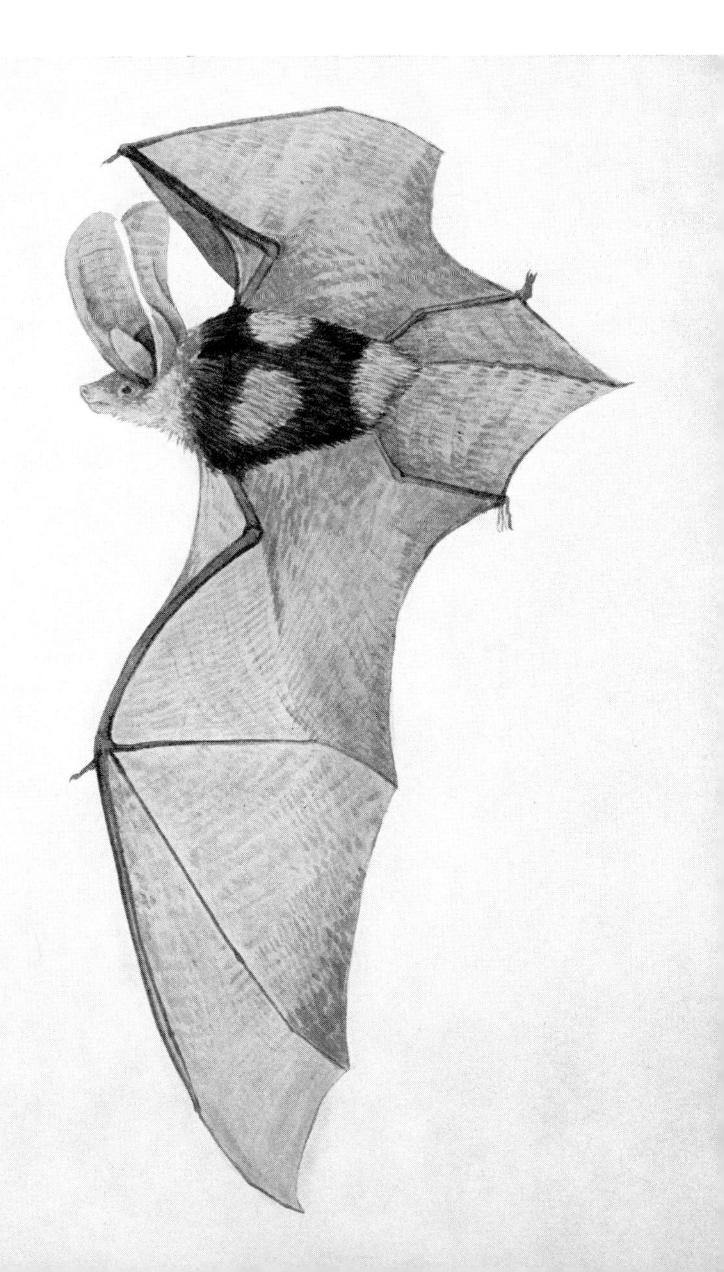

Pallid Bat. *Antrozous pallidus*

Description. A large bat (13—22 grams, wingspread 12 inches) with large, broad ears (about 30 mm. long). Pale tawny above and white below, the flight membranes also pale (fig. 75, p. 277).
Range. Central western Oregon, south-central British Columbia, eastern Oregon, central and northeastern Utah, and western and southeastern Colorado south to Cape district of Baja California and to Durango and Querétaro.
Occurrence in Monument. Recorded from: Indian Cove, Apr. 8 *; Cottonwood Spring, July 11 [1], 12 [1], 13 [1]; Pinto Wash Well, Aug. 18 [1], Sept. 29 [1].

A bat of this species was found dead floating in the water of a pothole in the gorge above Indian Cove on April 8. Otherwise it has been detected in some numbers at eastern stations in the Monument by parties from Long Beach State College. The pallid bats of this area belong to the desert race *Antrozous pallidus pallidus.*

RABBITS

Audubon Cottontail. *Sylvilagus audubonii*

Other name. Desert cottontail.
Description. A fairly small rabbit, with tail extensively white beneath, gray above; ears of moderate length (70—80 mm.) for a rabbit, sparsely haired on inner surface. Body pale brownish gray above with dusky tipping; rusty brown on back of head and on legs; under parts white.
Range. North-central California, central Nevada, north-central Montana, and southwestern North Dakota south to Cape district of Baja California and northern Sinaloa, and over the Mexican Plateau to northern Puebla.
Occurrence in Monument. Common locally. Recorded from: Little Morongo Canyon, Mar. 31; Black Rock Spring, Aug. 31, Sept. 4 *; Upper Covington Flat, June 26 [1]; Lower Covington Flat, June 12 [1], 28 [1], Aug. 20 *, 23, 25, 26, 27 *; Quail Spring, Jan. 21 *, 25, Apr. 22, May 19, July 1, 4 *, Sept. 9 *, Oct. 2, Dec. 20 *; Stubby Spring, Sept. 7; Lost Horse Valley, Feb. 10; Indian Cove, Apr. 7, Oct. 16; Twentynine Palms, July 13; White Tanks, Oct. 14; Cholla Garden, Apr. 10; Virginia Dale Mine, July 12, 14; Eagle Mountain, Oct. 18; Cottonwood Spring, Apr. 26, May 15, 18 *, July 8, Oct. 22.

Cottontails are seen almost always in the vicinity of extensive brush and rock tangles, which commonly occur along canyon bottoms and in washes. Here two or three may be flushed within a few hundred yards. In such places the variety and amount of succulent food is greatest, and hiding places in the rocks and heavier vegetation are close at hand. On at least three occasions we have seen them run into rock piles. Near oases the dense chrysothamnus bushes often afford shelter and doubtless food. In a wash in Lower Covington Flat on August 27, one was taken near a few opuntia cactuses of a flat-padded species. These cactuses had been severely gnawed down, apparently in large degree by the cottontails, which were obviously seeking moisture (fig. 80).

Fig. 79. Audubon cottontail foraging.

Cottontails, or their tracks, are often seen in sand or the fine gravel of the washes. This association is perhaps only incidental, but the species runs easily on loose, soft soil as well as on harder surfaces.

On April 7 at Indian Cove a very young cottontail ran from cover between bushes as we brushed against them. It was so young that it could run in the open only about 20 feet before stopping. It weighed only 60 grams. No parent or other young was seen at the point where the animal was flushed, at 11:45 A.M. On July 4 at Quail Spring a nursing female was collected in an area of thick bushes and catclaws. Half-grown rabbits weighing 459 and 479 grams were collected on May 18 and September 4; adults usually weigh about 800 grams. The breeding season is therefore fairly extended.

Life in desert areas is solved by cottontails much as by jack rabbits, by adhering to crepuscular or nocturnal feeding periods, using succulent vegetation for moisture sources, and staying in the shade of rocks or bushes during the heat of summer days. Indeed, burrows are also used by this species for shelter. The more limited vertical reach and cruising radius of cottontails compared with jack rabbits may make them more local in distribution in the desert, but, where proper vegetation and shelter permits, their numbers exceed those of jack rabbits.

The Audubon cottontails of the Monument belong to the desert race *Sylvilagus audubonii arizonae,* which occupies eastern California, western Arizona, Sonora, and the southern Great Basin.

Fig. 80. Opuntia cactus eaten by cottontails during a period of severe drought and heat; Lower Covington Flat, August 27, 1950.

Black-tailed Jack Rabbit. *Lepus californicus*

Description. A large, long-legged rabbit (hare) with dorsal surface of tail black and with large ears (130—160 mm.). Ashy gray above, the flanks and under surface of tail buffy, the belly whitish; margins and dorsal surface of ear tips black.
Range. Southeastern Washington, southern Idaho, northern Utah, southeastern Wyoming, southern South Dakota, and central Missouri south to the Cape district of Baja California, central Sonora, Aguascalientes, Hidalgo, and the Gulf coast of Texas.
Occurrence in Monument. Common and widespread. Recorded from: Morongo Valley, Dec. 31; Black Rock Spring, Sept. 4; 3 mi. S Warren's Well, Aug. 31; Lower Covington Flat, June 29 [1], Aug. 22, 23, 24, 26 *, Nov. 29 [1]; Quail Spring, Jan. 20, 25, May 19, July 1, 2; Stubby Spring, Sept. 5; Lost Horse Valley, Apr. 10; Key's View, July 15; Indian Cove, May 13—14; Pinyon Wells, Oct. 12—18; Cholla Garden, Apr. 10; Pinto Basin, Apr. 10, June 26 [1], July 15, Sept. 13; Virginia Dale Mine, Apr. 24; Cottonwood Spring, Oct. 13, 14.

The jack rabbits give the impression of abundance because of their conspicuousness, and yet actually the animals are widely spaced and not so numerous as many species of rats and mice. Moreover, the jack rabbits clearly do not achieve the large numbers in the Monument that they do in other ungrazed, undisturbed land of other desert and semiarid areas. Our sporadic visits to the Monument over a span of 15 years has given us no sure picture of annual fluctuations of striking order such as is known in other, more northern regions, although we thought numbers were larger than usual in 1953. Typically we have seen one to four animals in the course of a day

Fig. 81. Black-tailed jack rabbits on a creosote bush flat.

afield. On July 2, 1946, between Quail Spring and Twentynine Palms four were counted from the roadway. Three were noted on October 13 in late afternoon in the vicinity of Cottonwood Spring. At night at the Cholla Garden on April 10, 12 were seen along the road after 10 P.M. Otherwise, our notes record only one or two individuals at a time.

Like most rabbits, this species is seen chiefly in early morning or in the evening unless flushed from its shaded resting places in midday. Most foraging seems to be done in an hour or two at the beginning and end of the day. On October 14 at 5:30 A.M. a jack rabbit came toward the water at Cottonwood Spring. It stayed close to the edge of the vegetation and paused to stand erect on its hind legs to eat from a greenish yellow plant, the leafy parts of which had been cropped to just the height of a standing jack rabbit. All such bushes about the camp showed such a browse line. The rabbit also fed at ground level at the base of an indigo bush.

Fig. 82. Black-tailed jack rabbit crouched near its form in Lost Horse Valley in the early morning of April 10, 1951.

Jack rabbits range through all vegetation belts of the Monument. They do avoid the steep and very rocky slopes and are most to be expected on mesas and rolling hills in the Joshua tree areas, in the juniper woodlands, and on the creosote bush flats. We have seen them also in the sand dunes of Pinto Basin.

On April 10 at 7 A.M. in Lost Horse Valley among Joshua trees, we flushed a jack rabbit that had allowed us to approach cautiously within 5 feet (fig. 82). The rabbit panted noticeably and kept the belly and tail area down on the ground. The abdomen looked heavy, and we believe the animal was ready to give birth. When forced to run, it scooted at an even, fast gait, without any leaps or bounces, and with ears held down. Although the actions were unusual, the animal seemed to be vigorous and not sick. When stopped, the rabbit usually crouched with ears down, only occasionally elevating them. Finally the rabbit settled back into its "form" again and crouched there, head down. The "form" was under one side of a dense bush and was completely bare of litter from the bush. It was about 1 inch deep. At 11:30 A.M. the rabbit was still in the form and was flushed out, behaving as before. We had hoped to find the rabbit later with young, but were not successful.

Jack rabbits, as herbivores, obtain their water from fresh plant food entirely. At certain times on the desert the lack of succulent vegetation must limit local occurrence of rabbits and probably holds down numbers. Cactuses are known to be resorted to (Vorhies and Taylor, 1933) and must provide an essential last reserve of moist food during extremely dry periods. If kept on dry food in captivity, jack rabbits loose weight and obviously are not maintaining water balance. Rabbits avoid overheating in part at least by their behavioral adjustments; chiefly this consists of being quiet in midday and utilizing shade of a dense bush for a daytime resting place.

The black-tailed jack rabbits of the Monument belong to the widespread desert race *Lepus californicus deserticola.*

SQUIRRELS

Merriam chipmunk. *Eutamias merriami*

Description. A small (55–70 gram), squirrel-like mammal with stripes on both head and body. Light gray stripes above and below eye; back with four light stripes alternating with five dark stripes, the central dark stripe blackish; a poorly defined white patch behind ears; flanks tawny, under parts whitish. Tail long, with narrow brush, bordered laterally with black, the tips of the hairs buff; under surface of tail tawny medially; upper surface grizzled brown, with tendency to form buff streak on either side of dark center area. Summer pelage brighter than winter pelage and more sharply patterned, the light stripes nearly white and flanks much more tawny.

Range. Central and southwestern California, chiefly in coast ranges and Sierran foothills, and disjunctly in two areas in northern half of Baja California.

Occurrence in Monument. Common in piñon belt. Recorded from: Black Rock Spring, Apr. 3, Aug. 30, 31 *, Sept. 4 *; Upper Covington Flat, Aug. 23; Lower Covington Flat, Aug. 23 *, 24, 25 *, 26 *, 28 *; Smithwater Canyon, Apr. 24 [1]; Quail Spring, Jan. 26 *, Feb. 11, Apr. 22, May 19 *, Oct. 15, 27 *; Hidden Valley, Feb. 4 *, July 7 [1]; Lost Horse Valley, July 20 [1], 21 [1]; Stubby Spring, Sept. 6, 7; Barker's Dam, Feb. 18, Apr. 2 *; Indian Cove, Apr. 7, May 13 *; Pinyon Wells, Oct. 10 *, 12 *, 14 *, 15, 16; Eagle Mountain, May 16 *, Oct. 19, 20 *, 21; Cottonwood Spring, May 17 *.

Merriam chipmunks are typical of the piñon, juniper, and chaparral habitats along the crest of the Little San Bernardino Mountains. They do at times range through the Joshua tree "forests" but chiefly where there are scattered junipers and rocks. Two were seen at the lower edge of the piñons and rocks at Indian Cove, and one was found near Cottonwood Spring in scattered junipers. Such occurrences indicate that they can range across moderate desert barriers and populate such typical, isolated upland areas as that on Eagle Mountain. The greatest numbers are found in the piñons, however. For example, on a piñon-covered crest above Lower Covington Flat, 15 animals were counted in 2 hours of cruising along a ridge.

These chipmunks live in the rock and brush tangles, climbing and running readily on the rock surfaces and in the low trees. We saw them as high as 12 feet up in piñons. They depend for food largely on piñon nuts, acorns, manzanita seeds, and juniper berries. In the Pinyon Wells area in October, piñon cones were scattered widely where the chipmunks had gnawed them open for the nuts. We saw feeding stations on logs and on tops of large rocks, where cone fragments formed conspicuous little piles. Also in this area we saw chipmunks on October 12 up in the scrub oaks collecting acorns, and one taken on October 14 under a scrub oak had two unbroken halves of shelled acorns in each cheek pouch. On May 16 on Eagle Mountain a chipmunk we collected had juniper berries in the mouth, and the animal taken on May 17 near Cottonwood Spring had been feeding on juniper berries.

Chipmunks in this desert area are able to find shade readily in the rocks and among the trees and seem to use it during warm midday periods. Although not all of them by any means have water accessible, those near springs or tanks come to drink. We saw them at water near Lower Covington Flat on August 24, 25, 26, and 28. Here on the 26th one was watched working down a nearly vertical granite slope toward water in a sandy pit dug by a coyote. It either lost its footing on the steep face or intentionally dropped down to the sand, but it seemed badly frightened by its experience and ran hurriedly up the sand slope of the pit and disappeared in the rocks. The animal watched on the 24th moved through a group of seven mountain quail on its way to the water.

Chipmunks call especially when excited or disturbed, and sometimes their low-pitched bark may be continued for periods of 5 minutes or more. Squeaking by the observer will often bring them to view and sometimes induce them to start their high-pitched scolding notes. Except where they have become accustomed to people about camps, these chipmunks seem to be rather wild and difficult to approach closely.

On Eagle Mountain on October 19 at a pocket of wet sand where small mammals had been scratching for water, we found two-thirds of a tail of a chipmunk lying on the gravel. Apparently a chipmunk had been attacked there. Whether it was captured and killed or had merely lost its tail in the

Plate VI. Merriam chipmunk.

GMC

encounter, we could not ascertain. This find suggests again the general danger that animals face in coming to restricted water sources in the desert. It is here of course that predators concentrate their efforts.

The female chipmunks we took in August, October, and May did not have embryos. One taken on January 26 had 4 embryos, and one on April 22 had 4 large embryos. However, two females on April 2 had none. On May 19 a female was lactating. Thus the breeding season may start early, and the animals are obviously out and active in midwinter. A male on October 20 had scrotal testes, but they were small (about 4 mm. long).

Chipmunks, like antelope squirrels, undergo two molts each year, but there is not so great a difference in the denseness of the winter and summer pelages as in the antelope squirrels. Certainly chipmunks do not expose themselves to the sun in summer to the degree the antelope squirrels do and would not suffer so much from the possession of a moderately dense fur at that time. On April 22 an animal was one-fourth into new summer pelage, and on May 19 the molt of another was four-fifths complete. In general winter pelage seems to be held through April, and summer pelage is replaced in late August and early September.

D. H. Johnson (1943) has pointed out that the Merriam chipmunks of the Little San Bernardino Mountains are closer in characteristics to the race *Eutamias merriami merriami* than to *E. m. kernensis,* although they approach the latter in paleness. Our additional material from as far east as Eagle Mountain shows no increase in paleness farther into the desert. The race *kernensis* is stated to be smaller than *E. m. merriami,* but Johnson's figures (table 5) do not support this view and indeed indicate that no statistically valid separation on this basis is possible. The conclusion reached is that the animals of the Monument belong to the race *E. m. merriami,* which is typical of the San Bernardino Mountains, and that there is no substantial racial differentiation of the animals of the desert area.

Antelope Squirrel. *Ammospermophilus leucurus*

Other name. *Citellus leucurus.*
Description. A small ground squirrel with conspicuous white body stripes on each side. Tail short and characteristically held up over back to expose its white under surface; head and back gray to buff, the upper surface of the tail black; under parts whitish, the legs tawny.
Range. Extreme eastern part of northern California, southeastern Oregon, southwestern Idaho, northern Utah, and western Colorado south to north-central New Mexico and central Arizona, and through deserts of southern Nevada and California to Baja California, extending to the Cape.
Occurrence in Monument. Common throughout. Recorded from: Morongo Valley, May 2 *; Little Morongo Canyon, Apr. 1, 28; Black Rock Spring, July 6 *, Aug. 30 *; Upper Covington Flat, June 11[1], Aug. 23; Lower Covington Flat, Jan. 22[1], Apr. 8, 30[1], May 20[1], June 11[1], 28[1], Aug. 9[1], 24, 25 *, 26, 27 *, Oct. 10[1], 22; Quail Spring, Jan. 20, 21 *, 25, 27, Feb. 11, Apr. 10, 22, May 19–21, July 1, 2, 3 *, 4, Oct. 15, Dec. 20 *, 21 *, 22 *; Lost Horse Valley, July 20[1]; Stubby Spring, Sept. 7 *, 8 *; Key's View, July 3, 4, 15; Barker's Dam, Apr. 2 *; Indian Cove, Apr. 7, 8, May 13, 14; Fortynine Palms Canyon, July 24, 25; 4 mi. N

Twentynine Palms, July 24 *; 13 mi. SW Twentynine Palms, May 2 *; Pinyon Wells, Oct. 8, 10 *, 12; Pinto Basin, Apr. 5, July 15; Eagle Mountain, Oct. 21 *; Cottonwood Spring, Apr. 5, 26, May 15 *, July 8, 12 [1], 13 [1], Sept. 13, 15, Oct. 21—24.

Antelope squirrels occur from the highest sections of the Monument in the piñon woods down to the desert flats, as in Pinto Basin in the creosote bush association. They are seen singly, or two or three at a time, or in family groups, or in small concentrations about food sources, but not in colonies. Frequently they are heard giving their long, clear trill. Actually nowhere is there a dense population, but the widespread occurrence of the species through many habitats, from rocky areas to sandy flats, results in a total population of large magnitude. Moreover, their rapid movements and flickering white tail make them relatively conspicuous and augment the impression of abundance. Typifying actual occurrence are our notes of October 8 at Pinyon Wells, when 5 were seen on the plateau above the wells and 2 in the valley in the course of a morning. These 7 animals were in the Joshua tree belt, among junipers, and out on exposed rocks.

For food, these squirrels use a variety of plant material and will also feed on carrion along the roadways. Doubtless many kinds of seeds are taken, but succulent vegetation is evidently a requirement. Bartholomew and Hudson (1959) have shown that captives on a diet of sunflower seeds alone, and without drinking water, lose weight, dropping about 20 per cent in 9 days. They will then in 2 days recover their weight when given water. The species thus can withstand temporary dehydration but obviously cannot sustain water balance and good condition without moist food. Since these squirrels do not have drinking water in the desert, moist vegetable food must be essential. The investigations of Hudson (1962) also have shown that antelope squirrels can conserve water, especially under the stimulus of high temperatures, by exceptional concentration of the urine. In this respect they almost equal some species of kangaroo rats, although, unlike these rats, they are incapable of surviving entirely on metabolic water coupled with the limited moisture in seeds.

We have observed antelope squirrels climbing bushes and trees in foraging as well as working among sparse vegetation at the ground level. In July at Key's View they had eaten extensively of the moist fruits of the Joshua trees. This plant, then, through its seed pods and green seeds (fig. 84), probably is of much importance as a moisture source for these squirrels. At Quail Spring on April 22 one was taken as it perched a foot above ground eating flowers of a yellow composite. In September, 1950, at Cottonwood Spring these squirrels came to the feeding tray established for foxes, gathering dog food and table scraps placed there. Like the foxes, they had become tame in the presence of humans and spent much of their time about the tray watching for or chasing off others of their own kind.

Fig. 83. Antelope squirrel on the alert.

At this same station on September 15 a Cooper hawk made an unsuccessful strike at an antelope squirrel feeding in the campground.

These squirrels can apparently withstand a great deal of exposure to hot sun, as they are not infrequently seen in midday running far over the very warm desert gravel. In the hottest periods of the day their activity is reduced, to be sure, and they may be underground in their burrows part of the time or in the shade cooling off. Nevertheless, they do not conspicuously avoid the heat. Dawson (1955) and Hudson (1962) have shown that they can withstand significant elevation of body temperature, up to 108° F., without ill effects—an important element in their desert adaptation. The strikingly black dorsal skin of this species, although it would seem to function to "pick up" heat, could possibly also serve to shield the deeper tissues from the effects of ultraviolet light.

Our observations, like those of others (Hudson, 1962), show that antelope squirrels are active in midwinter and indicate that these squirrels do not hibernate.

Antelope squirrels breed in March in the Monument, as demonstrated by the fact that four females collected at Barker's Dam, 4500 feet, in April had embryos. On May 15 two females at Cottonwood Spring had no embryos, but one was lactating. Similarly, adult females taken in August, September, October, and December had no embryos. Young of the year have been seen

Fig. 84. Green seed pods of the Joshua tree, an important moist food source for antelope squirrels.

or collected on July 6 and 24, when they were about four-fifths grown. Doubtless young may be expected to range out of the burrows in June and even late May, as has been noted in the Providence Mountains to the northeast (Johnson, Bryant, and Miller, 1948: 355).

In this squirrel distinct summer and winter pelages are grown, and the sharp contrast in the two seems obviously adapted to the temperature differences in the seasons of the year. The winter pelage has extensive underfur, and, although not affording an extremely thick layer, it must serve materially to retard heat loss in the cold weather from October to early April. The summer pelage, on the other hand, consists almost entirely of sparse, coarse, short guard hairs, through which the black skin can often be seen. Such pelage must retard loss of body heat very little, thereby allowing cooling when the temperature gradient from body to air permits. Probably such pelage affords necessary protection of the skin in burrowing and other activities but has little use otherwise, although in certain situations it may be moderately concealing. The molt to summer pelage occurs in May, as shown by two animals taken on May 15 that had one-fourth to one-half the body area in new summer hair; an animal on May 2 showed no molt as yet. Summer pelage is carried through early September (specimens of September 7 and 8). On October 10 one specimen had the new winter pelage four-fifths complete, and on October 21 one was in complete winter fur. All others were in either full winter or summer pelage appropriate to the season. The timing of these molts thus seems clearly related to the principal changes in environmental temperatures. The heat stress period of June, July, and August coincides with the period of summer pelage.

The antelope squirrels of the area belong to the nominate race *Ammospermophilus leucurus leucurus,* which ranges over the southern Great Basin and the Mohave and Colorado deserts. We follow the well-supported conclusion of Bryant (1945) that on anatomical grounds the antelope ground squirrels merit generic separation from *Citellus.*

Beechey Ground Squirrel. *Citellus beecheyi*

Description. A large ground squirrel (500–600 gram) with moderately bushy tail equal to two-fifths of length of body. Mottled and grizzled pattern of gray and brown above with blackish triangular area between shoulders bordered laterally by whitish patches which extend posteriorly as imperfect lateral stripes; under parts buff. In heavily worn pelage, back is blackish and dark brown.

Range. Pacific coastal region from south-central Washington and western Oregon through California to northwestern Baja California.

Occurrence in Monument. Sparse resident at higher elevations. Recorded from: Black Rock Spring, July 6, Aug. 30; Upper Covington Flat, June 12 [1]; Lower Covington Flat, Apr. 8, May 15 [1], Aug. 24, 25 *, 26 *; Quail Spring, Jan. 20, 25, 27, Feb. 10 *, 11, Apr. 22, May 19–21, July 2, 3, 4 *; Barker's Dam, Feb. 18; Indian Cove, Apr. 8, Sept. 5; Fortynine Palms, Apr. 9 *; Pinyon Wells, Oct. 11 *, 16; Cottonwood Spring, July 2 [1].

Beechey ground squirrels occur on rocky slopes and among the open vegetation of the piñon belt east in the mountains of the Monument to the Pinyon Wells area. There is one record farther east at Cottonwood Spring. Unlike the situation on the coastal slopes of central and southern California, this squirrel does not build up to large numbers in the desert part of its range. Thus we typically see this species only occasionally about camps in the western section. Indeed, only twice have we noted two at a time, this at Lower Covington Flat on April 8 and at Quail Spring on April 22.

An animal collected at Pinyon Wells on October 11 was foraging in a sparse fringe of 6-inch-tall dry oat-like grass bordering a scrub oak patch near the base of a cliff. A squirrel taken at Quail Spring on February 10 had two seeds of catclaw (*Acacia greggii*) in its cheek pouches. The seeds of annual grasses are the principal food source for squirrels of this species, and the sparse and local distribution of such grasses in the desert mountains probably accounts for the sparseness of the populations of squirrels there. A squirrel taken in a steel trap set for carnivores had apparently been attracted by the bait of fish and scent.

The Beechey ground squirrels develop burrows in, or alongside, rocky outcrops and usually are seen running toward such retreats. Several times we have seen them within 50 yards of water and on another occasion saw their tracks in the mud of a wash, but we have no evidence that they seek or need drinking water, although elsewhere the species occasionally does so. Probably water

Fig. 85. Beechey ground squirrel among rocks.

derived and conserved from the metabolism of their seed diet is a large factor in keeping them in water balance. Moist plant food may be taken to some degree. They avoid the warmest part of the summer days by retreating to their burrows, where temperatures are probably moderate and humidity somewhat greater than on the surface in the open. In the morning they may be expected to come out to sun themselves and to dust bathe.

Our limited records for the Monument suggest no clear-cut period of estivation, for we have seen them out, as the dates listed show, in all parts of the year. This includes January and February at Quail Spring, when nighttime temperatures were below freezing. But the ground squirrels appeared here at midday on sunny slopes.

An adult female taken at Fortynine Palms on April 9 had seven embryos, each 24 mm. long. But one taken on August 24 had none. By July well-grown young of the year are in evidence. One such young, taken on July 4 at Quail Spring, had linear dimensions about two-thirds those of an adult.

The Beechey ground squirrels of the Monument belong to the race *Citellus beecheyi parvulus* of the Inyo and western Mohave desert districts and the desert slopes of the bordering mountains. The related rock squirrel (*Citellus variegatus*) does not occur closer than in the Providence Mountains on the eastern side of the Mohave Desert.

Round-tailed Ground Squirrel. *Citellus tereticaudus*

Description. A small (140–150 gram), nearly uniformly colored ground squirrel with reduced, scarcely visible ears and slender tail with short, harsh hair. Above buff or grayish; below whitish.
Range. Southeastern California, southern Nevada, and southwestern Arizona south to northeastern Baja California and in western Sonora south to Yaqui River area.
Occurrence in Monument. Recorded from: Pinto Basin, June 19[1]; 2 mi. S Pinto Wash Well, Mar. 18[1]; 3 mi. W Desert Center, Mar. 18[1].

This ground squirrel is inconspicuous and evidently is also scarce. Parties from Long Beach State College have encountered it twice in the Pinto Basin area in the creosote bush belt. The occurrence of June 19, 1962, was in the basin opposite the sand dunes, an area often visited by naturalists but where this species had not been detected heretofore.

The specimens of round-tailed ground squirrel belong to the race *Citellus tereticaudus tereticaudus,* which occurs in the Mohave and Colorado deserts west of the Colorado River and exclusive of the Coachella Valley.

POCKET GOPHERS

Botta Pocket Gopher. *Thomomys bottae*

Description. A stout, short-tailed, short-eared subterranean rodent, about 8 inches in total length when fully grown. Front feet with heavy digging claws and incisor teeth large and permanently exposed beyond lips; cheek pouches external and fur-lined. Color above pale

brown to light buffy brown in local races, whitish on sides and belly; tail sparsely haired.
Range. Southwestern Oregon, central and northeastern Nevada, northern Utah, and southern Colorado south to the tip of Baja California, northern Sinaloa, northwestern Chihuahua, southwestern Texas, and northern Coahuila.
Occurrence in Monument. Recorded from: Morongo Valley, Feb. 22 *; $1\frac{1}{2}$ mi. W summit Morongo Pass, Feb. 23 *; $1\frac{1}{2}$ mi. E summit Morongo Pass, Feb. 21 *, 22 *; Black Rock Spring, Aug. 30, Sept. 1 *; 4 mi. N Warren's Well, Jan. 1 *; Upper Covington Flat, Mar. 20[1], Aug. 27, Nov. 28[1]; Lower Covington Flat, Mar. 19[1], Aug. 26, 27, Oct. 24[1], Dec. 17[1]; Quail Spring, Feb. 10 *, 13 *, 17 *, Apr. 2 *, Dec. 22 *; Stubby Spring, Sept. 8 *, Feb. 10; Barker's Dam, Feb. 18, Apr. 4 *; Indian Cove, Apr. 7 *; 13 mi. SW Twentynine Palms, Apr. 29, 30 *, May 1 *, 2 *; Twentynine Palms, Feb. 20 *, 21 *; Pinyon Wells, Oct. 10, 14 *, 15 *, 16 *, 18 *; $3\frac{1}{2}$ mi. W Split Rock Tank, Feb. 18 *; White Tanks, Feb. 18; Virginia Dale Mine, Apr. 24; south side Pinto Basin, May 29[1]; Cottonwood Spring, July 8; 3 mi. S Cottonwood Spring, Feb. 19 *; 5 mi. S Cottonwood Spring, Feb. 19.

Pocket gophers occur widely through the Monument at all elevations regardless of vegetation belt. They seem to be absent only from extremely rocky terrain and from sand dune habitat. These latter types of ground condition doubtless make impossible or difficult the development of effective burrow systems and also limit the growth of plants, the roots or stems of which the gophers depend on for food. A reasonably constant supply of vegetable food, workable from the burrows and surfacing holes, must be present. In general, annual vegetation alone, which may develop little or not at all in certain years of poor rain and which in any event is only seasonal, would not seem to suffice to support pocket gophers in the desert.

Gopher sign in the form of surface mounds is seen most frequently near good plant growth. Thus at Pinyon Wells in October, gopher workings were

Fig. 86. Botta pocket gopher at mouth of its burrow.

Fig. 87. Pocket gopher mound, showing dark, slightly moist earth recently thrown out on the surface; Upper Covington Flat, April 9, 1960.

most common at upper levels where there was perennial bunch grass and bushes of *Eriogonum* and *Prunus*. Also here we noted that diggings were in the fine gravel of banks and washes and seldom among the larger rocks. Again at Black Rock Spring, gophers were in the sandy soil among manzanita bushes and in the adjoining washes. At Quail Spring gophers had workings in sandy loam and were feeding in the winter season on the available filaree (*Erodium cicutarium*), taking into their pouches the leafy parts as well as the root sections.

Fresh surface mounds are most often seen in periods following rains. During long dry intervals apparently the gophers stay in their deeper runs 5 to 20 inches under the surface and depend for food on roots intercepted there. New excavations leading to the surface probably can accomplish little in the extremely dry seasons, for usable fresh surface vegetation is scarce and at most the coming to the surface would suffice only to give an outlet for unneeded earth masses.

Fresh mounds, 8 to 14 inches across, almost always are conspicuous by the darker color of the soil in them, which results from their moisture content (fig. 87). Even when there may have been no rain for many weeks, the dampness of the earth thrown out is indicative of the fairly high humidity that must prevail in the burrows where the gophers are living. The brief exposure of the gopher at the surface for purposes of cutting plants seldom brings the animal all the way out of the burrow, and once active foraging ceases at a given location, the burrow is plugged with fresh dirt, deterring the entrance of predators and conserving moisture deeper down. The subterranean existence of the pocket gopher thus is well suited to desert conditions. Temperatures are relatively constant below the surface; at least no extremely high temperatures are experienced. And moistness of ground and air, although variable, must always be high enough to avoid serious threat of dehydration of the animal. The food is moist in large degree, even in the dry season, for the root systems of desert plants and other moisture-storing tissues, advantageous for plants and gophers alike, are located underground. Only the sparseness of vegetable food, on a sustained basis, and the unworkableness of certain soils in the desert terrain seem to impose limitations on rodents of this group.

On April 7 at Indian Cove, a young pocket gopher, about one-half grown, somewhat atypically was coming out of a dry, open hole at the base of a small shrub. No dirt mass had been pushed out here. Apparently the gopher was using an old burrow system of its species or some other rodent. The animal cut and snatched a fairly dry-appearing stalk of annual vegetation and pulled it down into the burrow. Later returning to the surface, it occasionally exposed its whole body in quick grasps for more food plants beyond the entrance. This was at midday in full sunlight, but the gopher was not conspicuous, except through its movement, as its pale buff dorsal color blended well with the gravelly, granitic soil about the burrow entrance.

Essentially no information has come to light concerning the breeding season of pocket gophers in the Monument. No embryos have been noted in February and April, when most of the specimens have been collected. Two pocket gophers were saved that were about one-half the body mass of adults, these on April 7 and September 1, the weight of the latter being recorded as 53 grams.

The racial differentiation of pocket gophers in southeastern California was reviewed in some detail by Chattin (1941), and we follow his appraisal of specimens in the main. Specimens from the Monument taken subsequent to his study do not modify the distributional pattern of the subspecies in any major way. At the west in the Morongo Valley the relatively dark-colored race *Thomomys bottae cabezonae* occurs, and Morongo Pass at its east end marks a sudden transition to the paler race *T. b. mohavensis*. Animals three miles apart on either side of the pass are rather sharply differentiated. The race *mohavensis* extends east then over most of the Monument and south to such stations as Black Rock Spring, Stubby Spring, and Pinyon Wells along the plateaus and the Little San Bernardino Mountains. It ranges on the north boundary through the Twentynine Palms area and presumably to the vicinity of Virginia Dale Mine, where workings have been seen but no specimens collected. In the southeastern section, the even paler race *T. b. rupestris* is known from the Cottonwood Spring area and presumably it occupies adjacent sections of the Monument. In view of real uncertainty concerning the conspecific relation of *Thomomys bottae* and *Thomomys umbrinus* (see Hall and Kelson, 1959), we continue to use the species name *bottae* for races of lowland western gophers.

POCKET MICE AND KANGAROO RATS

Little Pocket Mouse. *Perognathus longimembris*

Other name. Silky pocket mouse.
Description. A small (7–10 gram) mouse, with the short ears and external, fur-lined cheek pouches of the pocket mouse group. Pelage soft and silky, without conspicuous, spiny guard hairs; above buff with dusky tipping; under parts white; tail sparsely furred and weakly bicolored, slightly darker buff above than below.
Range. Southern California, southeastern Oregon, and northwestern Utah south to northern Baja California and western Sonora.
Occurrence in Monument. Fairly common to abundant, varying with the season and year. Recorded from: Black Rock Spring, Sept. 1; 3 mi. S Warren's Well, July 14, Aug. 31 *; Lower Covington Flat, Mar. 19[1], 20[1], June 30[1], Aug. 23 *, 24 *, 25 *, 28 *, 29 *; Quail Spring, Apr. 2 *; Lost Horse Valley, July 20[1], 21[1]; Stubby Spring, June 30[1], Sept. 6 *, 7 *, 8 *, 10 *, 11 *; Key's View, July 16 *; Indian Cove, Apr. 8; 4 mi. N Twentynine Palms, July 23 *, 24 *, 25; 13 mi. SW Twentynine Palms, Apr. 30 *, May 1 *, 2 *; 18 mi. SW Twentynine Palms, May 6 *; Squaw Tank, July 22[1]; Pinto Basin, Apr. 10; Virginia Dale Mine, July 14 *; New Dale, Apr. 17 *; 4 mi. NE Cottonwood Spring, May 22 *; 1½ mi. N Cottonwood Spring, May 10 *.

Little or silky pocket mice are noted for their variable abundance. First, it is apparent that they are active only in the warmer parts of the year, as we have had none enter our traps in the period from late September through early March, and most records are subsequent to mid-April. The animals presumably hibernate or at least stay below the surface in winter, where possibly they have food stored. Second, there are variations from year to year. In 1945 and 1946 we trapped many nights in favorable habitat in spring and summer months and yet obtained only three of these animals. Contrarily, in 1931, 1941, and 1950 we caught them in quantity. In this last year, catches of 8, 9, 10, 11, 11, and 14 per cent were recorded in lines of 60 to 100 traps each in suitable habitat. In 1931, one 8 per cent catch was recorded, and in 1941 a 19 per cent catch.

These mice have been recorded from the lower levels at about 1700 feet up to 5000 in the Little San Bernardino Mountains. This represents such diverse habitats as the creosote bush desert and the piñon woodland. Soil conditions range from sand and light or soft silt to gravel in the washes and on gentler slopes of the mountains. Trap lines where captures have been frequent have been among open juniper and burrobush, beside chrysothamnus and purshia bushes, and among open creosote bushes.

We have only limited evidence on breeding. However, judging from knowledge of the species in other, similar areas, such as the Providence Mountains (Johnson, Bryant, and Miller, 1948:360–361) breeding must occur in large part in April and May. We have one record of 4 embryos 10 mm. long on May 10, 1941, near Cottonwood Spring, and on July 23 and 24, 1931, gray-pelaged juveniles predominated in the catch of this species. Some eight females, fully grown and in adult pelage in late August and early September, were specifically recorded as without embryos.

These small mice, which are nocturnal and develop underground burrow systems, do not face a problem of high temperature in the desert. In fact, from evidence of their quiescence in winter, it is the cold winter night temperatures that seem to be detrimental and which they avoid, either by hibernation or by restriction of above-surface activity. Underground in winter, freezing temperatures would be avoided even a few inches below the surface. The water problem for the species probably is solved as in the related kangaroo rats (see p. 310) in which it is known that special physiologic mechanisms for water conservation exist so that the metabolic water derived from fairly dry seed food is sufficient to keep them in water balance indefinitely.

On July 16 at Key's View we found a fully grown pocket mouse of this species in the stomach of a speckled rattlesnake (*Crotalus mitchelli*).

All the little pocket mice of the Monument seem to belong to the widespread race of the Mohave Desert, *Perognathus longimembris longimembris*. Even in the Cottonwood Spring area, specimens seem to show no approach to the pale, small race of the lowlands of the Coachella Valley immediately to the south.

Long-tailed Pocket Mouse. *Perognathus formosus*

Description. A large (16–24 gram) pocket mouse. Above grayish brown, grizzled, the long hairs of rump not spiny and stiff; under parts cream to white, fairly sharply set off from back. Tail long, with long erect blackish hairs on dorsal surface of distal half which form a fairly conspicuous terminal tuft.

Range. Northwestern, central, and northeastern Nevada and northwestern Utah south through eastern California, northwestern Arizona, and northeastern Baja California to central Baja California.

Occurrence in Monument. Fairly common in a few localities. Recorded from: mouth of Fortynine Palms Canyon, July 24 *, 25 *; 1 mi. S Twentynine Palms, Apr. 26 *; south side Pinto Basin, Mar. 26 [1], Oct. 22 [1]; north side Eagle Mountain, Oct. 20 *; 3 mi. S Cottonwood Spring, Feb. 19 *.

This large gray pocket mouse has been trapped only at the five places listed. In each instance in our experience, the animals were in or near rocky washes of canyon mouths or of bajadas. The soil conditions at the trap sites have been noted as hard-packed gravel, stony bench, and well-packed, stable alluvium above adjoining wash bottoms. Vegetation in these areas is generally sparse, creosote bushes and cactuses often being present.

The gravel or rocky ground favored by this mouse is also used, although not exclusively, by the more common *Perognathus fallax*, and *fallax* has been taken in the same trap lines on the north side of Eagle Mountain and in the mouth of Fortynine Palms canyon, where its numbers were greater than those of *formosus*. At the latter station 55 traps yielded 5 *formosus* and 7 *fallax*, and on the next night 115 traps caught 6 *formosus* and 9 *fallax*. Apparently this species is active in winter, unlike the little pocket mouse.

None of the long-tailed pocket mice collected had embryos.

In racial characters the long-tailed pocket mice of this area fall with those of the Mohave Desert, to which the name *Perognathus formosus mohavensis* (Huey, 1938:35–36) has been applied. The animals from the northern border of the Monument seem consistently to show the dark color and size prevalent in this race, and the same holds for two taken on the north side of Eagle Mountain. However, of three animals from Cottonwood Spring, one is dark, another pale and very small (possibly a runt), and the third of somewhat intermediate color. This inadequate sample from Cottonwood Spring suggests intergradation with *Perognathus formosus mesembrinus* of the Coachella Valley, a form characteristic of the southern border of this valley and of the desert slopes of San Diego County, in which latter region its pale coloration is best developed.

San Diego Pocket Mouse. *Perognathus fallax*

Description. A large (15–25 gram) pocket mouse. Above grayish brown, grizzled, the long hairs of rump stiff and spiny but the flanks and anterior back soft and moderately silky; under parts white, a cream or light buff border zone on flanks usually fairly well defined. Tail long, with short hair basally, the distal half with elongate dusky hairs (fig. 88).
Range. Southwestern California and western desert border of California south through western Baja California to Turtle Bay.
Occurrence in Monument. Common and widespread, extending east to Cottonwood Spring area. Recorded from: Little Morongo Canyon, Apr. 2 *; Black Rock Spring, Sept. 1 *, 4 *;

Fig. 88. Little pocket mouse (left) and San Diego pocket mouse (right).

Upper Covington Flat, Apr. 4 [1]; Lower Covington Flat, June 27 [1], 28 [1], Aug. 24 *, 25 *, 29 *, Oct. 10 [1], 11 [1], 24 [1], 25 [1]; Smithwater Canyon, June 27 [1], 28 [1], Aug. 28 *; Quail Spring, Jan. 26 *, Apr. 2 *, July 3 *, 4 *; Lost Horse Valley, July 19 [1], 20 [1], 21 [1]; Stubby Spring, July 22 [1], Sept. 6 *, 8 *, 9 *, 11 *; Indian Cove, Apr. 23 [1], May 13 *; mouth Fortynine Palms Canyon, July 24 *, 25 *; Twentynine Palms, July 13 *; 13 mi. SW Twentynine Palms, Apr. 30 *, May 2 *; Squaw Tank, July 22 [1]; Pinyon Wells, Oct. 13 *; Pinto Basin, Oct. 22 [1]; N side Eagle Mountain, Oct. 19 *, 20 *, 21 *; 4 mi. NE Cottonwood Spring, May 22 *; Cottonwood Spring, May 15 *, 17 *, 18 *, July 8 *, 9 *, 11 [1], 12 [1], Sept. 16 [1], Oct. 22.

This is the common, large gray pocket mouse of the Monument, a species that centers in the arid coastal and desert-border areas to the west but which ranges east through the mountains of the Monument as far as Twentynine Palms and the Eagle Mountain–Cottonwood Spring district. We have not found it farther east nor in the flats far from rocky or gravelly ground. The greatest numbers seem to be at the middle and upper levels, in the yucca "forests" and in the open brushlands near or in the piñon–juniper belt, but independently, we judge, of these tree species. Although often in or near rocks or coarse gravel, this species, unlike the other larger pocket mice, is not dependent on such terrain and ranges out into silt and fine sand although not into dunes.

Our notations on locations of capture show the following representative details: At Twentynine Palms on July 13 near the oasis, 3 were taken in 118 traps near the mesquites among low scattered bushes where fine sand and silt prevailed on the alluvial flats. In Little Morongo Canyon on April 2, 3 were taken in a side canyon on steep gravelly slopes and among rock rubble. On July 24 and 25 in the mouth of Fortynine Palms Canyon, this species was common in the rocky wash (for proportional catch, see p. 303). At a point 13 miles southwest of Twentynine Palms, 3 were taken under *Prunus fasciculata* where there were freshly dug burrows. On August 24 at Lower Covington Flat, captures were made on fine gravel and sand in the vicinity of small piñons, chrysothamnus bushes, junipers, purshia, and yuccas. On September 4 on gravel and hard ground on slopes among piñons and oaks at Black Rock Spring, 5 were taken in 50 traps. At Cottonwood Spring, 4 were taken on ridges of bare granitic gravel among well-spaced bushes. An exceptional capture was made in the willow thickets at the oasis in Smithwater Canyon below Lower Covington Flat, on August 28, but the surrounding wash and hill slope consisted of typical dry, open habitat for this species. Also unusual was capture of one of this species inside a cabin at Quail Spring on January 26.

These pocket mice often have dry seeds in their external cheek pouches when caught. In addition, one caught on April 30 had green grass seeds, and another caught on May 15 had two large nutlike seeds over $\frac{1}{2}$ cm. in diameter in the pouches.

This common pocket mouse is presumed to have water-conserving mechanisms like those of other heteromyid rodents such as *Dipodomys merriami* (see p. 310) and *Perognathus baileyi* (Schmidt-Nielsen, *et al.*, 1948), the latter a member of the same subgenus as *P. fallax*. In these heteromyids dry food can be used indefinitely, and water balance is maintained on it without drinking water through conservation of metabolic water. *Perognathus baileyi*, for example, has shown especially high capacity to concentrate the urine and excrete salt with passage of very small amounts of water.

Temperatures in winter in the desert do not preclude the activity of a species such as *P. fallax*, although they may influence the amount of movement on the surface during cold spells. We have taken a few of this species in October and January, and the other large kinds of pocket mice similarly have been taken in winter. In summer, *fallax* avoids the severe high temperatures of the daytime by the natural insulation of its subsurface burrows.

The San Diego pocket mice have been caught in fairly large numbers, and yet we have only one record of a pregnant female, on June 27. Adult females taken in the following months have been specifically recorded as without embryos: May, 2; July, 5; August, 1; and September, 1. Juveniles in gray pelage have been captured on May 18, July 3, July 4, 13, 24, and 25; those on the last two dates had partly molted into postjuvenal pelage. Because of the appearance of these young, it seems probable that breeding occurs chiefly in March and April.

The mice of this species in the Monument belong to the pale, eastern or desert race of the species, *Perognathus fallax pallidus*, that occurs on the western borders of both the Mohave and the Colorado deserts.

Spiny Pocket Mouse. *Perognathus spinatus*

Description. A large (15–22 gram) pocket mouse. Above grizzled gray, the flanks and rump with elongate stiff, spiny hairs; pelage of flanks and entire back coarse, not silky; under parts cream to white, fairly sharply set off from back. Tail long with very short, sparse hair basally, the distal half with elongate dusky hairs.

Range. Deserts of southeastern California south throughout northeastern, east-central, and southern Baja California, including adjacent offshore islands.

Occurrence in Monument. Encountered in only a few localities at low elevations, once commonly. Recorded from: Fortynine Palms, Apr. 9 *; mouth of Fortynine Palms Canyon, July 24 *, 25 *; Cottonwood Spring, July 8 *; 2 and 3 mi. S Cottonwood Spring, Jan. 28 [1], Feb. 19 *.

This species of pocket mouse has been taken only in two general areas, and on only one occasion was it found commonly. In July, 1931, it and the two other species of gray pocket mice were taken in numbers in the rocky wash of the mouth of Fortynine Palms Canyon. Two *spinatus* were taken in 55 traps on July 24, and 10 in 115 traps on July 25. On April 9 a single animal was caught up in the Canyon, at 3000 feet elevation, in a clump of sedge at the oasis, but typical rocky desert habitat was close by.

Apparently this species occurs only in the central and eastern sections of the Monument and reaches there the northern limits of its range. To the west and east it reaches Palm Springs and the vicinity of Needles. It is primarily a low desert type inhabiting the rocky mesas and hills and tolerating the extensive hot and dry deserts of Baja California and the lower Colorado River valley.

The series of specimens taken in Fortynine Palms Canyon contains four animals molting from the fine gray pelage of the juvenal stage to adult coarse pelage. One of these immatures weighed only 9.9 grams.

The spiny pocket mice of the Monument correspond well with the race *Perognathus spinatus spinatus* of the Colorado River valley rather than with the northwestern race that reaches the Palm Springs area.

Chisel-toothed Kangaroo Rat. *Dipodomys microps*

Description. A middle-sized (50–65 gram) species of kangaroo rat with 5 toes on the hind foot. Markings as in other kangaroo rats (see p. 307). The lower incisor teeth broad and flat anteriorly, not rounded or awl-like. Coloration dorsally slightly darker brown, less pinkish buff than in *Dipodomys merriami,* and dark tail stripe fully developed at base of tail; soles of hind feet prominently black.
Range. Southeastern Oregon and northwestern Utah south to southeastern California and northwestern Arizona.
Occurrence in Monument. Rare. Taken twice at only one locality. Recorded from: Stubby Spring, Sept. 6 *, 8 *.

This species of kangaroo rat was found most unexpectedly near Stubby Spring, where it is far south of its previously known range in northern and western San Bernardino County, California, at Clark Mountain and Victorville. The fact that it has not been trapped in any other part of the Monument, in areas that often seem similar, suggests that its numbers are small or much localized. Probably we have not yet determined fully the features of the environment which it requires here at its southern periphery. In general it is a species typical of sagebrush and shadscale areas of the Upper Sonoran levels of the Great Basin. Probably the low levels of the Monument are unsuitable for it.

The animals taken near Stubby Spring were trapped in a flat approximately on the divide at 4750 feet in the low bush and juniper association. The soil was whitish to reddish sandy gravel. The bushes were 2 to 3 feet high with open alleyways of 2 to 20 feet. After the first animal was captured on September 6, the particular flat was heavily trapped (60 to 120 traps) on several succeeding nights and only one more was obtained. Both were adult females with no evidence of recent breeding.

The racial affinities of the scattered southernmost samples of this species are not readily determined because of lack of series of adults. Hall and Dale (1939) in their review of this species refer the one adult from Victorville to *Dipodomys microps microps,* a race typical of the Owens Valley of Inyo County, but point out that it is a specimen with broken skull. The pale color and small size of this animal do justify such a tentative assignment.

The two females from the Monument, on the other hand, seem best referred, again tentatively, to *D. m. occidentalis,* the closest point of capture heretofore being Clark Mountain on the north side of the Mohave Desert. This referral to *occidentalis* is based on the skull dimensions and shape, which is less narrow than in *D. m. microps.* Examination of Hall and Dale's table of measurements convinces us that diagnostic differences in measurements for these races are with one exception poorly founded, and even most of the average differences are statistically invalid. However, greatest breadth of the skull of adult females ranges from 22.5 to 23.7 mm. in 12 *occidentalis* and from 21.3 to 22.3 in 6 *microps,* suggesting little or no overlap. The two specimens from the Monument are 23.0 and 23.6, thus well within the range of *occidentalis.* Fresh pelages of *microps* are paler than those of *occidentalis. D. m. levipes* of the Panamint Mountains is even larger and generally darker than *occidentalis.* The two specimens from the Monument are somewhat paler than most *occidentalis* but darker than *D. m. microps.*

Merriam Kangaroo Rat. *Dipodomys merriami*

Description. A small (35–47 gram) species of kangaroo rat, with 4 toes only on hind foot. Markings typical of group, with buff above set off sharply against white under parts and with white stripe on each flank extending to base of tail above dark area of hind leg; face marked with white patches on upper lips bordered above with dusky; also white spots above each eye and behind each ear. Tail greatly elongate, with dusky above and below, except at base, and white lateral stripes, the tip with a conspicuous dusky tuft. Ears small and rounded.

Range. Northwestern and southern Nevada, southwestern Utah, northwestern and central Arizona, central New Mexico, and western Texas south through southern California, Baja California, Sonora, and northern and eastern Durango to Aguascalientes and central San Luis Potosí.

Occurrence in Monument. Abundant and widespread. Recorded from: Little Morongo Canyon, Apr. 2 *; Black Rock Spring, July 6, Sept. 1 *, 4; 4 mi. N to 3 mi. S Warren's Well, Jan. 1 *, Aug. 31 *, Dec. 28; Upper Covington Flat, Apr. 4[1], June 27[1]; Lower Covington Flat, June 29[1], 30[1], July 26[1], Aug. 29 *, Oct. 10[1], 24[1], 25[1], Nov. 7[1]; Quail Spring, Jan. 22 *, Apr. 2 *, May 19 *, 20 *, July 2 *, 4, 5 *, Dec. 19 *, 20 *, 21 *, 22 *; Stubby Spring, Sept. 7 *, 8 *, 11; Lost Horse Valley, July 19[1], 20[1], Oct. 21 *; Key's View, July 16 *; Indian Cove, May 13 *; mouth of Fortynine Palms Canyon, July 24; 4 mi. N Twentynine Palms, July 23 *; 1 mi. S Twentynine Palms, Apr. 26 *, July 13 *, 23, 24, 26; 13 mi. SW Twentynine Palms, May 2 *; 1½ mi. NE Pinyon Wells, Oct. 13 *; 1½ mi. W White Tanks, Oct. 18; Pinto Basin, Apr. 6, 11, June 30[1], July 7 *, 10[1], 11, 15 *, Oct. 22[1]; Virginia Dale Mine, Apr. 24 *, July 14 *; north side Eagle Mountain, Feb. 19[1], Oct. 19 *, 20, 21 *; Cottonwood Spring (1½ mi. N to 3 mi. S), Feb. 19 *, Apr. 5, May 10 *, 17 *, July 11[1], 12[1], Oct. 22 *, 23 *.

The Merriam kangaroo rat is the most widespread and abundant species of the kangaroo rat–pocket mouse group in the Monument, and it serves as the prime example among the rodents of successful adaptation and adjustment to desert conditions. Much is known about its solution of the moisture problem through experimental testing, so that this species, as a well-known type, permits making some assumptions about how several other kinds of rodents survive in the desert.

The species shows a wide tolerance for conditions of soil and vegetation, and it ranges from the lowest parts of the Monument, as in Pinto Basin (1700 feet), up to at least 5000 feet at Key's View. Dense vegetation, such as chaparral, and cliffs and slopes nearly solidly covered with rocks are not tolerated, but vegetation that is sparse with openings three feet or more between shrubs is entirely acceptable, and much scattered rock may be present. Soil on which it has been caught ranges from loose sand of large dunes to coarse gravel and to hard-packed, gravelly slopes of well-eroded, compacted ridges. Always present is some diggable soil, consisting usually of only moderately firm gravels, alluvial slopes, and sandy lenses in washes, large sand dunes also may be occupied.

The better catches concerning which data on soil and plant associations were noted are as follows: On October 13 on gently sloping creosote bush flats below Pinyon Wells, 8 kangaroo rats were taken in 22 traps on gravelly soil. Another 8 were taken in 28 traps nearby on similar surface but free of creosote bushes. Both areas had *Ephedra* and *Beleporone* plants, but among the creosote bushes there was more dry grass. At Twentynine Palms on July 13 on a clear, moonlit night, with little wind, 6 were taken in 58 traps set in open brush south of the mesquite thickets; the soil was well-packed, fine sand and silt. On July 14 at Virginia Dale Mine, 5 were caught in 60 traps set in almost barren, ripple-marked sand in which a person would sink an inch and a half in walking over the surface, and 5 were captured in 39 traps set in cross washes, some in fine sand and some on gravel near burrows. In Pinto Basin on flats of loose sand and gravel in the creosote bush association, 10 were caught in 60 traps on July 15, and in the dune area there, 10 were taken in 30 traps on July 7.

This abundant species, active on nights throughout the year, must afford a basic, dependable food supply for carnivores. We have definite evidence that kit foxes eat kangaroo rats, and on April 11 a badger taken in Pinto Basin had eaten a young Merriam kangaroo rat. This rat had tiny milk incisors and a hind foot 27 mm. long; judged from Chew and Butterworth's (1959) data on growth of this species, it was only 12 days old, which is an age before that when they normally leave the underground nests, although young can hop at this point in development. Probably the badger dug out this young one from a nest or nesting burrow. On October 18 at White Tanks, a rat of this species was found in the digestive tract of a gopher snake. Almost certainly coyotes capture many kangaroo rats, and probably gray foxes and sidewinders take them.

The breeding season extends from March until August. Pregnant females have been trapped on April 2 (2 individuals), May 2, May 20, and July 6 and 13. Embryo counts were: 2, 2, 5, 2, 2, and 2, respectively. The young of April 2 must have been conceived in March, and the young one taken by the badger on April 11 was born about March 30. In July, 22 adult females

Fig. 89. Desert kangaroo rat (above) and Merriam kangaroo rat (below).

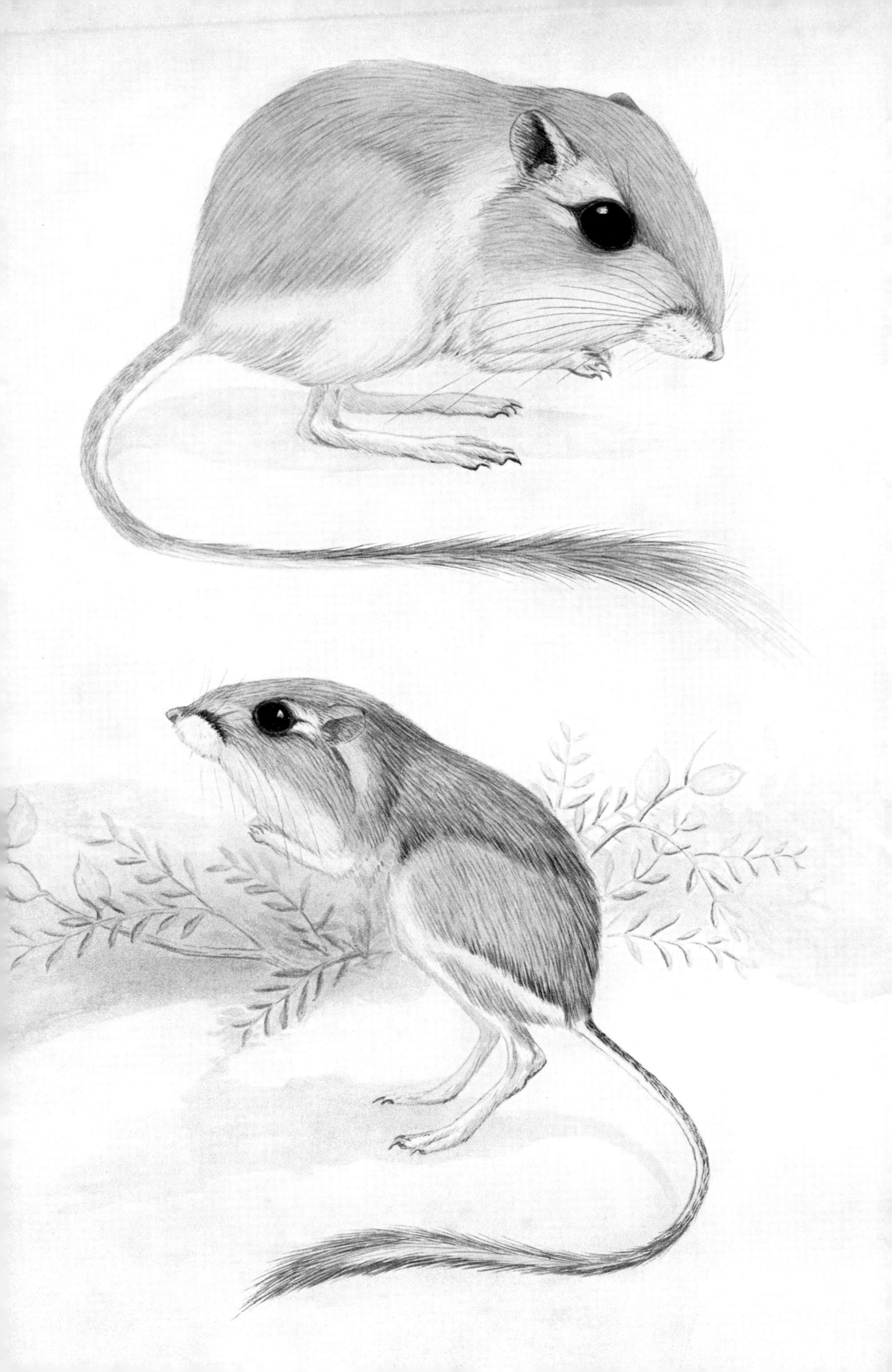

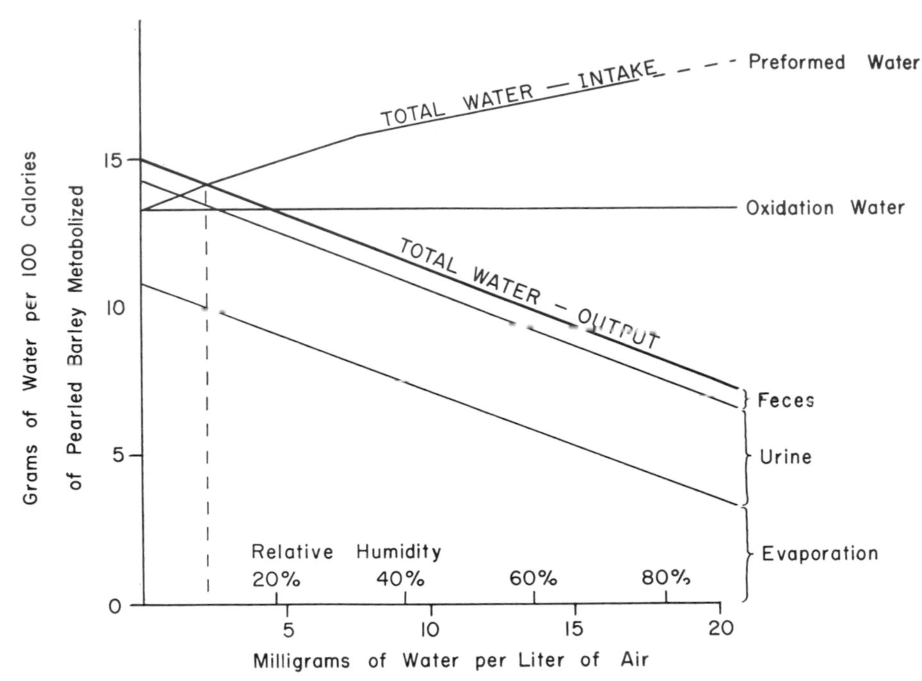

Fig. 90. Diagram of the water balance of the Merriam kangaroo rat, showing sources and use. For explanation see text. Figure redrafted from Schmidt-Nielsen and Schmidt-Nielsen (1952) through the courtesy of these authors.

on dates from July 2 to 16 were specifically noted to be without embryos and only one was obviously lactating. Therefore a large proportion of the females of the population must have stopped breeding by that month. However, 11 males in this same period had scrotal testes, ranging from 3 to 14 mm. long. One juvenile taken on September 7 had a hind foot 33 mm. long, indicating a birth in August.

Merriam kangaroo rats maintain water balance on dry seed food without resort to drinking water. The metabolic water derived from use of this food, namely the water resulting from oxidation, is carefully conserved in several different ways, physiologic and behavioral. The water balance is shown graphically by the Schmidt-Nielsens (1952) in their figure reproduced here (fig. 90). This represents what happens to the water derived from 100 calories of pearl barley (25-gram dry weight) that the rats were fed at various levels of atmospheric humidity. When the atmospheric humidity is high, the food absorbs moisture and the loss from lung and surface evaporation of the animal drops, as does that from the urine and feces. Total water output and total intake are in balance when the air has 2 to 3 mg. of water per liter. Absolute humidity of this sort is critical rather than relative humidity, since evaporation rate is a function of the former at the constantly warm respiratory surfaces. When the air is extremely dry, below 2 mg., the animals lose weight and are not keeping up the necessary water supply. But the Schmidt-

Nielsens find that the characteristically plugged burrows, in which the rats spend a large part of the 24-hour period, have an atmosperic humidity of 7 to 15 mg., so that they can gain water under those conditions. The nocturnal activity of course has an advantage in permitting the animals to avoid high surface temperatures and thus also avoid heat stress in which they would have to use water for heat regulation. Moreover, they are probably less readily captured by carnivores at night than they would be by day.

When the kangaroo rat's body temperature approaches lethal value at 105° F., it then and only then shows an emergency cooling mechanism. This consists of a copious flow of saliva, which is spread about the lower jaw and throat, and which by evaporation can for a short time keep the body temperature below that of the surroundings. The species has no sweat glands for cooling.

The key to the achievement of water balance at low atmospheric humidity consists then of water conservation. Behaviorally the animals can and do limit the time outside their burrows, exposing themselves for only short periods, if necessary, to humidity levels below 2 mg. They keep the humidity somewhat elevated in the burrows by plugging them, and they avoid heat stress and water-depleting, cooling devices until close to an emergency level. They sleep with their noses buried in the fur so that a small pocket of moist air accumulates around the nose and little moisture is lost (Lidicker, 1960). Physiologically the low balance point is achieved by water-conservation measures of four types. (1) The feces are dried before extrusion so that they contain only about a fifth of the water normally lost by this means in the laboratory rat (*Rattus norvegicus*). (2) The urine is about twice as concentrated before discharge as that of the laboratory rat; to make a different comparison, in this kangaroo rat the urea concentration of the urine can be as high as 24 per cent, whereas in man it can be concentrated only to the 6 per cent level. (3) The loss of water by evaporation, which in this species occurs principally from the surface of the lungs and the respiratory passages, is reduced in completely dry air to roughly 60 per cent of that lost by the laboratory rat and to about the same proportion compared with man's loss from the lungs alone (thus not counting loss from human perspiration). The mechanism for this conservation of water evaporated from the lungs is not fully clear. Chew (1961) has pointed out that the linings of the nasal passages in kangaroo rats are maintained at the surprisingly low level of 75° F. and that these cool surfaces would condense and thus salvage considerable moisture from the warmer, saturated air coming from the lungs. (4) The animals are able especially to lower their metabolic rate under arid conditions, thus reducing the rate of respiratory water loss, as Howell and Gersh (1935) have shown.

According to the Schmidt-Nielsens (1953), the temperatures that kangaroo rats encounter in their burrows in the Arizona desert in summer range from

75° to 88° F., and the nighttime temperatures aboveground from 60 to 75°. Thus high temperatures of critical levels, in the high 90's, are not encountered. Actually low temperatures in winter aboveground are probably more troublesome to these rats than high temperatures, as the animals often come out when the air is down to 32° F. Then the metabolic rate would have to be increased to offset loss of body heat. This, however, would have an incidental affect of increasing the supply of metabolic water. A rich seed supply would provide adequate food for such a situation.

We may think, then, of kangaroo rats emerging at night to forage on abundant seed supplies, low in moisture though seldom fully dry. In so doing they meet moderate to cool temperatures and move freely and at times rapidly over the sandy or loose gravelly surfaces between well-spaced plants. Quick travel and retreat from danger are greatly facilitated by their bipedal hopping on elongated hind legs, the large feet when necessary serving as sand shoes. The small forefeet, when not used in slow, quadripedal creeping, aid in gathering seeds and in placing them in the external cheek pouches to be carried underground, in part for storage. Their nightly forays, not by any means entailing nightlong exposure, may often be at low humidities, yet at those above the low critical level for the water balance of the species. But should the humidity fall below the critical level, a retreat to the burrow would enable the animals to regain water balance.

The Merriam kangaroo rats of the Monument belong to the race *Dipodomys merriami merriami* (*D. m. simiolus* is a synonym), according to the latest monographic treatment of this group by Lidicker (1960). However, in the area of the Monument, Lidicker shows that there is much variability. "A few specimens are noticeably redder" than those to the north; that is, there is a reduction of the dusky hair tips and a generally paler coloration. This represents a trend toward the race *arenivagus* of the southwest side of the Salton Sea. Lidicker postulates that these two races were once separated by the flooding of the Salton Sea–Coachella Valley basin. But "with the recent breakdown of this barrier, the two . . . have intergraded broadly and some influences of *arenivagus* can be detected throughout the Coachella Valley and in the Mohave Desert as far north as the southern boundary of San Bernardino County" (p. 191).

Desert Kangaroo Rat. *Dipodomys deserti*

Description. A large (80—130 gram), pale kangaroo rat with 4 toes on the hind foot. Pattern of body markings characteristic of the group (see p. 307), but reduced in contrast by reason of paleness of ochraceous buff of dorsal area. Tail white-tipped, the under surface of tail without dark stripe (fig. 89).

Range. Western and southern Nevada south through southeastern California, southwestern Arizona, and northeastern Baja California to northwestern Sonora.

Occurrence in Monument. Locally fairly common. Recorded from: 4 mi. N Twentynine Palms, July 22, 23 *; 16 mi. E Twentynine Palms, Feb. 5 *; Pinto Basin, July 7 *, 11, 15 *, Dec. 8 [1]; 1 mi. S Virginia Dale Mine, July 14 *; Pinto Wash Well, Nov. 24 [1].

Fig. 91. Typical open burrow of a desert kangaroo rat in Pinto Basin near the base of a smoke tree.

This conspicuous, sand-dwelling species is restricted to the low parts of the Monument where deep, soft soil is present. Thus along the north edge of the Monument it may be expected in the Twentynine Palms basin, and it is also present in Pinto Basin. Probably there is not enough deep, open sand-dune country for it elsewhere at the middle or higher levels.

The large burrows of this giant kangaroo rat are conspicuous and characteristic. The open mouths of the burrows, 4 inches in diameter, frequently run into a bank or mound (fig. 91) of sand held firmly as a small hillock by a creosote bush or clump of other sparse vegetation characteristic of the habitat. The burrows are often clustered, each animal having several entryways, although only one or two seem to be in active use at a time and show trailways leading to them. The active burrows do not show plugs at the entryways, although deeper in there may be such.

In Pinto Basin in early April, 1951, the "colonies" of these rats about the large sand dune seemed to be deserted. Burrows were not active, being caved in and showing cobwebs and debris blocking the entrances. The preceding four dry years had greatly reduced the growth of small annual seed-producing plants, and we thought it likely that a die-out of the desert kangaroo rats had taken place. Such must occur occasionally in unfavorable climatic cycles in these extreme desert areas. In 1960 occupied burrows were again seen at this locality.

This species usually could not be caught in numbers even when special sets were made at the mouths of their burrows. The population is probably not so dense as one might casually infer from the multiple burrow systems. In 60 traps set in Pinto Basin on July 15 in suitable habitat, only two of this species were taken while 10 *Dipodomys merriami* were caught. Near Virginia Dale Mine, only one desert kangaroo rat was caught in 75 traps. The best percentage was 3 in 25 traps at Pinto Basin on July 11.

Near Twentynine Palms desert kangaroo rats have been reported a number of times about dwellings, where they learn to come for food. On July 22, 1931, Benson watched them coming for bread thrown out for them.

None of the rats taken by us showed any sign of breeding.

The animals of this section of the desert belong to the nominate race *Dipodomys deserti deserti,* described from near Hesperia in the Mohave River valley.

WHITE-FOOTED MICE AND ALLIES

Western Harvest Mouse. *Reithrodontomys megalotis*

Description. A very small (8—13 gram) mouse, without external cheek pouches, the tail slender and sparsely haired, and with deep longitudinal groove present on each upper incisor. Above mixed dark brown and light buff, grading to buff on sides; under parts and feet white; reddish hairs present on ears. Tail sharply bicolored, brown above and whitish below.

Range. Southern interior British Columbia, southern Alberta, central North Dakota, and southwestern Wisconsin south to northern Baja California, northern Sonora, the Mexican Plateau to Oaxaca, western Texas, Kansas, and northern Arkansas.

Occurrence in Monument. Rare. Recorded from: Morongo Valley, Feb. 22 *; Little Morongo Canyon, Apr. 3 *.

This small mouse is normally found in grassy or sedge areas in the vicinity of water in the desert regions. Most of the Joshua Tree Monument is unsuited for it because of the lack of such conditions or the isolation of those small grassy areas that do occur. We have taken it in Little Morongo Canyon, where there is permanent water. One male was trapped there on April 3 among scattered sedge growing in a dense stand of baccharis and chrysothamnus on soil white with alkali on a bench above but near the stream. Miss Alexander trapped two in grass at the edge of a tule patch in Morongo Valley.

The specimens belong to the interior, somewhat pale desert race of the species, *Reithrodontomys megalotis megalotis.*

Canyon Mouse. *Peromyscus crinitus*

Description. A member of the white-footed mouse group with tail 90 per cent or more that of head and body. Dorsal stripe of tail present but not sharply defined. Fur relatively long and lax. Pectoral mammary glands absent. Color above pale ochraceous gray or buff with dusky-tipped hairs. Under parts white.

Range. Central and eastern Oregon, southwestern Idaho, northern Utah, and western Colorado south through eastern California and Great Basin to northeastern Baja California, northwestern Sonora, western and northern Arizona, and northwestern New Mexico.

Occurrence in Monument. Common in mountainous terrain. Recorded from: 1 to 4 mi. N Warren's Well, Jan. 1 *, Dec. 28 *; Lower Covington Flat, June 28 [1], Oct. 8 [1], 10 [1], 11 [1], 24 [1]; Quail Spring, Jan. 23 *, Apr. 2 *, May 20 *, Dec. 22 *; Lost Horse Valley, July 19 [1], 21 [1]; Stubby Spring, Sept. 6 *; Indian Cove, Apr. 23 [1]; mouth Fortynine Palms Canyon, July 25 *; 13 mi. SW Twentynine Palms, Apr. 30 *, May 1 *, 2 *; Squaw Tank, July 22 [1]; Virginia Dale Mine, Apr. 23 *; N side Eagle Mountain, Oct. 20 *; Cottonwood Spring, July 9 *; 3 mi. S Cottonwood Spring, Feb. 19 *.

This species of *Peromyscus* is a rock-dwelling form that has been taken in all instances in rocky arroyos or on rocky canyon slopes or near enough to such places that the animal could readily have ranged out from a home base in the rocks. We have never caught it out on the flats in creosote bush habitat, where *Peromyscus eremicus*, a very similar-appearing species, is common. Elevations of locations range from 1900 feet to 4800 feet. Miss Alexander's record of a trap line on the plateau 13 miles southwest of Twentynine Palms, at 4500 feet, describes a setting typical of that in which this mouse occurs. Here in the vicinity of a camp among huge granite boulders, traps were set under junipers and shrubs on a rocky slope near the base of some steep rock outcrops; 46 traps here on April 30 yielded 6 *crinitus* and no other members of the genus. Six more were taken in similar situations on May 1 and May 2. Similarly at Quail Spring on May 20 one was trapped on a rocky outcrop near the spring.

Most breeding seems to occur in April and May. Five of the 8 females taken in the period from April 30 to May 2 were pregnant, and the other 3 were lactating. Embryo counts were: April 30, 5 and 4; May 1, 6; May 2, 5 and 3. Embryos ranged from 8 to 25 mm. long. Also, one gray-pelaged juvenile was trapped on May 1. Breeding apparently does occur later, also, as two males on July 9 had scrotal testes 8 mm. in length. No evidence of breeding was noted in August, September, and October, nor in winter.

The canyon mice of the Mohave and Colorado deserts of California belong to the race *Peromyscus crinitus stephensi*.

Cactus Mouse. *Peromyscus eremicus*

Description. A fairly large white-footed mouse (adult length normally 180 mm. or over), with long, sparsely haired tail (distinctly longer than head and body). Tail without obvious dorsal stripe, upper surface only moderately darker than lower surface. Fur normal, not long or lax. Pectoral mammary glands absent. Upper body surface pale ochraceous gray or buff with dusky-tipped hairs; under parts white (fig. 92).

Range. Southern California, southern Nevada, southwestern Utah, central Arizona, southern New Mexico, and western Texas south to Cape district of Baja California, central Sinaloa, central Durango, northern and eastern Zacatecas, and southern San Luis Potosí.

Occurrence in Monument. Common. Recorded from: Little Morongo Canyon, Apr. 1 *, 3 *; 3 mi. S Warren's Well, Aug. 31 *; Black Rock Spring, Aug. 30 *, Sept. 1 *; Lower Covington Flat, Jan. 8 [1], 29 [1], Feb. 19 [1], May 1 [1], 7 [1], June 28 [1], Aug. 23 *, 24, 28 *, Oct. 8 [1], 9 [1], Nov. 7 [1]; Quail Spring, Jan. 21 *, 22 *, 23 *, Apr. 2 *, May 19 *, 20 *, July 2 *; Lost Horse Valley, July 21 [1]; Stubby Spring, Sept. 11 *; Indian Cove, Apr. 23 [1], May 13 *; Squaw Tank, July 22 [1]; Pinyon Wells, Oct. 9 *, 13 *; White Tanks, Feb. 19 [1], 25 [1]; north side Eagle Mountain, Oct. 19 *, 20 *; 4 mi. NE Cottonwood Spring, May 22 *; Cottonwood Spring, May 17 *, July 7 *, 8 *, 22 *, 23 *, Oct. 22 [1].

Cactus mice may be expected in all parts of the Monument except in the chaparral and denser piñons and on very rocky slopes. We have not taken them far out in the large flats or basins, but they may nevertheless occur there. The greatest numbers are found on the mesas and washes along the bases of the mountains, where yuccas and cactuses are common. Since occasionally they are caught near rocks, rocky ground is apparently to a degree invaded or tolerated but is not sought.

At Lower Covington Flat on August 23, two were taken among 60 traps along a sandy draw with scattered borders of desert willow and chrysothamnus. At the north base of Eagle Mountain, 3300 feet, on October 19, 30 traps on granite gravel, but among plentiful *Yucca mohavensis*, yielded 3 cactus mice. Four miles northeast of Cottonwood Spring on May 22, 4 were caught under junipers in a sandy wash and on a slope near a divide between the mountains and outlying hills. One of the best catches was on October 9, when 6 were taken in 97 traps in the wash below Pinyon Wells. Here desert willow, cholla, and small yuccas are common along the lower slopes and in the open canyon mouth.

Cactus mice recently have been studied by Murie (1961) with a view to ascertaining their metabolic characteristics and tolerance of high temperatures. He found that this species when compared with *Peromyscus maniculatus* has a lower metabolic rate (of an order of 10 to 20 per cent). This, it seems to us, is possibly an adaptation to the scarcer food resources prevailing at times in the characteristically desert habitat of the species. Also, this mouse tolerates higher temperatures before putting into play the emergency cooling behavior of wetting the face and throat with saliva. This emergency reaction of course uses up water reserves rapidly. *Peromyscus eremicus* generally is a more placid animal and avoids heavy activity that takes energy and tends to elevate body temperature.

White-footed mice as a group do not burrow but find daytime retreats under vegetation or in crannies and holes developed by other species. Thus they are somewhat more subject to high daytime temperature than the burrowing kangaroo rats and pocket mice. The evaporative water loss of the canyon mouse, and presumably of this species also, is low, at the same level as in the Merriam kangaroo rat (Schmidt-Nielsens, 1952), but apparently the conservation of water through the urine and feces is not so great as in kangaroo rats. Lindeborg (1952) tested *P. eremicus* in comparison with *maniculatus* and found that the cactus mouse in captivity normally drinks 1.8 cc. of water daily whereas the deer mouse uses 3.2 to 3.8, depending on race. With progressively reduced water supply, *maniculatus* begins to die out, after weight loss, at the 0.2 cc.-daily water-intake level, whereas *eremicus* tends to stabilize weight and survive under this treatment. *Peromyscus eremicus,* for example, survived an average of 41 days after water was down to 0.4 cc. daily, whereas *maniculatus* survived 24 to 29 days. Unfortunately, Lindeborg does not specify the water content of the food supplied and we do not know the make-up of the diet of wild cactus mice. Presumably much dry seed material is used, but it is also evident that green and more moist seeds and other moist plant materials are taken. *Peromyscus eremicus* probably requires some water content in its diet and may not be supposed to maintain water balance with such great reliance on metabolic oxidation as do kangaroo rats. However, the cactus mouse apparently has greater tolerances of low water intake and of high temperatures than does any other member of the genus *Peromyscus* thus far tested, and it may broadly be characterized as a desert-adapted white-footed mouse.

On August 30 at Black Rock Spring a screech owl (*Otus asio*) had captured and eaten a juvenile of this species of mouse (see p. 86).

The cactus mouse seems to show evidence of a protracted breeding period. A lactating female was taken on May 20, and on May 22 four gray or still partly gray juveniles were taken which must have been born in late winter. Gray-pelaged juveniles also have been taken on July 8, August 23 and 30, and September 1. Pregnant females were trapped on April 1 (4 embryos), July 2 (3 embryos), and July 8 (2 embryos), and a female that had just given birth to young was taken on October 13.

The cactus mice of the Monument belong to the race *Peromyscus eremicus eremicus,* which ranges throughout the Mohave and Colorado deserts.

Deer Mouse. *Peromyscus maniculatus*

Description. A small (total length less than 180 mm.) member of white-footed mouse group with ears less than 20 mm. from notch and tail less than 85 mm. and less than 90 per cent of length of head and body. Tail with distinct, narrow dorsal blackish stripe. Pectoral mammary glands present. Upper parts varying from mixed brown and black to ochraceous or reddish; under parts white.

Range. Central Yukon and central Mackenzie east to northern Quebec and extreme northern Labrador and south to Cape district of Baja California and along the Mexican Plateau to northern Oaxaca; in the east, south to central Texas on the Gulf coast and to southern Arkansas, Tennessee, northern Georgia, and western South Carolina.

Occurrence in Monument. Uncommon but widespread. Recorded from: 4 mi. N Warren's Well, Jan. 1 *; Upper Covington Flat, Aug. 28 [1]; Lower Covington Flat, Jan. 8 [1], Feb. 19 [1], Mar. 20 [1], Apr. 2 [1], Aug. 23 *; Smithwater Canyon, May 1 [1]; Quail Spring, May 19 *; Stubby Spring, Sept. 8 *; 13 mi. SW Twentynine Palms, Apr. 30 *, May 2 *, 6 *; 1½ mi. N Cottonwood Spring, May 10 *.

The deer mouse is a member of the *Peromyscus* group which is scarce in desert country. Even in the piñon and juniper belt we have taken it infrequently. In all the trapping we have done since 1945, only three have been caught. However, earlier, in 1941, 13 miles southwest of Twentynine Palms at 4500 feet Miss Alexander took five in juniper habitat, and also in that year two were taken near Cottonwood Spring in an area of scattered junipers. One of these was a lactating female on May 10, and on May 2, at the first station, a female had 4 embryos. The parties from Long Beach State College took moderate numbers in the Covington Flat area in 1960 and 1961.

None of our records is for lower than 3200 feet, which means that all were within the lowest border of occasional juniper growth. Nevertheless, the species does not seem to be clearly associated with any one soil or vegetation type. Apparently it is much less successful than the prevailing desert members of this group of mice, the cactus mouse and the canyon mouse. As indicated elsewhere (p. 316), the deer mouse is less adapted than these species to dryness and to high temperatures, as shown through experimental tests. A few deer mice evidently can find adequate food and retreats locally in the Monument at the middle and upper levels.

The deer mice of the deserts of California belong to the race *Peromyscus maniculatus sonoriensis.*

Brush Mouse. *Peromyscus boylii*

Description. A fairly large member of white-footed mouse group with tail about equal to head and body and sparsely haired and coarsely scaled. Ears moderately large, 20–23 mm. long from notch. Color above brownish with pronounced buff on sides; below, white.

Range. Northern California, northern Utah, southern Colorado, and southeastern Kansas south to northern Baja California and through mainland México to Guatemala, Honduras, and El Salvador; absent from most of desert areas and lower elevations of Great Basin.

Occurrence in Monument. Recorded from: Lost Horse Valley, July 20[1], 21[1].

The brush mouse was found by a party from Long Beach State College in a canyon near Lost Horse Ranger Station in July, 1961. Three specimens were taken. Identity of these has been verified on the basis of cranial and dental characters by Seth B. Benson, to whom we are indebted. The species is similar in general appearance to the cactus mouse and the piñon mouse and could easily be overlooked. However, it has not been taken by our parties at the several stations in the upper elevations of the Monument, where it might occur in bushy areas and where we have trapped fairly extensively.

The brush mice of this area belong to the race *Peromyscus boylii rowleyi,* which is present in coastal southern California and disjunctly in the uplands of the southern Great Basin and southern Rocky Mountain system.

Piñon Mouse. *Peromyscus truei*

Description. A large (adult length normally 190 mm. or over), white-footed mouse with long, hair-covered tail (longer than head and body) with distinct dorsal, dark stripe. Ears large, 21–24 mm. long from notch. Color above tawny-olive, mixed with blackish; below, white.

Range. Southwestern and central Oregon, northern Nevada, northern Utah, and central and southeastern Colorado south to northern Baja California, southern Nevada, central and southeastern Arizona, and Méxican Plateau to central Michoacán, Morelos, and northern Oaxaca; disjunctly in Cape district of Baja California.

Occurrence in Monument. Common in piñon belt of western section. Recorded from: Black Rock Spring, July 6 *; Upper Covington Flat, June 27[1]; Lower Covington Flat, Feb. 2[1], 19[1], Mar. 18[1], 20[1], Apr. 2[1], June 27[1], 28[1], 29[1], Aug. 23 *, 24 *, 25 *, Oct. 10[1]; Smith-

Fig. 92. Desert wood rat (above), cactus mouse (center left), piñon mouse (center right), and grasshopper mouse (below).

water Canyon, June 27 [1], 28 [1]; Quail Spring, Apr. 2 *, 22 *, May 19 *, 20 *, July 2 *, 5 *, Dec. 19 *, 21 *; Stubby Spring, July 22 [1], Sept. 6 *, 7 *, 8 *, 10 *; Key's View, July 15, 16 *; Lost Horse Valley, Mar. 9 *, July 21 [1]; Indian Cove, May 13 *; 13 mi. SW Twentynine Palms, Apr. 30 *, May 1 *.

The piñon mouse occurs in the piñon and juniper woodland and among yuccas. The lowest place we have taken it is at Indian Cove, where one was caught near desert willows at the canyon mouth. This is not far distant from scattered piñons on rocky slopes. More typical is the situation at Key's View, 5000 feet, where 3 were taken in 57 traps on July 15 in an area of rocks and junipers with close-set bushes. At Quail Spring on May 20, 10 were taken in 57 traps among tree yuccas, catclaw, and chrysothamnus on relatively flat ground. Apparently a necessary condition for this species is substantial woody plant growth providing downed logs or low trunks and leaf litter, as from piñons and junipers, on part of the ground surface. The activity of these mice must center about the bases of piñons and junipers in many instances.

This mouse, by adhering to heavier plant growth than its relatives of the *Peromyscus* group, and by staying at higher elevations, must encounter less severe daytime temperatures in its hiding places.

The only evidence of breeding which we obtained was from a female taken on April 30 in the junipers at 4500 feet 13 miles southwest of Twentynine Palms. This animal had four embryos. Absence of embryos was specifically recorded in adult females taken on May 19 and 20, July 16 (2), and September 10.

The piñon mice of the Monument are of the race *Peromyscus truei chlorus*, which was named originally from Lost Horse Mine about two miles east of Key's View. This is a race that ranges through the mountains of southern California, from the San Gabriel Range south to northern San Diego County (see Hoffmeister, 1951) and east to the type locality in the Monument. We were not successful in finding the species southeast of this point at Pinyon Wells.

Southern Grasshopper Mouse. *Onychomys torridus*

Description. A species with superficial resemblance to the white-footed mice but with distinctly short, heavy tail, the tail measuring only slightly more than half that of head and body. Ears of moderate size, as in the deer mouse. Above, light pinkish cinnamon, gray in juveniles and subadults; below white. Tail bicolored basally, but tip white above and below (fig. 92).

Range. Central California, central Nevada, southwestern Utah, central Arizona, and central New Mexico south to northern Baja California, northern Sinaloa, Aguascalientes, and southern San Luis Potosí.

Occurrence in Monument. Sparse but widespread population. Recorded from: Morongo Pass; 3 mi. S Warren's Well, Aug. 31 *; Lower Covington Flat, Mar. 18 [1], 20 [1], May 1 [1], June 28 [1], 29 [1], Aug. 23 *, 29 *, Oct. 10 [1], Nov. 29 [1]; Smithwater Canyon, May 1 [1]; Quail Spring, Apr. 2 *; Lost Horse Valley, July 20 [1]; Stubby Spring, Sept. 6 *, 7 *, 8 *; 1 mi. NW Fortynine Palms, Apr. 1 *; Pinyon Wells, Oct. 13 *; 1½ mi. N Cottonwood Spring, May 10 *; 2–4 mi. S Cottonwood Spring, Jan. 23 [1], Nov 4 [1].

This rodent, as an insect eater, has available moisture sources, which place it in a very different category from related mice that are seed-eaters. The Schmidt-Nielsens (1952) report that mice of this group drink water and eat juicy vegetation in captivity and will suffer if these are not supplied. Their insect food is estimated by them to be 60 to 85 per cent water, and in a wild state with extensive use of insects the water problem should be amply solved by this diet. Being nocturnal, the grasshopper mice are not subjected to heat stress and are thus not called on to use water in keeping cool.

Grasshopper mice have been captured essentially throughout the altitudinal range represented in the Monument and in the gravelly or sandy flats of the creosote bush association, Joshua tree areas, and piñon–juniper woodland, but never on rocky or steep terrain.

Grasshopper mice were taken only occasionally, many trapping efforts yielding none at all. On August 31, 3 miles south of Warren's Well, two were taken in a line of 97 traps. Near Stubby Spring in 100 traps set among low bushes and junipers, two were taken on September 6 and one on September 7. Two were caught in 60 traps there on September 8, also. These latter represent maximum numbers we encountered.

Evidence of breeding consists of 4 embryos recorded on March 18 and 5 embryos on May 1 in females taken by parties from Long Beach State College. Gray juveniles in the 11- to 16-gram size range have been caught on April 1, August 23, 29, and 31 (2), and September 6 (2). These data indicate an extended breeding season. The mouse taken on April 1 weighed 16.6 grams, which contrasted with adult weights of 20 grams or over; this juvenile was in gray pelage, and yet had testes 8 by 14 mm., suggesting that it was reaching breeding condition.

The southern grasshopper mice of the Monument belong to the race *Onychomys torridus pulcher,* which ranges over the Mohave and Colorado deserts and which was described originally from Morongo Pass.

WOOD RATS

Desert Wood Rat. *Neotoma lepida*

Description. A fairly large rodent (adult 26—32 cm. long), with prominent, almost naked ears and furred, bicolored tail, the tail approximately three-fourths the length of head and body. Soles of feet naked. Above buffy gray; under parts white or pale buff, the hairs on throat plumbeous basally; hair on feet entirely white (fig. 92).

Range. Central California, southeastern Oregon, southwestern Idaho, northern Utah, and extreme west-central Colorado south to Cape district of Baja California, northwestern Sonora, and southwestern Arizona.

Occurrence in Monument. Common throughout. Recorded from: 4 mi. N Warren's Well, Jan. 1 *; Upper Covington Flat, Aug. 19[1]; Lower Covington Flat, Jan. 8[1], June 29[1], Oct. 8[1], 10[1], 11[1], 24[1], Nov. 29[1]; Quail Spring, Dec. 19 *, 22 *; Stubby Spring, Sept. 6, Oct. 15; Key's View, July 15; Lost Horse Valley, Mar. 9 *, July 19[1], 20[1]; Barker's Dam, Apr. 2 *; Indian Cove, Apr. 23[1]; Jumbo Rocks, Oct. 15; Fortynine Palms, Apr. 8; mouth Fortynine

Palms Canyon, July 25 *; 4 mi. N Twentynine Palms, July 25 *; 13 mi. SW Twentynine Palms, May 1 *; Squaw Tank, June 16[1], July 22[1]; Pinyon Wells, Oct. 13 *, 15 *, 16 *; White Tanks, Feb. 25[1]; Cholla Garden, Apr. 10; Pinto Basin, Jan. 29[1], May 6[1], Oct. 21[1], 22[1]; Eagle Mountain, Oct. 19, 20 *; 5 mi. NE Cottonwood Spring, Feb. 18[1]; Cottonwood Spring, May 15 *, 18 *, July 8 *, 11[1], 12[1], Sept. 16[1].

This wood rat occurs in all sections of the Monument and from the lower flats up to the piñon belt. The species tends to concentrate around rock outcrops and bases of larger plants, such as cactuses and yuccas. The animals have their burrows and houses especially in such places. The houses are often slight accumulations, poorly formed, but disposed about the entrance to a burrow or along a rocky crevice in which they have a retreat. On Eagle Mountain, for example, the nests were commonly seen beneath boulders, among catclaw bushes, and in clumps of *Yucca mohavensis.* One desert wood rat was trapped at its house beneath a large rock. At Pinyon Wells three were taken along rocky ledges and one was seen by day as it came within 10 feet, hurrying to its nest at the base of a scrub oak. At this station also desert wood rats had set up quarters in a pile of about 15 juniper fence posts. Again, north of Warren's Well, nests were found and rats trapped at them in crannies of large rock piles. But in more open gravel or sandy terrain, wood rat nests are piled at the bases of chollas on flat ground, as shown in figure 93. At such times the extensive use of the most available material for houses

Fig. 93. Nest of desert wood rat constructed of cholla burrs and sticks at the Cholla Garden area.

Fig. 94. Diagram of nest chambers of a desert wood rat in the base and trunks of a Mohave yucca at Jumbo Rocks.

leads to a formidable pile of the spiny shed cholla joints. The rats show remarkable ability to handle these spine clusters, arrange them, and enter through passageways among them to their underground burrows. Houses of spines such as these must serve as a partial deterrent at least to coyotes and other predators that would not, at least rapidly, tear into them to get at the rats.

The debris piled into nests is of great variety, representing whatever is available locally but also somewhat reflecting what may be utilized for food: juniper cuttings, dry sticks, cholla sections, tree yucca fruiting pods, yucca leaves, and clippings of small bushes. An occupied nest at Jumbo Rocks was built in the base of a *Yucca mohavensis.* An entry hole at ground level went beneath a live yucca stalk and had vertical branch tunnels upward into two live stalks and one dead one. These vertical tunnels were in both soft-decayed and live "heart" tissues and extended up as much as 2 feet. The ground-level burrow was as much as 1 foot below the soil surface at one place (fig. 94). About the base of the yucca stalks was a small accumulation of sticks and leaves. The rat had a nest chamber at the side of one of the vertical tunnels in the dead yucca stalk, about a foot above ground level. The rat was found in another side tunnel and when disturbed ran from this place up on the outside of an adjacent stalk to a juniper tree. It then ran up a nearly vertical cliff face and into a crevice 10 feet above the nest in the yucca.

Desert wood rats are omnivorous, eating many types of plant materials, including yucca pods, parts of cactus stems and fruits, bark and green twigs, and berries. They have been attracted also by meat and fish bait used on traps. Obviously their varied diet includes many items in which there is fair to good water content. This species does not then, as do the strictly seed-eating desert rodents, depend heavily on metabolic water and the rigorous conservation of it. Wood rats of this or related species (*Neotoma albigula*) in the desert have been found to be intolerant of a diet of dry grain alone, without water, dying in 4 to 9 days (Schmidt-Nielsens, 1952). Cactus with 80 to 90 per cent water content may constitute 43.8 per cent of the annual food of *albigula* and may amount to 90 per cent in the driest periods of the year (Vorhies and Taylor, 1940). Nest temperatures inside crevices or in the typical subsurface runs of nests on flat ground probably do not exceed 92° F. in the daytime and would not demand use of water to hold down body temperature.

On October 19 at midday near a spring on the north side of Eagle Mountain, a wood rat was disturbed as we broke through some dense *Zauschneria* bushes. This induced it to give its alarm rattle, made by vibrating its tail against the vegetation.

The desert wood rats of the Monument belong to the race *Neotoma lepida lepida,* which is of widespread occurrence in the Mohave and Colorado deserts.

Dusky-footed Wood Rat. *Neotoma fuscipes*

Description. A large wood rat (adults 33–46 cm. long) with somewhat darker, browner dorsal coloration in adults than that of the similar desert wood rat (see p. 321). Feet with parts of dorsal or lateral surfaces dusky rather than pure white. Rostrum long and relatively narrow (this and diagnostic dental characteristics evident in prepared specimens of skulls).

Range. Willamette Valley and southwestern coast of Oregon south through California, west of Sierra Nevada, to northwestern Baja California. Ranges east only to borders of Mohave and Colorado deserts.

Occurrence in Monument. Present in higher levels of western section. Recorded from: Little Morongo Canyon, Apr. 1 *; Black Rock Spring, Aug. 30, Sept. 1 *; Lower Covington Flat, June 27 [1], 29 [1], Oct. 10 [1], 11 [1]; Quail Spring, Dec. 20 *.

This species of wood rat is not readily distinguished from the desert wood rat except by skull characters and the features of the skin and dimensions of fully adult animals. It has accordingly generally been overlooked among the occurrences of the better known desert wood rat but has now been verified as present in the piñon belt and the heavier brushy areas of the western section of the Monument east to Quail Spring. It is apparently fairly common in that section.

This species tends generally to build larger stick nests than *Neotoma lepida,* and the numerous nests of this kind at Little Morongo Canyon, Black Rock Spring, and Lower Covington Flat probably represent the efforts of this species. The subadult animal trapped at Little Morongo Canyon was taken in an area where large nests were situated in dense mesquite clumps. At Black Rock Spring, near the spring, several large occupied nests were seen in a mesquite clump, one of which was 3 feet high. On August 30 after dark, wood rats were tail-rattling in this thicket, and on September 1 two rats of this species were trapped at these nests.

The dusky-footed wood rat has not been recorded east of the San Bernardino Mountains and San Gorgonio Pass heretofore. The few specimens we have at hand seem to accord racially with those of that area, which, as Hooper emphasized (1938), are intermediate between *Neotoma fuscipes simplex* and *N. f. macrotis* to the south, although best allocated on the basis of all characteristics to the former. The animals of the Monument may similarly be designated as *simplex* pending analysis of a better series of adults from the area.

VOLES

California Vole. *Microtus californicus*

Other name. California meadow mouse.
Description. A mouse of moderate size (160—190 mm. long in adults) with aspect of the vole group, namely, short tail (40—60 mm.) and broad body with fluffy hair tending to obscure the ears, which are in turn hair-covered. Color brown with coarse overhairs dark brown to black; below gray. Tail brown above and gray below.
Range. Central western Oregon south through California to northwestern Baja California; extends east locally into Inyo district and Mohave desert.
Occurrence in Monument. Locally common at western border of area. Recorded from: Morongo Valley, Feb. 22 *; Little Morongo Canyon, Apr. 1 *, 2 *, 3 *.

The voles or meadow mice require green plant food and at the arid borders of their range are confined to marshes or wet meadows, where moisture is always present and green grasses or sedges are available. The occurrence of this species in the Monument is then, as expected, confined to the western border area, where surface water is present at all times in Morongo Valley and Little Morongo Canyon.

On February 22, 1935, Miss Alexander and Miss Kellogg trapped five of these mice in a swampy place at the eastern, lower side of Morongo Valley, catching the animals in runways in the grass at the edge of a patch of tules. At this time a swamp covered several acres here. Fresh grass cuttings were in evidence, as usual, in the active runways of these mice.

In Little Morongo Canyon we took three mice along the small stream. One was caught beneath a pile of dead brush, 3 inches above the water level, where there was some indication of a runway, and others were taken in sedge on the stream bank, one of them by daylight just after the traps were set. The animals caught were all full-grown. Of four females taken in February, two were lactating and a third had four embryos. The three females caught in April were not pregnant.

The specimens from the Monument are referable to the race *Microtus californicus sanctidiegi* of adjacent parts of coastal southern California, although they seem to average less rich, bright brown in dorsal color.

CARNIVORES

Coyote. *Canis latrans*

Description. In size and appearance similar to a large, slender dog, the body about 3 feet long and the tail 1 foot long. Ears pointed and erect. Coloration grayish buff above, mixed with black on back; below whitish. Tail moderately bushy and with dark tip.
Range. Kenai Peninsula and extreme northern Alaska, central Mackenzie, central Saskatchewan, southern Manitoba, central Ontario, and extreme southern Quebec south through México to Costa Rica. In the eastern United States south to western Louisiana, central Arkansas, southern Illinois, and extreme southeastern Michigan.
Occurrence in Monument. Common. Recorded from: Morongo Pass, Nov. *; Little Morongo Canyon, Apr. 2, 3; Black Rock Spring, Aug. 31; Upper Covington Flat, Aug. 24, 27; Lower Covington Flat, Apr. 8, 9, Aug. 23, 25, 26 *; Quail Spring, Jan. 24, Feb. 11, May 19, July 1, Oct. 20 *; Stubby Spring, Sept. 9, 12 *, Oct. 15; Key's View, July 3, 4; Indian Cove, Apr. 8, Sept. 5, Oct. 15, 16; Twentynine Palms, July 12, 13; Pinyon Wells, Oct. 8, 11, 12, 13 *, 14 *, 16; Split Rock Tank, Apr. 9; White Tanks, Sept. 12; 3 mi. E White Tanks, Feb. 18; Pinto Basin, Apr. 6; Eagle Mountain, May 16, Oct. 20; Cottonwood Spring, Apr. 5, Oct. 22.

Coyotes range through all parts of the Monument from the lowest flats and sand dunes to the piñon-covered crests. In the upper levels they are probably no more abundant than the gray foxes, but their calling at night and their tracks and droppings, which are not readily confused with any other wild predator species, make them the best known and most often recorded carnivore in all parts of the area.

The yammering, high-pitched call of a coyote often deceives a novice into thinking that more than one animal is producing the sound. But of course on occasions calls coming from two or three different directions, simultaneously, clearly show that several animals can engage in barking at once and that to a degree at least they are responsive to one another's calls. Dawn is a time when we most often heard them, but we have recorded calling in all periods of the night; our early evening records are numerous but perhaps reflect our greater alertness at that time. We have heard calling in all parts of the year.

Tracks are seen best along sandy washes and in game trails, as also on the sand dunes. The coyotes probably do travel such places by preference, and yet the coursing over other surfaces is not rare although often undetectable because of the nature of the soil or rock. Tracks compared with those of kit foxes in the sand dunes of Pinto Basin showed a gait of 12 inches for the coyote, compared with 8 inches for the fox. This is when, in both species, the animals were placing the hind foot in the track of the front foot.

Fig. 95. Coyote and pups beneath a piñon tree.

Coyotes, although using moist animal food, frequently seek drinking water and exert great effort to obtain it by digging. We note their tracks about those springs and tanks that are undisturbed by people, but also we find them excavating in damp sand. In the canyon leading from Upper Covington Flat on April 9 and August 27 there were coyote excavations. In August they had gone down 40 inches in soft sand and had removed or disturbed the sand over an area 12 by 15 feet in diameter, forming a cone-shaped pit (fig. 2). In the morning there was a little pool of water 6 inches across and 3 inches deep in the bottom. The hole was in a narrow part of the stream course cut in a granitic formation. On April 9 two holes were being kept open in this canyon, one 30 inches deep and the other only 15 inches deep with a pool 12 inches across, the water obviously being at higher ground level at this season. On October 16 at Pinyon Wells coyotes had dug holes 3 feet down, below rock dikes in the canyon, in futile search for water. The coyote's exploitation of subsurface water regularly serves the interests of other vertebrate species in the area such as quail, doves, and house finches that come to drink.

Digging for food by coyotes is often noted also, especially in their pursuit of kangaroo rats. To our embarrassment they completely excavated a 3-foot-deep garbage dump we had carefully covered upon leaving our camp site above Stubby Spring.

Coyotes are versatile feeders, although rodents are doubtless the principal element in their diet. In coyote droppings examined at Pinyon Wells we identified bones or hair of wood rats (*Neotoma*), antelope squirrels (*Ammospermophilus leucurus*), and deer (possibly from carrion). Also in the feces were manzanita and juniper berries, the latter in large numbers. The coyotes were apparently gathering dried or ripe juniper berries that had fallen to the ground, as the pits of the berries were naturally brown, not green like those hanging in the trees. At this station coyotes and gray foxes were showing a black diarrhea. We could not be sure of the dietary cause of this, although possibly an excess intake of juniper berries could have produced it. At Cottonwood Spring on April 5 coyote scats contained rabbit and wood rat remains and also part of a shell of a desert tortoise. On October 20 at the north base of Eagle Mountain, a coyote had visited a line of mouse traps, robbing them of our catch. At Black Rock Spring on September 4 a coyote had apparently caught a mountain quail (see p. 69).

At traps set for coyotes, the animals have been attracted by fish bait and scent made from fish as well as by rabbits used as bait and by carcasses of other coyotes. These carnivores are characteristically cautious about traps and often manage to avoid capture. On April 6 at Pinto Basin the fresh tracks of a coyote showed it had circled the remnants of bait at a site from which a trap had been removed; the animal stayed 15 feet away from it but obviously was aware of it. Yet at other vacated trap sites on this same line coyotes came in and took the bait. One had urinated at the site, and two had defecated. Apparently three coyotes at least visited this old line of a mile and a half extent on this night. In Little Morongo Canyon on April 3, four of eleven steel traps were sprung by coyotes but no animal was caught.

At Pinyon Wells in October, of three sets of coyote traps on October 11, one was uncovered by a coyote and another was sprung but held only the toe of the animal. On October 13 and 14 coyotes were caught here at the mouth of a canyon in the gravel bottom. On October 16 a coyote pulled the carcass of one of these specimens from the bush where it was hung and dragged it across the trap, springing it. A young coyote caught on August 31 at Black Rock Spring pulled out of the trap and escaped as we came up to it. Thus coyotes proved generally wary and capable of avoiding capture by various means.

Occasionally coyotes are seen by day. On February 18 two ran across the road 3 miles east of White Tanks and trotted up a low hill less than 100 yards away, where we watched them for about 5 minutes. On August 24 at about 7 A.M. on Upper Covington Flat, a full-sized coyote loped off into the wash ahead of us. Again on May 16, at the north base of Eagle Mountain an animal ran out of the canyon bottom ahead of us; its pelage was notably shabby.

The young are born in the spring and by late summer are scattered from the dens. We caught young of the year on August 26 and 31. The animals saved as complete specimens were all adults taken on August 26, September 12, and October 13 and 14 and were largely or entirely in newly grown fur. Males on the first two dates weighed 20½ and 21 pounds, respectively.

The coyotes of this area belong to the desert race *Canis latrans mearnsi.*

Kit Fox. *Vulpes macrotis*

Description. A small (3–5 lb.), delicately built fox, with exceptionally large ears. Color nearly uniform sandy gray, lacking distinctly blackish central area and contrasting tawny sides of the gray fox. Tail with black tip. Hairs inside ears heavy and white. Soles of feet with long fur.

Range. Southwestern deserts from southeastern Oregon, southwestern Idaho, and northern Utah south to Cape district of Baja California and to central Sonora, northeastern Durango, and southeastern Coahuila. Extends west to San Joaquin Valley of California and east to central and southeastern New Mexico.

Occurrence in Monument. Fairly common in lower basins and flats. Recorded specifically from: Morongo Pass, Nov. 23 *; Queen Valley, Feb. 10[1]; 4 mi. S Twentynine Palms[1]; Pinto Basin, Apr. 5 *, 6, July 15.

The kit fox is a species of very open and usually flat terrain. Many casual reports of foxes from the Monument are said to be of this species but in fact may relate to the more common fox of the area, the gray fox of the middle and upper levels.

Our own detailed acquaintance with this species in the Monument has been made in Pinto Basin in the creosote bush flats and sand dunes. Here on the night of April 4–5 we had set a line of 18 traps extending about a mile and a half through sandy and gravelly washes and the outlying parts of the large sand dune. Traps were baited with canned fish and rotted fish scent. Two foxes were caught at points about a half-mile apart. These tiny foxes growled as we approached, but they were not very formidable. They weighed only 3½ and 3 lbs., adult male and female, respectively. The traps were removed after taking the animals, and the next day the line was followed to search for fresh signs of carnivores; a rain had obliterated old signs and made new tracks distinct. On the morning of the 6th we noted that another kit fox had visited a now vacant trap site, taken the fish bait, and left droppings at the place. At another trap site a kit fox had passed within 15 feet and had missed the bait. At still another site in the dunes a fox had also been by but missed the bait. There had been at least three coyotes along this line also. Thus the two species of carnivores were clearly ranging the same area, and within two nights at least four and perhaps five different kit foxes had crossed or visited the mile and a half of trap line.

The kit foxes showed two types of gaits, one in which all four feet struck in line but not on top of one another, and the second where the hind foot was placed in the track of the front. In this latter type the tracks were found to be 8 inches apart. On dry sand the kit fox tracks showed no pad or claw marks at all when walking, as the feet have heavy hair which makes a kind of sand shoe. In dragging a trap, of course, the toes are spread far apart and claw marks are then plain.

One of the kit foxes taken had an empty stomach, but the stomach of the other contained hair, including the tips of the tails of two kangaroo rats. Also noted was an insect leg and wings. In the intestine was a scorpion claw. Neither animal had any fat, and we were impressed with the slim, laterally compressed bodies. These are indeed delicate creatures, able to move lightly over the sand and dig and enter small holes in the sand. In this connection, they also show adaptation of the ears in the development of heavy, stiff fur inside the pinnae, which must help exclude sand.

The kit foxes were not in breeding condition. The female was an old animal to judge from her heavily worn molars and canines; she was not pregnant nor lactating, but was in good pelage. The male, an adult but not an old animal, had testes 15 mm. long, exclusive of the epididymis.

The kit foxes we trapped were examined extensively for parasites. They were quite free of fleas and other external parasites, but, as is frequently true of carnivores, had tapeworms and nematodes in the intestine and a very small species of tapeworm in the liver. The latter proved to be a hitherto unknown genus which has been described by Voge (1952) as *Mesogyna hepatica.*

Fig. 96. Kit fox on the sand flats of Pinto Basin with freshly caught kangaroo rat.

The kit foxes of the desert areas of eastern California belong to the race *Vulpes macrotis arsipus.*

Gray Fox. *Urocyon cinereoargenteus*

Description. A medium-sized fox (5–8 lbs.), blackish gray on the upper parts, with rich reddish brown on flanks. White on throat and belly; stripe along upper side of tail and tail tip black; muzzle blackish, except for white at each side of nose. Skull distinctively marked by well-spaced lyrate temporal ridges rather than closely approximated central ridges or a single ridge as in other local canids.
Range. Northern Oregon, central Nevada, central Utah, northern Colorado, northern Texas, eastern North Dakota, northern Minnesota, northern Michigan, southern Ontario, and southern Maine south through México and Central America to central Colombia and northern Venezuela in northern South America.
Occurrence in Monument. Common. Recorded from: Black Rock Spring, Sept. 3, 4 *; Lost Horse Mine, Mar. 9 *; Fortynine Palms, Apr. 7; Pinyon Wells, Oct. 8, 9 *, 17 *; Split Rock Tank, Oct. 29; Cottonwood Spring, Sept. 12, 15, 23 *.

The gray fox is the most abundant species of fox in the Monument insofar as our evidence shows. It is the only kind we have proved to be present in the chaparral and piñon–juniper habitats, and it descends in the mountains and canyons to Cottonwood Spring at 3000 feet. Indeed, the species may be expected lower, as it occurs regularly in and near the Colorado River valley. The presumption is very strong that all fox sign seen in the canyons of the upper parts of the Monument relates to this species rather than to the kit fox, which favors low or open terrain. Our records of certain occurrence based on gray foxes seen or trapped are all in manzanita brush, piñon–juniper country, Joshua tree and scrub oak growth, or in canyons with patches of fairly dense low scrub or palm and mesquite thickets.

Occasionally gray foxes are seen by day. Thus on October 29 at Split Rock Tank one crossed the road by the camp site, and on October 8 near Pinyon Wells one was seen in a manzanita patch in late afternoon. In September of 1950 at Cottonwood Spring, gray foxes had been fed by former campers near the spring and the steep slope rising above it on the west. On our arrival on September 12, the foxes were very tame. That afternoon before sundown between 4 and 5 P.M., four of them came out to food placed within 30 feet of our camp (fig. 98). One smaller, young-appearing animal was timid toward the other foxes and watched the brush back of him for the presence of the others more than he watched us. The animals ran about a great deal, reappearing quickly at some distance and visiting other parts of the general campground. We noted that they moved the ears separately and frequently and were constantly on the alert for sounds. They seemed particularly startled by the rustling of paper and other high-pitched sounds. When the foxes retreated up the rocky slope back of the spring, we noted they moved over an almost vertical rock slope 10 feet high with a smooth gliding motion. No special jumps or effort to find toe holds were evident.

Fig. 97. Gray fox in Queen Valley.

On September 15, the foxes were slower in coming out, appearing only after the sun had left the feeding area, and 30 to 45 minutes later than usual. Also, only two were seen on this date. After dark we heard them growling, but we could occasionally shine their eyes with a flashlight, as we did also at Black Rock Spring at 8:30 P.M.

The foxes at Cottonwood Spring had been attracted by dog food and table scraps originally. But on the 15th they readily took bird bodies. The foxes we have trapped have been baited with rabbits in three instances and with a coyote body and scent made from spoiled fish in two other instances. Animals taken in traps have shown no stomach contents other than the bait, and thus we have had no other direct evidence to assist in judging food habits. However, we found droppings of foxes that contained rabbit and wood rat bones, so that apparently these mammals form a considerable part of the diet of the gray foxes.

The fox taken at Pinyon Wells on October 9, a male, was extremely fat. A female at Black Rock Spring on September 4 was an old animal with badly worn teeth; she weighed 5½ pounds and was not notably fat but was in good summer fur. Neither this female nor the other taken on September 23 showed any breeding activity.

The gray foxes in southern California are separable into the desert race *Urocyon cinereoargenteus scottii* and the coastal race *U. c. californicus.* Comment on the characters of these races has been offered in later years by Grinnell, Dixon, and Linsdale (1937, 2:430–436), Hall (1946:241), and Johnson, Bryant, and Miller (1948:348). In an effort to evaluate the features of size, we have restudied the original measurements in these works and find that differences in external dimensions taken from fresh animals are in no instance statistically reliable. Even casual inspection of the dimensions in the tables of Grinnell, Dixon, and Linsdale show essentially the same ranges of values in each race. We have separately calculated the ratio of tail length to body length in each specimen and find no basis, either, for the claim of relatively longer tail in *scottii.* However, *scottii* and *californicus* are distinct from *townsendii* in relatively longer tail and lesser general size.

In skull dimensions *scottii,* as represented in the Colorado River valley, is more slender on the average. This is revealed in zygomatic breadth and especially in width of rostrum, where the average difference is significant. This was a feature early pointed out by Grinnell and Swarth (1913:373). The three male foxes from Joshua Tree Monument, an area not heretofore represented in collections available to others, show rostral widths of 18.0, 18.0, and 19.0 mm., and zygomatic breadths of 62.3, 64.6, and 66.4. Comparison of these values with data in the tables presented by Grinnell, Dixon, and Linsdale shows that in rostral width the animals conform best with the average for *californicus* and that in zygomatic breadth they are equivocal. No firm conclusion in this regard can be drawn from the small sample, but the data on the one perhaps most important dimension favor *californicus.*

In coloration, we find that the fresh winter pelage represented in four of our specimens is a close match for winter pelages of *californicus* from the eastern border areas of coastal California, whence *californicus* was named, and are appreciably darker than winter-taken animals from the Providence Mountains and the Colorado River valley as representative of *scottii.* This circumstance and the slight evidence from dimensions of skull lead us to place the gray foxes of the Monument in *californicus.* This finding admittedly is not supported by strong evidence, both because our sample is small and the races overlap greatly in characters. However, it is not discordant with our findings on the geographic affinities of other upland vertebrates in the Monument, such as California thrashers, scrub jays, and titmice, in which the populations in the Little San Bernardino Mountains represent extensions of coastal types out onto this desert spur, either unmodified, or if modified with obviously close affinities to the coastal populations.

The one gray fox from the far eastern part of the Monument, a female from Cottonwood Spring, although dark like *californicus,* has a small and slender skull like *scottii.* The other female from the western station at Black Rock Spring was in summer pelage on September 4, and thus was not readily comparable in respect to race, but its skull is of the dimensions typical of females

Fig. 98. Gray fox at sunset foraging at the campground at Cottonwood Spring, September 12, 1950.

of *californicus.* The specimen from Lost Horse Mine is a pick-up skeleton of uncertain age.

Raccoon. *Procyon lotor*

Description. A well-known carnivore type, the size of a small, chunky dog (9 to 16 pounds), with black mask through the eyes bordered above and below by white, the fairly bushy tail ringed with black and pale brown; body grayish brown.
Range. Southern British Columbia and northeastern Alberta east across southern Canada to southern Quebec and Nova Scotia and south to Panamá.
Occurrence in Monument. Recorded with certainty only from Little Morongo Canyon, Mar. 31–Apr. 2.

Raccoons are generally restricted to the vicinity of water except as vagrants, and records of them are lacking for great sections of the southwestern deserts of North America. We have sure evidence of the presence of raccoons only along the permanent small stream in Little Morongo Canyon. Here on March 31 raccoon droppings were found filled with fragments of the exoskeletons of crayfish. These were associated with the very characteristic 5-toed, long-heeled track of the hind foot of this mammal. On April 2 fresh raccoon tracks were seen in the mud by the stream superimposed on our own tracks of the previous day. Apparently the crayfish and amphibians in this stream were an attractive food source.

Badger. *Taxidea taxus*

Description. A heavy and flat-bodied carnivore with short ears and tail and long powerful claws on front feet. Head and neck marked with median white central stripe, bordered with brown, and by black patches behind and below eye and at bases of ears; body grizzled, but whitish to clay color beneath.

Range. Southeastern British Columbia and northern Alberta southeast to southern Ontario and Great Lakes area and south to the tip of Baja California, southern Sonora, northern Tamaulipas, and over the Mexican highlands to western Puebla.
Occurrence in Monument. Sparse resident. Recorded from: Upper Covington Flat, Aug. 27; Indian Cove, Apr. 8 *; Pinyon Wells, Oct. 8; Pinto Basin, Apr. 11 *.

Badgers are not numerous in the Monument, although conditions for them would seem to be satisfactory in the Joshua tree "forests" and on the creosote bush flats. We have occasionally seen their characteristic diggings—wide and flattened entryways excavated as the animals have gone down after rodents. Diggings were seen on the broad plain of Upper Covington Flat among Joshua trees on August 27, and fresh holes were noted in an open canyon at Pinyon Wells on October 8. Here also at 4500 feet a part of a badger skull was found.

On April 8 a thoroughly weathered badger skull was picked up on the flat below Indian Cove in an area of fine gravel where there were scattered workings of kangaroo rats.

In Pinto Basin a badger was seen trotting along a wash bordered by desert willows, turning to one side and then the other every few feet and sniffing at the ground. When thus hunting and before it saw the observer, it was moving at about a man's walking speed. When alarmed, it turned back, periodically looking behind at the disturbance; this was on April 11 at 9 o'clock on a cool, overcast morning. The animal was a very old male, yet a surprisingly small-sized individual. It was fairly fat and in good condition, but the skin of the face was heavy and wrinkled as in an old dog. The badger had eaten a *Dipsosaurus* lizard and a young Merriam kangaroo rat (see p. 308), the latter so young as probably to have been dug from its burrow. The lizard also might have been dug out in the early morning, as it was partly digested and this type of lizard was only beginning to appear at the surface about 9 A.M.

The badgers of the Monument belong to the small desert race *Taxidea taxus berlandieri.*

Mountain Lion. *Felis concolor*

Other names. Puma; cougar.
Description. A large cat (100–160 pounds), of slender build, with long legs and long tail (over 2 feet long). Color uniform tawny or grayish to reddish brown above, whitish beneath.
Range. All of the Americas, from northern British Columbia and south-central and southeastern Canada south to Patagonia.
Occurrence in Monument. Rare. One record, based on direct evidence, at Lower Covington Flat, Apr. 8.

Mountain lions are evidently present in the western part of the Monument, where there is a moderate-sized deer population on which they may be expected to prey. We have had several second- or third-hand reports of lions seen, but evidence we could ourselves evaluate was gained only once from a set of tracks. One of these animals had crossed the wash draining Lower Covington Flat, traveling the wind-smoothed sand and fine gravel from one ridge to another, where the outlet canyon narrows to the north. The tracks were observed on April 8 and it was estimated that the lion had crossed several days before, but so recently that no other confusing or superimposed tracks of other animals were present. The lion had moved rapidly in great bounds, the feet striking close together, the hind feet partly on the front-foot tracks; each cluster of tracks was 10 feet apart with completely unmarked surface between. The animal had covered 50 yards and then gone up a steep ridge to the north. One well-formed track was found above the wash, but those in the wash were caved in because of the soft sand and showed no detail; the impressions in the sand were 5 to 6 inches in diameter.

Bobcat. *Lynx rufus*

Other name. Wildcat.
Description. A short-bodied cat with relatively long legs and very short tail. Ears tufted and hair of lower sides of face elongate. Spotted pattern of coloration, the spots more conspicuous, black on white, on lower sides and belly; back black and gray-brown; margins, tips, and bases of ears and a spot near tip of tail black.
Range. Southern parts of the Canadian provinces from British Columbia to Nova Scotia and south throughout the United States to northern Oaxaca and southern Veracruz in México.
Occurrence in Monument. Sparse resident. Recorded from: Upper Covington Flat, Apr. 9; Quail Spring, Oct. 20 *; Stubby Spring, Sept. 5 *; Pinyon Wells, Oct. 13; Cottonwood Spring, Oct. 22.

Bobcats are possibly more common than our few records indicate. However, the brushlands favored by this cat are of limited extent, and the rocky canyons, although suitable, may in many instances be too barren of plant cover to attract this species. At Upper Covington Flat, Quail Spring, Pinyon Wells, and Cottonwood Spring we have picked up fragmentary skeletal remains and a skull; the dates listed above only indicate the times we found remains.

Our one significant encounter with this species was on September 5 at Stubby Spring. At sundown on this day, Russell had walked quietly to within 70 yards of the spring, watching for deer, and observed a group of 10 mountain quail walk to the water and then on leaving retrace their route up along a trail. A bobcat had seen them at a distance and ran downslope ahead of them and crossed the trail they were on. The cat crouched and then scuttled forward a few feet to hide behind a 3-foot-high rock at the side of the trail the quail were using. As the quail approached, one hopped and walked to the top of this rock. The cat and the quail jumped in the air above the rock, but the cat missed and the bird was free. The cat then rushed at some of the other quail that were 6 feet away, but they also flew and it missed them. The

Fig. 99. Bobcat feeding on a chipmunk.

bobcat then went down the trail in a slow walk. It showed no agitation over the recent episode. It then stopped at the spring, where it was collected. The animal was a female, very fat, and weighed 10½ pounds. The stomach contained a small amount of hair and one tail vertebra of a rodent, probably that of a wood rat.

The single skin of a bobcat, the one referred to above, is insufficient to establish the racial affinity of the cats of this area. The specimen is surprisingly dark-colored on the back and richly tawny on the legs. However, although the pelage is partly new, the full development of typical winter fur was not achieved and part of the dark dorsal color might have become obscured by later growth of long, light-tipped hair. The restriction of the black areas of the ears and tail conform best to the condition in *Lynx rufus baileyi,* the pale desert race (see Grinnell, Dixon, and Linsdale, 1937:602). Animals from both the deserts and bordering mountains of southern California have previously been referred to that race.

DEER AND SHEEP

Mule Deer. *Odocoileus hemionus*

Other names. Black-tailed deer; burro deer.
Description. A deer with long ears and small, narrow tail, black at tip and on upper surface. Upper surface and flanks of body gray (winter) or brown (summer), the fawns spotted; rump area, posterior abdomen and inner sides of legs whitish. Antlers of male dichotomously forked, shed annually.
Range. Southern Yukon, southern Mackenzie, north-central Saskatchewan, and southern Manitoba south to Cape district of Baja California, central Sonora, eastern Durango, and northern Zacatecas.
Occurrence in Monument. Sparse resident. Recorded from: Black Rock Spring, Apr. 3, Aug. 31, Sept. 1, 2, 3; Lower Covington Flat, Apr. 9; Quail Spring, Jan. 22, 26, Oct. 14, 15; Stubby Spring, Apr. 10, 11 *, Sept. 11; Pinyon Wells, Oct. 8, 9, 10, 12, 13, 14, 15.

Deer focus in the higher parts of the Little San Bernardino Mountains in the western section of the Monument. The animals are detected principally near or within cruising radius of springs, especially in the summer and fall. The area where deer are present is in the piñon belt, but the association of significance is that with the scattered chaparral patches of this belt, especially the open clumps of scrub oak (*Quercus dumosa*) and mountain mahogany (*Cercocarpus*).

The deer population of the Little San Bernardino Mountains is not hunted, except very peripherally or illegally, as the area of occupancy is almost entirely within the Monument. Longhurst, Leopold, and Dasmann (1952) estimate that this region falls generally in the abundance category of 1 deer per square mile with a local spot mapped at 2–4 deer per square mile. Their estimates were formed from scattered reports, partly ours, from knowledge of vegetation conditions, and particularly from background information of censuses of deer populations elsewhere. We generally agree that there are a few square miles, perhaps 10 to 15, along the axis of the mountains from Black Rock Spring to Stubby Spring where the higher figure is correct. Our observations at or near the principal water sources which bear this out are as follows.

At Black Rock Spring in late August and early September deer were visiting the spring, partly at night, to drink. The surrounding small canyons with scrub oak and manzanita patches form good cover. On August 31 at 5:30 A.M. a large doe ran from the spring as we approached, and later in the day what was probably the same animal was seen at the water. At 6 P.M. on September 1 as we drove to the spring, a buck was standing near it; it was a mature animal, but we could not count the points of its antlers. At 8:30 A.M. on September 3 a doe came to drink and was watched for 30 minutes from a hiding place in the cover. At times she was within 15 feet of the observer. To the southeast within a mile of the spring were many deer trails, and among

Fig. 100. Buck mule deer in the velvet in the Black Rock Spring area.

manzanita and piñons, in the shade, were several deer beds with leaves and piñon needles in them. A half-mile to the west, a set of shed, 4-point antlers was found near an oak thicket; the two members of the matching set were within 50 feet of one another. Again on April 3 at this point fresh tracks and droppings were seen, indicating that the animals were browsing on scrub oak, but there were fewer tracks than in September and less concentration of them about the spring.

In Lower Covington Flat and the springs below it in Smithwater Canyon, deer were present although we did not actually see them. On April 9, wet, fresh droppings were found and three cast antlers representing three different individuals were picked up in the area of scattered oak brush, manzanita, and piñon on the southern crest of the basin; these were from 3- and 4-point bucks. On August 26 in the willow thicket at the spring, parts of a dried deer skeleton were found.

The adjoining area about Quail Spring had a group of deer working the piñon–juniper and oak slopes. In January fresh tracks were seen about two water holes in a side canyon to the southeast. On October 14 at this locality a half-grown deer was seen, and on the next day three does and a half-grown individual were seen here at 8:50 A.M. In the night preceding at 3:00 A.M. one of us was wakened by a crash of antlers and in the next hour heard this twice more; a low blowing sound was also given. When a light was turned on, a deer was heard to bound away. At 5:45 A.M. that morning the first bucks were seen, 4 different individuals. One had 4-point antlers, another 3, and one 2. Two bucks came within about 8 feet of one another, and one lowered its head and snorted. The others walked by up to the spring but did not drink, hesitating, apparently, to take their attention off the campers. Thus in this period there were evidently at least 8 animals near the spring and the rutting season was in full force.

In the vicinity of Stubby Spring, deer were seen on the ridge and mesas to the north but not about the spring proper, which was utilized heavily, perhaps exclusively, by mountain sheep. On September 11, 1950, in a shallow valley draining north from the divide, five deer were seen, two does, two fawns of the year, and one of undetermined age, but clearly not a buck. These animals were moving as a group and had been working in a wash where rains of a week before had already stimulated green growth in the perennial grass clumps and had brought a few lupines into flower. Purshia bushes in this wash had been grazed by deer, so that they were low mats. On April 11, a specimen of a male deer was taken within 300 yards of this spot among scattered oak brush. It was an old buck, the antlers of the season just starting to form; the "velvet" knobs extended only an inch above the burr. The animal was estimated to weigh about 130 pounds. It was without subcutaneous fat, but the muscles were not in poor condition. The hair of the muzzle was well-worn from the animal's browsing on harsh vegetation.

It was in the full gray winter pelage with conspicuously sooty chest and anterior belly. Its first upper premolar on one side had been worn down so heavily as to have become almost useless, and little was left but the roots. Judging from Taber and Dasmann's (1958) information on age in relation to tooth wear, this buck was in the 8- to 10-year-old class.

In the Pinyon Wells area deer were actually seen only once, although in October fresh tracks were seen on top of our foot and tire tracks several times. In general the deer tracks lasted a long time and probably very few animals were present to make them. On October 14 along the ridge of the divide back of Pinyon Wells, fresh deer tracks revealed the following story. Two deer were traversing a trail single file, then side by side. One thereafter went to the ridge top and the other continued southwest along a hill slope to a point where it made a bed, the fresh-kicked sand showing at the edge of the bed. The animal sooner or later continued on 48 feet to another bed, where marks from swishing of its tail could be seen, made as the animal lay in the "form." Here at the edge of the bed there was a pile of fresh droppings. The deer left this bed to run downhill to the northwest.

The food of deer in the Little San Bernardino Mountains probably consists in large degree of scrub oak and purshia, as already noted, beside some green grass and annuals when available. Mountain mahogany and manzanita may be used, but we have not specifically recorded this. The animal taken above Stubby Spring had Joshua tree blossoms, grass, and liliaceous plants in the stomach beside great masses of material that may have been browsed oak foliage. A Joshua tree near this point had been browsed heavily at deer height, and the yucca spines were clipped back thoroughly as though by a mower.

Longhurst, Leopold, and Dasmann (1952:45) have pointed out that for the deer range in the southern coastal arid lands of California the most critical season is summer. "Populations may be limited . . . by the amount of *succulent forage available during drought periods* and by water supplies rather than by winter forage." We believe this applies particularly to the deer of the Little San Bernardino Mountains. A large mammal such as a deer can sustain temporary water shortage and heating better than a small species, and yet on a sustained basis, with respect to water, it is equally vulnerable and moreover cannot seek relief underground. However, deer, by staying close to drinking water during periods of heat stress, by resting in the shade in midday, and by adhering to the higher elevations in the desert mountains, alleviate the water and heating problem. Probably the environment is no more severe for them in these respects than in parts of the inner coast ranges of coastal southern California.

The mule deer is represented by a low desert race, *Odocoileus hemionus eremicus* in the bottomlands and adjacent desert washes of the lower Colorado River valley. This is not the form occurring in the Little San Bernardino

Mountains, but it may range into the east end of the Monument. Our evidence on this score is unsatisfactory. Tracks were recorded once near Cottonwood Spring, but in retrospect we cannot feel wholly confident that these were of deer rather than sheep. The Coxcomb Mountains, which lie partly in the eastern end of the Monument, are reported to be reached by *eremicus*, the burro deer (Grinnell, 1933:208), apparently on the second-hand evidence obtained by McLean (1930:119). We do not know of any specimens verified from this area to show that *eremicus* is involved.

The specimen we obtained in the Little San Bernardino Mountains has all the diagnostic features of color of the race *Odocoileus hemionus fuliginatus* as given by Cowan (1936:229) in contradistinction to *O. h. californicus*. These are dark color throughout, especially a well-defined blackish middorsal line; broad black dorsal tail stripe extending from tip to base; and dark spots on the lower lip restricted to lateral areas, not meeting in the midventral line. The skull characters consist chiefly of average differences between these two races, and the single Joshua Tree specimen cannot contribute significantly to a racial allocation on this basis. The qualitative feature of the vomer's being closely applied to the basisphenoid is present and thereby relates the skull to *fuliginatus*. The animal, we believe, must therefore be classed with *fuliginatus* despite the fact that Cowan gives the range of this form north only to the San Jacinto and Santa Rosa mountains. The deer of the San Bernardino Mountains are mapped as *O. h. californicus* by Cowan (1936:234), but exact localities and specimens from that range are not given by him. Thus, although one might suppose that the animals in the Little San Bernardino Mountains would be *californicus* because of the somewhat uncertain mapping of that form in the adjacent San Bernardinos, there is actually contrary evidence in our specimen. The more southern race *fuliginatus* could well range north on the west edge of the desert across the eastern part of San Gorgonio Pass to the Monument area. A more or less intermediate or intergrading population is actually to be expected in all of western Riverside County and southwestern San Bernardino County.

Mountain Sheep. *Ovis canadensis*

Other names. Bighorn; desert bighorn.
Description. A stocky, fairly short-legged ungulate (body the size of a deer but legs much shorter), with permanent unbranched horns, recurved ventrally and laterally. Horns massive in adult males; variously developed in young males and in females, in which they may be slender and only slightly curved. Above gray or smoky brown, the rump area white except for narrow black central stripe leading to the short tail; muzzle, ear linings, and posterior belly whitish or light gray.
Range. Originally inner mountainous regions from southern British Columbia, southwestern Alberta, northern Montana, and southwestern North Dakota south to southern Baja California, central Sonora, central Chihuahua, and central Coahuila.
Occurrence in Monument. Common for an ungulate type of mammal. Recorded from: Little Morongo Canyon, Apr. 3; Black Rock Spring, Aug. 30, 31, Sept. 4; Twentynine Palms Highway east of Joshua Tree (road kill; newspaper photograph and report);

Fig. 101. A social group of mountain sheep.

Lower Covington Flat, Aug. 23, 24 *; Stubby Spring, Sept. 6, 8, 10, 12, Oct. 15, Nov. 1; Indian Cove and Rattlesnake Canyon *, Apr. 8; Fortynine Palms; Pine Spring; Pinyon Wells, Oct. 13, 14; White Tanks, Sept. 12; 1 mi. S Virginia Dale Mine; Eagle Mountain, May 16, Oct. 19; Cottonwood Spring *, Apr. 5, Oct. 21–24; Lost Palm Canyon, Sept. 14; Coxcomb Mountains. (Specimens saved consist only of pick-up skulls or skeletons.)

The mountain sheep symbolizes the protection of wildlife afforded by the Monument. This animal can withstand little or no hunting pressure, and it suffers in most parts of its range from competition from grazing. These sheep require undisturbed conditions, and especially in desert areas access to drinking water during parts of the year. The springs and oases of the Monument which are not situated by public campgrounds provide this facility and are the centers for groups of mountain sheep. Most notable is Stubby Spring.

Mountain sheep are gregarious at all seasons, and this circumstance, combined with their habit of stopping to stare at the source of disturbance after running a short distance, has made them very vulnerable to hunters in unprotected areas. The sheep of the Monument, because of protection for several decades, probably have reached a level of numbers reflecting what the desert forage resources of the region can support. Buechner (1960) concluded with respect to the desert range of sheep in general that "forage supplies appeared sufficient for the number of bighorn sheep present, and it is believed that shortage of water, rather than food, limits the number. . . . Competition may arise if the numbers of deer increase in the vicinity of water developments." In this latter connection, we have at times gained the impression that deer and sheep do not use the same water holes in the Monument. For example, Stubby Spring is used largely if not exclusively by sheep although deer are in the near vicinity, and deer chiefly were seen at Black Rock Spring. On the other hand, both species have come to the water there and in Smithwater Canyon. Several water sources in the lower, rockier mountains are used only by sheep, as the adjoining areas are not suitable for deer occupancy. Such seems to be true of springs in Rattlesnake Canyon, at Fortynine Palms, and in and about Eagle Mountain.

The estimated numbers of sheep in the Monument cited by Buechner (1960:60) are 200 in 1950 and 150 in 1955, based on reports provided him by Park Service personnel. We are not aware of how these figures were derived and doubt that any rigorous censusing methods formed the basis for the number for 1950. Indeed, in part the figures may reflect our own casual estimates reported to Park headquarters in that year. We are inclined to doubt that there is real evidence for the decline to be inferred from the figures, and we suggest that the total number for the Monument area is placed too low. In the some 200 square miles of the better-known mountain sheep areas of the Monument, centering on Barker's Dam, there are probably 150 to 200 animals, since there are probably 6 or more watering stations here with groups of 6 to 40 animals focused at each. But also there are bands in and about Pinyon Wells, Black Rock Spring, and the Cottonwood, Eagle, Pinto, and Coxcomb mountains outside this central sector which may well bring the total for the Monument to 300 or 400.

In our field work we have been fortunate to be able to observe sheep at close range on several occasions. A synopsis of certain of our experiences may convey something of their activities to those seeking to learn about these attractive animals.

In the vicinity of Black Rock Spring in early September on the day after a heavy rain, sheep tracks were seen in soft fine sand. The tracks, as well as some that were sketched at Pinyon Wells on October 14, were blunt-toed and parallel-sided and reflected a short gait, with a tendency to toe out (see fig. 102). At another location, on the flats north of Indian Cove, a large ewe and a ram ran at a steady pace across in front of us. The tracks were then checked and found to have a maximum of 30 inches between prints.

At Lower Covington Flat, single animals were twice seen on August 23. One of these was a ram that ran almost leisurely along a canyon slope opposite the observer, crossing the canyon head eventually, tumbling sizable rocks as it clattered along. At this station, below Covington Well on the 24th, a young sheep ran up a canyon wall from a cottonwood thicket where there was water. Here also a skull of a sheep was found, as well as a deer skeleton. We also checked sheep skull fragments in Little Morongo Canyon, where fresh tracks were seen on April 3.

Our most extended observations were at Stubby Spring in September, 1950. Here there is a soft muddy area about 25 feet across. On September 6 this was heavily trampled and grazed by sheep. Here and surrounding the surface water is about a half-acre of chrysothamnus bushes and sedge. The sedges were clipped down by the sheep to within $\frac{1}{2}$ inch of the ground except where they were protected under or among the bushes. One of the sheep was actually seen feeding on the sedge. The area had the aspect of a domestic sheep corral except that it did not have the odor. The following spring, on April 9, the sedges and grasses had greatly recovered, and no sheep had been trampling the area for some time. Probably through the early spring succulent forage was widespread and water was not especially sought.

The animals come to Stubby Spring usually in bands of 3 to 7 at a time. As many as 18 in a day have been reported to us, but these were not seen simultaneously and there may have been some duplication of individuals. On September 8, we saw 7 sheep come to water at 11:45 A.M. They were very tame, as is usual at this spring, and approached within 25 yards while we were in the open talking in normal tones. The group consisted of two adult males, two males of a younger age class, two males of the year, and one adult ewe. The adult female led the group in all instances, and the sheep often bunched together within a 10 x 10 foot space. One of the large rams made a weak attempt to mount the ewe, at the same time giving a grunt. Only the ewe was

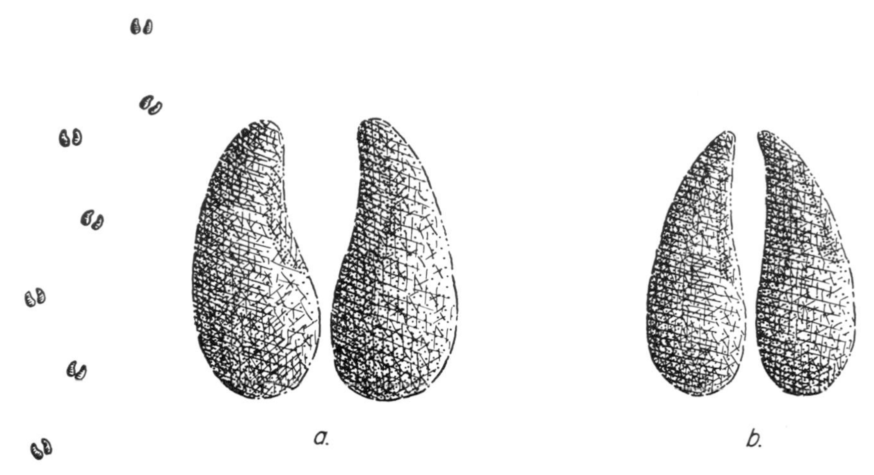

Fig. 102. Diagrams of tracks of mountain sheep (*a*) and deer (*b*), showing the relatively blunt toe impressions of sheep. The trackway of a sheep at the left shows the characteristic short gait and tendency to toe out.

in poor condition, with some of her ribs showing. We were impressed with the way the color of the animals matched the shaded sections of the rock outcrops against which they stood.

On September 10 we returned to the spring and watched a group come in that must have consisted in the main of different individuals from those there on the 8th. We stayed quietly and in partial hiding from 8 to 11 A.M. A group was frightened from the area, but soon after we had hidden, these or others returned. We heard a rumbling belch and a large ram came into view across the small canyon, working up the crest of rock on a lateral ridge. Here he stood, often with his forefeet on the edge of the rock. This animal was in sleek gray brown pelage and seemed to be in prime condition and mature, the testes appearing to be about 8 inches long. Soon other animals were heard disturbing the rocks on our side of the canyon, and a ewe and following lamb were moving toward the water tank in the canyon bottom 30 feet below us. They became frightened and retreated 50 yards, and the ram quickly moved across to join them, at once nosed the ewe's genital area and lifted one forefoot, but made no further effort to mount. The ram seemed less wary than before and more concerned with the ewe than with us. The ewe was lighter gray than the ram and not sleek; occasionally a rib could be detected, but she was not seriously thin. The lamb was about half the height of the ewe and had horns about 2 inches long; hers were 7 inches long.

The lamb stayed always within 20 feet of the female and occasionally it ran at her from 2 or 3 feet away, ramming her in the udder from straight behind. This was a very rough, vigorous punch, and the ewe would lose her footing and have to shift about. Nursing contact was never made, however; there were 5 attempts in 2 hours.

Fig. 103. Mountain sheep near Stubby Spring; left to right, lamb, ewe, ram, and second lamb, September 10, 1950.

The ram now led the party toward the spring. At a distance of 20 feet from us they were frightened by our efforts to take photographs, the ram snorting and bouncing off. Another ewe and her lamb joined the waiting group and the ram sniffed this ewe, but again went no farther in an effort to mount than to raise the forefoot. Once the ram nosed the hind quarters of the lambs but did not raise his foot.

The five sheep began to browse, lambs and adults alike, eating scrub oak leaves at their head height. The ram then fed on leaves of *Yucca mohavensis,* pulling off entire leaves at head height and chewing each down, the white basal part waving about, and the whole stalk being crushed like a stalk of celery. He ate at least three of these.

The five animals now clustered together on a rock spur, often within a fifteen-foot circle. The second lamb, although more adventurous in straying away than the first, attempted, as had the first one, to nurse from its mother; it was a young male. The animals occasionally were alarmed by

Fig. 104. Ewe and ram on rock outcrop above Stubby Spring, where they were waiting to come to water, September 10, 1950.

our movements with the cameras, but when we froze, they stared at us for a short time, then seemed soon to forget the source of the disturbance.

In the Indian Cove—Rattlesnake Canyon area our only direct observation of sheep again indicated the social nature of the animals. On April 8 below the desert willows in the mouth of the wash, two sheep crossed 70 yards ahead of us toward a rocky spur. The lead animal was a large ram. Soon two more followed by 60 yards, the leader a ewe after which came a large ram.

On Eagle Mountain and in Lost Palm Canyon and in the washes near Cottonwood Spring sheep signs are abundant, especially well up on the mountain slopes. At Lost Palm Canyon on September 14 at a deep water hole among the palms at 10 A.M., a clatter of rocks was heard on the slope above. A large ram was sighted on a steep knife edge of rock 30 feet away. It gave a guttural snort, ran up the very steep slope, but then stopped to look back halfway to the crest and again on the crest.

On May 16 on the east shoulder of Eagle Mountain, two sheep were sighted 100 yards ahead of us in a basin of scattered piñon and junipers. We sat quietly and the animals approached downwind. One gave a low, rumbling blat. Occasionally they browsed on scrub oak, and we could hear them forcing their way through tight places in the rigid brush. They came with evident curiosity directly toward us to within 30 feet and finally to 20 feet.

One began circling us, flaring its nostrils very widely; their dark interior looked like a double-barreled shotgun aimed at us. The animal circling us evidently got our scent and walked back rapidly to the other and both ran 50 yards, stopping to look back, but soon continued across the basin upwind. Apparently they sighted us about the time we saw them, but this only aroused their interest and no deep fear. Not until our odor was perceived at close range did they become alarmed. Of course we had remained completely motionless. These animals were fully grown but young males.

The mountain sheep of the Monument are of the race *Ovis canadensis nelsoni,* typical of the Inyo district and the Mohave Desert and their borders.

SEVEN

AMPHIBIANS

TOADS AND FROGS

Western Toad. *Bufo boreas*

Description. Adult $2\frac{1}{2}$–5 inches long. Differs from the red-spotted toad (see p. 351) in being of larger size and less flattened form; parotoids less rounded, and cranial crests weak or absent; white middorsal stripe present. Dorsal pattern of blackish blotches on a pale background; warts brownish; below cream-colored. No vocal sac. Tadpole: resembles tadpole of red-spotted toad but tends toward finer dark flecking on dorsal fin and has shorter upper labial tooth rows.
Range. Southern coast of Alaska south to northern Baja California and southern Colorado. Absent from Colorado Desert and more arid parts of Nevada and Utah.
Occurrence in Monument. Occurs only in western border area. Recorded from: Little Morongo Canyon, Mar. 30 *, Apr. 3 *, May 1 *.

On March 30, 1951, western toads were found at permanent water in the upper part of Little Morongo Canyon. Natural dikes and sharp rock spurs from the sides of the canyon back up water to form pools and sedgy meadows. Other plants in the area were mule fat, mesquite, arrowweed, and Washington palms.

On May 1, 1952, at the Berl Duncan ranch farther up canyon, toads were found in a meadow near several large earthern reservoirs. Mesquite and cottonwoods were present along the canyon bottom. The stream border below the ponds was grown to willows, mule fat, sedge, and tules. Other amphibians in the area were the California and Pacific treefrogs and the bullfrog. Tadpoles of the western toad, ranging to over 1 inch in total length, were found in the stream in flowing water 1 foot wide and 1 inch deep; water temperature at 4:45 P.M. was 72° F., and the air $\frac{1}{4}$ inch above damp sand near the stream, 71° F. The chirruping of male toads was heard.

Eggs are laid in a double row within a long double-layered cylinder of jelly. Ova measure 1.5 to 1.7 mm. in diameter and are black above and pale below.

Seven adult males and an adult female from Little Morongo Canyon are similar in form and color to toads from the San Bernardino Mountains to the west (*Bufo boreas halophilus*). They differ, however, in having larger dark blotches on the lower sides. From the side, especially when the animal is in the light phase, they appear boldly marked with contrasting dark and light color. The markings are best seen from near ground level and would appear to be directed toward animals operating close to the ground. Since toads are attracted to other individuals by movement, the markings may aid detection during nocturnal breeding activity.

The toads in the Morongo Valley area seem to be at present isolated from the nearest populations to the west in the San Bernardino Mountains. However, only a slight increase in rainfall would be required to reëstablish connections which may have existed during, or perhaps since, the last pluvial period.

Red-spotted Toad. *Bufo punctatus*

Other name. Desert toad.
Description. Adult 1½–2½ inches long. When viewed from above, parotoid glands about same size and shape as upper eyelids. Warts reddish orange or buff, usually set in small black blotches which give spotted appearance; below whitish to pale beige, unspotted or with some black spotting, especially on chest. Juveniles usually with bright orange or orange-brown warts and more ventral spotting than adults; undersides of feet yellow. Male averages smaller and darker than female, has dusky brown or grayish olive throat which in life may be buff to yellow-orange anteriorly; usually has brownish pad on thumb and first finger. Vocal sac round when inflated. Tadpole: Black or sooty above; small, evenly distributed dark flecks on tail fins; iris and blotches on belly bronze in life; older larvae often mottled dorsally. Edges of mouth disc with pronounced lateral indentations; no labial papillae on lower lip.
Range. Southern Nevada, Utah, Colorado, and Kansas south to Cape district of Baja California and Guanajuato and from central Oklahoma and east-central Texas to desert slopes of mountains of southern California.
Occurrence in Monument. To be expected at all water holes. Recorded from: Black Rock Spring, July 6; Quail Spring, June 30 *, July 1, 2 *, 3 *, 4, 5, Sept. 8, 10 *, Oct. 14; Stubby Spring, Sept. 6 *, 7 *, 10 *, 12 *; Indian Cove, Apr. 7 *, 8, 11 *, 30, May 1, 8 *, 12, 13 *, 14, Sept. 5 *, Oct. 12; White Tanks, Sept. 12 *; Cottonwood Spring, April 29 *, May 14, 15 *, June 22, 23, July 7 *, 8, 9, Sept. 12, 14 *, Oct. 21–24, 22, 23; Lost Palm Canyon, Sept. 14 *, Oct. 22 *, 23 *.

In the Monument this toad is primarily an inhabitant of rocky canyons and gullies where there are springs or seepages, or where pools form after storms. Clean sand or fine gravel are often present, since favorable breeding sites are seldom found in the steeper scoured portions of the canyons. Good examples of canyon habitat are to be found at Indian Cove (fig. 106) and Lost Palm Canyon. Because of rapid runoff from rains in the canyons, losses of eggs and larvae are probably great there, and thus more favorable habitat

Fig. 105. Red-spotted toad, approximately twice natural size. Photograph taken in Tahquitz Canyon, Riverside County, by Nathan W. Cohen.

Fig. 106. Breeding habitat of red-spotted toads and California treefrogs in rocky pools at Indian Cove, April 7, 1960.

would seem to be spring areas less subject to flooding, as at Quail and Cottonwood springs. Although found mostly in rocky areas, in some parts of its range this toad ventures far out on river bottoms or plains to breed in temporary pools.

In a number of structural and behavioral attributes the red-spotted toad is especially well suited for desert life, and yet like all amphibians it is sharply limited by water. Standing water is required for the duration of the tadpole period and sufficiently high air humidity and soil moisture are needed to prevent desiccation of metamorphosed individuals. Under the desert conditions of uncertain water supply, water shortage is probably an important decimating factor, particularly for eggs and tadpoles. After the dry winter of 1945–1946, half-grown larvae were found at Black Rock Spring (July 6) and Cottonwood Spring (July 7, 1946) in nearly dry pools. It appeared they would not transform in time.

Recently transformed young and adults during the height of the breeding season may be seen in the daytime, sometimes in full sunlight. Tadpoles expose themselves freely to the full desert sun. When algal growth is present, they may take refuge under it, but it is not uncommon to see them in shallow clear pools over coarse sand and gravel where there is little plant cover. On May 1, 1952, in a pool at Indian Cove, five pairs of toads were found in amplexus in bright sunlight at 10:45 A.M. However, nocturnal and crepuscular activity are the rule in adults, when conditions of humidity and temperature are generally more favorable.

The flattened form of this toad aids it in penetrating deeply into crevices and in climbing. Both young and adults evidently spend the winter and dry periods in fissures or other openings at damp locations in the vicinity of the breeding sites. The flat body, with its low center of gravity, helps locomotion over steep rock surfaces. Adults readily ascend the vertical rock walls (3 to 4 feet high) of the Quail Spring tank. However, they are unable to climb the polished walls of deep water-scoured potholes. On May 8, 1953, three carcasses were found floating in a pothole in the otherwise dry stream bottom granite at Indian Cove. The toads appeared to have been dead for one or two days. On October 12, 1953, the hole, then dry, was reexamined, and five mummified individuals were found.

The presence of these toads in areas where the ground is usually moist but air humidity low seems to have placed a premium on the development of the "seat patch," a ventral region of thin skin characteristic of toads, through which they can absorb water from the ground. It is conspicuous in *Bufo punctatus*. The "seat patch" is broadly in contact with the ground when the toad is at rest.

When red-spotted toads are encountered in the Monument, they are usually engaged in some phase of breeding activity (fig. 107). Although in some especially favorable areas large numbers may congregate for breeding, counts

Fig. 107. Red-spotted toads in amplexus in the water at Indian Cove, April 7, 1960. A cluster of eggs may be seen below the adults on the granite floor of the pool.

of adults (mostly males) in breeding aggregations in the Monument (Cottonwood and Quail springs) ranged from 5 to 15, averaging 9. Numbers were slightly higher at Quail Spring. Assuming a balanced sex ratio and considering the limited habitat available, a total of 30 to 40 adults at each of these spring areas seems a reasonable estimate. Turner (1959) estimated population density at Cow Creek in Death Valley as roughly 8 adult toads per acre.

As is generally true with tailless amphibians, a disproportionate number of males and females is found at any one time at the breeding sites. Often only males may be seen. The females apparently do not stay long after depositing their eggs, and in some species the movements of individual females are not well synchronized with one another. Temperature conditions under which breeding occurs in the Monument are unknown, but Turner (1959), at Death Valley, concluded that breeding activity was roughly correlated with mean daily temperatures of about 70° F. or above.

Trilling usually occurs after dark and sporadically in the daytime during the height of the breeding season. It was heard from April to July. The voice is a sustained bird- or cricket-like trill pitched approximately two octaves above middle C. Pitch may vary several whole notes among individuals. Small males tend to have high voices. The trill is clear and rapid, on one pitch, or with a slight rise toward the end. Sometimes discordant or wavery trills are heard when males scuffle in attempts at amplexus, and occasional individuals give a wheezy call. The duration of a single trill averaged 6 seconds (range 2 to 17 seconds), in 30 calls given by three individuals. Intervals between calls of a series were highly variable but were not recorded as less than 5 seconds.

Males sit erect when trilling and seem to display considerable tolerance for one another, for it is not uncommon to see two of them trilling within 6 inches. In a species so limited as to breeding sites, territorial behavior when breeding may be weak. Trilling does not prevent an amplexus, for males at the height of the breeding season will clasp other males. But the trill probably contributes to release. At Indian Cove, on April 30, 1952, a male was watched attempting amplexus with another male. The seized individual fell on his back with the other toad on top. When the prostrate toad trilled, the other one promptly withdrew.

In rock-bound pools with deep water, individuals called from the water's edge where they could rest their hindquarters on the bottom. Others called from stations on the bank. The vocal sac appears white and spherical when inflated. Although females do not trill when disturbed, they may vibrate the chest and upper abdomen.

In the Monument the red-spotted toad apparently breeds whenever it can. Evidence for this is based on the condition of the reproductive organs in adult females, the size range of individuals in the populations, variation in larvae, and dates of capture of recently transformed toads. Gravid females, with ova ready for deposition (1.6 mm. or larger in diameter), have been found over a period of five months—April 30, 1952 (Indian Cove), July 7, 1946, and September 14, 1950 (Cottonwood Spring and Lost Palm Canyon). A recently spent female was found on May 8, 1953 (Indian Cove).

Eggs were found at Indian Cove on May 1, 1952, in a pool of clear water, roughly 12 x 4 feet and 1 foot deep. They were most abundant in the shallows at a depth of 1 to 3 inches where there was an accumulation of sediment, and they were scattered over the bottom, singly or in groups, in a layer one egg thick. Clusters varied from 2 to 20. A female caught on May 8, 1953, at Indian Cove, oviposited in the container in which she was carried to camp. Typically, *punctatus* eggs are laid singly, in short strings, or in loose, flat clusters. There is a single gelatinous envelope (3.2 to 3.6 mm. in diameter) which may become coated with silt. The egg proper, black above and pale below, measures 1.0 to 1.6 mm. in diameter.

Tadpoles within the same pool may show a considerable range in size, suggesting that they may be hatched from eggs laid at different times. At Indian Cove on May 8, 1953, the range in head–body length (preserved specimens) was 7.3 to 11.7 mm. and development ranged from those without limb buds to those with legs 2.5 mm. in length. At the end of the season, on September 10, 1950, at Quail Spring, the range was 7.1 to 13.0, and at Lost Palm Canyon on October 22, 1945, 6.0 to 11.0 mm. At both sites transforming individuals were found. On September 5, 1950, recently hatched tadpoles (2.9 mm. head–body length) were found at Indian Cove.

Recently transformed toads, 10 to 14 mm. in head–body length, have been found over a period of about four months on the following dates: July 3, 1946 (Quail Spring), September 5, 1950 (Indian Cove), September 12, 1950 (White Tanks), October 14, 1953 (Quail Spring), October 22—23, 1945 (Cottonwood Spring and Lost Palm Canyon). A small toad, 17.3 mm. head–body length, found at Indian Cove on April 8, 1951, seemed too advanced to have hatched from eggs laid even as early as mid-February. It probably had transformed in the fall of 1950 and had wintered over. Recently transformed toads may be found in the open, under stones, in niches among rocks and roots, or in mud cracks near water.

Nine young toads collected on October 22 and 23, 1945, at Lost Palm Canyon ranged from 10.2 to 23.4 mm. in head–body length and were graded in size. We have found no evidence for distinct size groups as reported by Johnson, Bryant, and Miller (1948). Indeed, remeasurement of the Providence Mountain (Pachalka Spring) toads studied at an earlier time has failed to yield a convincingly bimodal curve.

California Treefrog. *Hyla californiae*

Other names. Canyon treefrog; formerly *Hyla arenicolor*.
Description. Adult 1 to 2 inches long (fig. 108). A round adhesive disc at tip of each toe; skin slightly rough. Ground color blackish or dark brown (dark phase) to pale gray, never green; no black eye stripe; above gray usually spotted or blotched with dusky; below whitish or cream with yellow or orange on hind legs and in axillae. Male smaller than female, with dusky throat. Vocal sac round when inflated. Tadpole: maximum size about 2 inches in total length; labial papillae on lower lip; eyes do not extend to outline of head when viewed from above; light gray dorsally; throat gray with coarse mottling at sides; tail irregularly mottled with dusky.
Range. Occurs disjunctly in mountains from central San Luis Obispo County of California south to near Rosario, Baja California.
Occurrence in Monument. Present in vicinity of water in canyons. Recorded from: Little Morongo Canyon, Mar. 30 *, 31, Apr. 1, 2 *, May 1; Smithwater Canyon, Apr. 10 *; Indian Cove, Mar. 30–Apr. 3, Apr. 7 *, 8, 11 *, 30, May 1, 8 *, 13 *, 15, Sept. 5 *, Oct. 12; Forty-nine Palms, Apr. 7 *, May 7 *, 8, 16, Oct. 12.

In the Monument, this frog frequents palm oases, springs, and intermittent stream courses in rocky canyons. Typical habitat includes clean rock surfaces, crevices, shade, and, during the breeding season, quiet, clear water.

We failed to find this frog at Quail Spring, where the red-spotted toad was abundant. Plants noted at collection sites were Washington palm, maidenhair fern, mule fat, cattail, and willow.

The frogs are usually found in niches in the rocks within a few feet of water. When at rest the limbs are drawn in under the body helping to minimize loss of water by evaporation through the skin. Color usually harmonizes with the background, and to human eyes the animals are often well camouflaged.

California treefrogs spend little time in the water except during the breeding season. Even then pairs in amplexus may be seen out on rock surfaces nearby. When frightened into the water, these frogs often swim to the other side and climb to a rock niche. Occasionally they seek shelter among debris on the pond bottom.

Many individuals may use the same shelter. On October 12, 1953, 16 adults were found within a distance of $3\frac{1}{2}$ feet, near the opening of a crevice. All were within one jump of water. All but three faced in the same direction, many in contact with one another.

During winter and in dry weather, these frogs withdraw for long periods into the depths of moist retreats. They seem to prefer to rest on rock and have not often been found under stones that rest on soil. Association with rocks is so marked that it is notable that on May 1, 1952, at Little Morongo Canyon, an adult was found after dark perched $1\frac{1}{2}$ feet above the marshy stream bottom on a slanting tule stalk, over 30 feet from the nearest rock wall of the canyon.

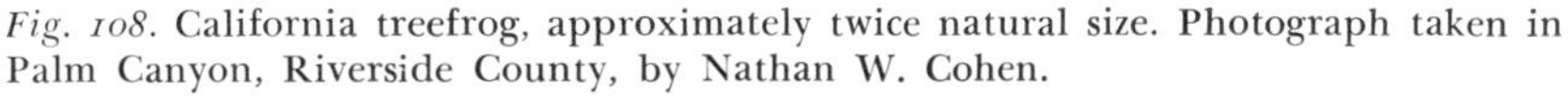

Fig. 108. California treefrog, approximately twice natural size. Photograph taken in Palm Canyon, Riverside County, by Nathan W. Cohen.

Although temperatures may soar very high in dry areas in the sun where treefrogs occur, in the immediate vicinity of the canyon pools moderate temperatures prevail. Body temperatures of the animals are usually below that of the air because of evaporation of water from the skin.

The quacking note of the California treefrog is readily distinguished from the melodious bird-like trill of *Bufo punctatus*, with which it is often associated. At Little Morongo Canyon near the western boundary, *Hyla californiae* may be heard in company with *Hyla regilla*. Although there is some similarity between the calls of these two frogs, there is no difficulty in distinguishing them once they are heard together. The voice of *californiae* is hoarser and lower in pitch; *regilla* has a two-parted call, with the second element terminated with a rising inflection, whereas *californiae* usually gives a single "quack," with little or no inflection. However, at Fortynine Palms Canyon, on May 7, 1953, a group of five California treefrogs was heard calling, all but one producing a two-parted call. The voices of the individual frogs differed in pitch about two whole notes. The duration of a single call was $\frac{1}{2}$ to $\frac{3}{4}$ second. Intervals of silence between chorusing varied from 2 to 10 seconds (average 3 to 4 seconds). There appeared to be no leader, since choruses were started up again by different individuals. Usually two or more frogs participated, and sometimes all called in syncopation. After about 5 minutes of observation, the chorus was interrupted by a blatant duck-like quacking note, heard on previous occasions. The significance of this sound is unknown. There were at least three other chorusing groups at rock-bound pools along a 200-foot section of the canyon bottom.

On March 30, 1951, at Little Morongo Canyon, calling began at dusk, about 6:30 P.M. Approximately 10 individuals of each treefrog species along 50 yards of stream participated. They called from shallow water among grass, algae, and water cress. After 9:00 P.M. *californiae* was seldom heard, perhaps because of falling temperature. Both species may be heard in the daytime.

The California treefrog appears to be less wary than the Pacific treefrog. Individuals may allow close approach, evidently relying on background matching for protection. Sometimes they can be prodded before they will move. At Little Morongo Canyon they called even when a light was shone on them, whereas *regilla* would not.

Breeding was noted at Fortynine Palms on April 7, 1951. Many clasping pairs were seen about the larger, deeper pools in 6 to 12 inches of water, amongst vegetation, and out of water on the rocks. The female of one pair extruded a few eggs when handled, but no eggs were found in the pools. Calling was heard all morning in the shade of the palms. The next day a pair was found in amplexus at Indian Cove. Daytime calling was heard again. On April 10, at Smithwater Canyon, at 4200 feet elevation, a pair was found in amplexus. On May 8, 1953, at Fortynine Palms Canyon, a pair was seen in amplexus, facing upward on the side of a boulder behind a grass clump about 6 inches above the water. Eleven individuals were seen in an area 4 x 7 feet on an 80° rock slope above a pool.

Eggs were found on May 8, 1953, at Indian Cove, in clear water in the deepest part (depth 2½ inches) of a shallow granite basin (5 x 4 feet) in the canyon bottom. The eggs rested on a layer of silt in the vicinity of an overhanging granite ledge, where they were shaded except at midday. Bits of decomposed granite and silt adhered to the jelly capsules. One egg hatched when it was lifted out of the water. Water temperature at 9:55 A.M. was 54° F.

On May 7, 1953, tadpoles of two sizes were found at Fortynine Palms Canyon. They averaged approximately 5 and 15 mm. in head–body length. The smaller ones were dark brown, the larger ones grayish. The latter blended well with the bottom of the pool. The next day at Indian Cove, several sizes of tadpoles were noted with those of *Bufo punctatus*. The mature tadpoles of *californiae* were readily distinguished by their light gray color and pointed tail.

Probably since sometime following the last pluvial period there has been little or no contact between most of the populations of *californiae* in the Monument and those of other arid portions of the range of the species. Although separated by an air line distance of only 3 miles, the frogs at Fortynine Palms and Indian Cove appear to be isolated from one another. Minor differences in color seem to exist. Frogs at Indian Cove generally are tanner above and on the average show more guanin on the rump and undersides of the femora than those from Fortynine Palms. A sample of 11 adults from each locality suggested that there might also be differences in dorsal spotting. A count was made of the number of spots on the back in an area delimited on each side by a line extending from the posterior eye corner through the sacral eminence on that side to the vent, and in front by a line connecting the anterior eye corners. Any spot on or within these boundaries was counted. Animals from Indian Cove ranged from 15 to 49 (average 29) spots, and those from Fortynine Palms 16 to 44 (average 22) spots. Six living adults from Fortynine Palms Canyon were also compared with 9 living adults from Tahquitz Canyon near Palm Springs, Riverside County, an air line distance to the southwest of 34 miles. The animals from Fortynine Palms differed from the latter as follows: dorsal spots darker, smaller, and of sharper outline, iris lighter, throat darker in males, less guanin on underside of hind limbs, yellow color of hind limbs and groin brighter and more contrasting.

Fig. 109. The small stream in Little Morongo Canyon where male California and Pacific treefrogs were stationed for calling on March 30, 1951.

Pacific Treefrog. *Hyla regilla*

Description. Adult 1 to 2 inches long. Resembles California treefrog but skin smoother. Web of hind foot less fully developed (see fig. 110). Above green, brown, or rust to nearly black, differing in different individuals and with changing color phase; usually shows dark blotches; color change from the dark to light phase may occur in a few minutes. A black eye stripe extends from nostril to behind eardrum, but is difficult to see when frog is in dark phase. Males average smaller than females and have olive or dusky throats. Vocal sac round when inflated. Tadpole: maximum size about $1\frac{3}{4}$ inches; differs from California treefrog in having eyes set farther apart, extending to outline of head as viewed from above; throat often yellowish or cream without coarse blotching.
Range. Southern British Columbia south to Cape district of Baja California and from Pacific coast to extreme western Montana and eastern Nevada.
Occurrence in Monument. Recorded only from Little Morongo Canyon, Mar. 30 *, Apr. 1, 2 *, 3.

This is the frog most often heard and seen on the Pacific Coast, where it occurs in a variety of habitats in grassland, woodland, and forest from sea level to 11,000 feet in the Sierra Nevada of California. West of the Cascade-Sierran mountain system the range is nearly continuous, but in the arid lands to the south and east isolated populations occur, perhaps relicts of the last ice age. The one known population in the Monument in Little Morongo Canyon may be such a relict.

The call is a loud two-parted note, *kreck-ek,* the last syllable with rising

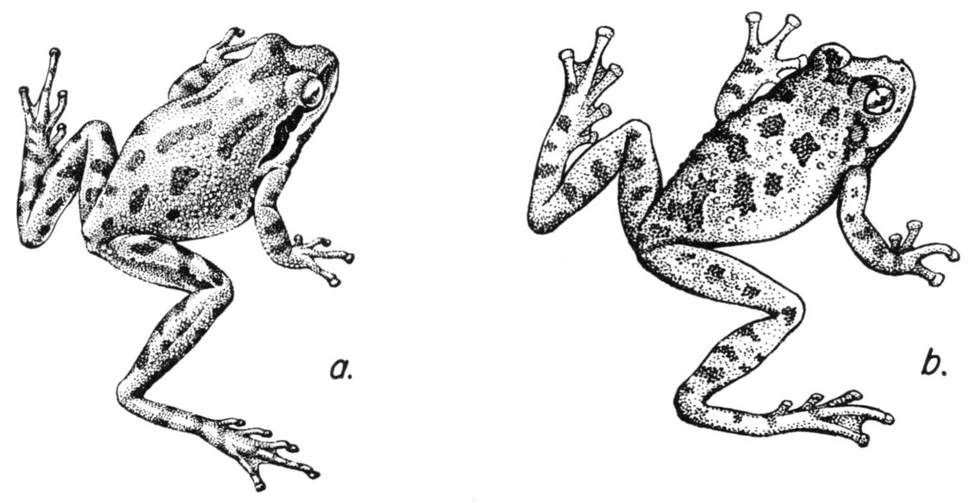

Fig. 110. Comparison of Pacific treefrog (*a*) and California treefrog (*b*).

inflection. Calls are often uttered in rapid sequence, about one a second. The characteristic voice enabled us at once to recognize this species in the stream in Little Morongo Canyon, where it was interspersed with calling males of *Hyla californiae* on March 30 (see p. 358).

The eggs of this frog are deposited in a loose, often irregular, cluster. There are two jelly envelopes, and the ova, brown above and yellowish below, measure 1.2 to 1.7 mm. in diameter.

Bullfrog. *Rana catesbeiana*

Description. Adults range up to 8 inches in length. Eardrum large and conspicuous, with a ridge extending around its rear margin from behind the eye. Above dark olive to greenish; green color especially evident on head; dusky spots on back and cross bars and spots on hind limbs; below whitish, often with a tinge of yellowish on chin and hindquarters. Male has yellow throat and eardrum is larger than eye. Tadpole: maximum size about $5\frac{1}{2}$ inches; labial papillae confined to the sides of the mouth disc; sides of disc indented; above olive-green speckled with black; below whitish to yellowish.
Range. Widespread east of Rocky Mountains. Introduced at many localities in western United States.
Occurrence in Monument. Recorded only in Little Morongo Canyon, Apr. 25, May 1.

At the Berl Duncan ranch at the head of Little Morongo Canyon, bullfrogs have been introduced along with perch, bluegill, crappie, and bass. They were reared in several large artificial ponds. On a visit to the ranch on May 1, 1952, bullfrogs were frequently heard calling during the night.

The eggs of this species float at the water surface adherent to one another in a roughly disc-shaped raft often several feet in diameter; 10,000 to 20,000 eggs may be laid by one female. Ova are surrounded by a single jelly layer, measure 1.2 to 1.7 mm. in diameter, and are black above and white or cream below.

EIGHT

REPTILES

TORTOISES

Desert Tortoise. *Gopherus agassizi*

Description. Adults range up to 1 foot in shell length. Carapace high and dome-shaped with scutes bearing prominent growth ridges. Shell yellow-brown to dark brown. Limbs elephantoid and tail short. Male averages somewhat larger than female, with concave plastron and longer gular horn and tail. Hatchlings are about 1½ inches in shell length, have a transverse furrow across the plastron between the limbs, and an "umbilical" scar. The shell may remain flexible until the fifth or sixth year of life.
Range. Mohave and Colorado deserts from southern Nevada and southwestern Utah south to southern Sonora and from desert slopes of mountains of southern California to southern base of Arizona plateau and western slope of Sierra Madre Occidental.
Occurrence in Monument. Widely distributed except on the steep slopes and tops of the more rugged mountains. Recorded from: 3 mi. S Warren's Well, Aug. 31; 10 mi. W Twentynine Palms, May 12 *; Indian Cove, May 8, 13 *, 14, Sept. 5; 3 mi. S Twentynine Palms, May 8; 3 mi. E Twentynine Palms, Apr. 25 *; Pinto Basin, Apr. 6, 12, 26, 30, July 6, 7 *, 11 *, Sept. 11, 13, 15 *, Oct. 13, 14; Virginia Dale Mine, Apr. 24, July 12, 15 *; 1 mi. S Virginia Dale Mine, July 14; Cottonwood Spring, Feb. 7, 22*, Apr. 12, 17 *, July 8 *.

Tortoises are most common in the vicinity of dunes and washes wherever they can find soil conditions suitable for construction of burrows and the plant growth upon which they feed. Although occasionally found among rocks, where they may go in search of food, they prefer the easier footing of open terrain. It is not uncommon to encounter them on roadways. They may range widely from their den sites. Woodbury and Hardy (1948) found that individual ranges in Utah varied from 10 to 100 acres. On its excursions a tortoise may spend the night in the temporary shelter of a bush, rock, or opening in the ground.

Two kinds of burrows are constructed: temporary summer holes, 3 or 4 feet long, usually dug in banks or beneath bushes, and more permanent ones, usually constructed in the banks of washes. These may be 8 to 15 feet long, occasionally 30 feet, and are usually horizontal. Burrows have an arched roof and are wider than high. Measurements of the openings of three burrows were as follows: 8″ x 4½″, 6″ x 4″ and 12″ x 4″. Burrows are common in the Pinto Basin, where they may be found in the sides of sand hummocks, beneath bushes, and in the vicinity of kangaroo rat burrows, which the tortoises probably modify for their own use.

Tortoise signs in the form of tracks, droppings, and skeletal remains are found more often than the animals themselves. A resting tortoise can be easily overlooked because of its resemblance to a stone. In sandy areas tortoises can be found by following their tracks. Direction of movement is indicated by heaped-up sand at the rear of the footprints. Tracks indicate that they commonly follow the easiest course, even avoiding small plants that would easily be crushed by their weight but which would, nevertheless, impede their progress.

Tortoises eat grasses (brome and mesquite grass), the blossoms of desert composites (*Encelia* and others), and probably almost any plant when food is scarce. On July 12, one was found near the Virginia Dale Mine eating short, dry grass.

When opportunity affords, as after a thundershower, tortoises may drink deeply. A captive increased its weight 41 per cent after a single long drink (L. Miller, 1932*a*).

When picked up, the animal may release a copious amount of a nearly

Fig. 111. A young desert tortoise about 6 inches long.

odorless, tea-colored urine, especially if it is turned on its back. The bladder has a large capacity, which apparently is important to the tortoise in conserving and storing water.

Activity occurs in the daytime in spring, early summer, and fall, and sometimes at night in summer when daytime temperatures are high. Cold weather is spent underground in dens. Low thermal tolerance permits these reptiles to be active early in the morning and on overcast days. On July 11, one was found abroad at 6:40 A.M. and on the 15th another, sunning at the mouth of its burrow, at 5:10 A.M., just after sunrise. Fourteen cloacal temperatures taken over the years, of individuals abroad in the field, averaged 86° F. with a range from 63° to 97°.

Mating occurs in spring and summer but has not been seen by us in the field. A captive male from Pinto Basin was placed with a pet female in Berkeley on October 20, 1953. He promptly began courting, placing himself in front of her whenever she turned aside, and approaching her with his shell held high and head bobbing. He frequently bit her legs and shell and lunged at her with his gular horn, retracting his head at the moment of impact. The blows were struck at the front of her shell, and his gular horn often entered the space between her head and forelimb. In attempting copulation, he had difficulty in maintaining position. He mounted from the rear and placed his forefeet on top of her shell, but his feet tended to slip off when she tried to crawl away. Whenever she stopped, he assumed a more erect position, extended his hind legs slightly and then suddenly flexed them, bringing the rear margin of his shell against the ground with considerable force. His tail was protected by the recess between the anal shields. The "pumping" action, executed with mouth agape and accompanied by a hissing sound each time his shell struck the ground, continued as long as the female was quiet. With each movement, the male swung his tail inward, endeavoring to push aside her tail in efforts to expose her vent. However, she kept her tail curled firmly downward. The short finger-like copulatory organ, which is everted through the vent opening, was not seen.

Robert Heckley (in correspondence) frequently observed nesting of captive tortoises near Barstow in the Mohave Desert. An urn-shaped hole was dug in firm soil with the hind legs. The hole extended downward and forward, in the direction in which the tortoise faced when digging. The brood chamber was an inch or so below the surface. The eggs were directed forward into this chamber with the hind feet. The tortoise usually urinated copiously on the earth plug used to close the nest. Nesting occurred in open areas where there were no bushes.

The eggs are white, oval in shape, hard-shelled, and about 45 mm. long. Two to 9 (often 5) constitute a clutch. Most laying apparently occurs in June. Eggs deposited in captivity may require about 3 months for development.

An infant tortoise found at 9:30 A.M., February 7, 1947, by Loye Miller in the canyon south of Cottonwood Spring was far from water. It had not travelled a great distance because its claws were unworn and the stalk of the yolk sac was still evident. It must have been hatched the previous October.

Although well protected by their shell and heavily armoured legs, tortoises are not immune to attack. A large, scarred adult male was found on October 14, 1953, near the east end of the dunes in Pinto Basin. The horny covering of the gular horn was missing, and the scales of the front surfaces of the forelegs were sheared off. Bone showed at the end of the horn and between the scutes along the anterior margin of the carapace. Perhaps a coyote had attempted to chew its way in. The tortoise was seen again in the same area on April 26, 1955, about a year and a half later. There had been no change in the appearance of the damaged parts.

GECKOS

Banded Gecko. *Coleonyx variegatus*

Description. Adult 2½–3 inches in snout–vent length. A lizard with smooth, delicate skin and large, pale yellow eyes with movable eyelids and vertical pupil. Tail usually conspicuously enlarged at base. Ground color above pinkish or pale yellow, spotted or banded with reddish brown; below usually plain white or pinkish. Adult males have preanal pores, postanal swellings on underside of tail, and spurs on each side of tail at its base.

Range. Southern Nevada and southwestern Utah south to northwestern México and from coastal southern California and western slope of southern Sierra Nevada to central Arizona and southwestern New Mexico.

Occurrence in Monument. To be expected throughout; probably most common at lower elevations. Recorded from: Twentynine Palms, Apr. 28 *; Pinyon Wells, Oct. 15, 16 *, 17 *; Pinto Basin from near Cholla Garden to vicinity of Cottonwood Spring, Apr. 10 *; Cottonwood Spring, Apr. 28, May 1, Sept. 13 *.

The banded gecko, even though of delicate, almost fetal appearance, ranges widely in the hottest, driest country in North America from below sea level in the Salton Basin and other desert sinks to around 5000 feet in the piñon–juniper belt. In the desert it frequents terrain grown to creosote bush, Joshua tree, mesquite, catclaw, cactus, and other xeric plants. Although it seems to prefer rocky environments, it ranges through a variety of ground conditions to completely sandy ground, and in the Monument it occurs among the rocks of the Little San Bernardino Mountains and down onto the sandy floor of Pinto Basin.

Like most geckos this species is nocturnal and sensitive to high temperatures. Thus it often seeks deep rock crevices and rodent burrows as daytime retreats. It is almost impossible to find geckos in such places, but they can be observed when they take refuge under cap rocks, or stop for a time under boards, fallen yucca stems, road signs, rock flakes, dried cow dung, and other litter. We found most individuals under cap rocks rather than beneath rocks resting on soil.

Fig. 112. Banded gecko, one and one-quarter times natural size. Photograph taken by Nathan W. Cohen.

The period of activity in southern California seems to be from March to October, with a peak in May, as judged from numbers of individuals found on roadways at night (Klauber, 1945). On a given night the first ones may be abroad before it is fully dark, but peak activity is not reached until about two hours after sunset. The banded gecko seems to be the most cold-tolerant of our desert lizards, individuals having been found active on highways at night at air temperatures of 60° and 62° F., even during strong winds. However, it prefers higher air temperatures, and Klauber reports that 40 per cent of those he found on highways occurred at air temperatures of 80° to 84° F. (range 60° to 97° F.). We have 14 cloacal temperatures of individuals active on the road at night. The records were obtained in April, from 7:30 P.M. to 12:55 A.M., and averaged 76° F. (range 70°–82° F.). In addition, one record was obtained of an individual in a crevice. Its temperature at 9:15 A.M. was 88° F.

One usually finds these geckos singly, but they sometimes occur in pairs, and on one occasion we found four under the same rock. When first exposed or touched, they may squeak and stand stiff-legged with back swayed and tail elevated and arched. The tail may be moved in a series of undulations. Once when one of these lizards was dropped in an attempt at capture, it lay where it fell, on its back without moving, until it was picked up.

When encountered at night, geckos sometimes are found crawling with the tail curled forward, spitz dog-fashion. The position of the tail, and accompanying writhing movements, may help focus a predator's attention on the tail which, if seized, can be shed quickly at its constricted base. When cast, the tail thrashes about violently and probably further distracts the lizard's enemies.

Geckos feed on insects, spiders, and other arthropods that are abroad at night.

In mating the male holds the female by her neck skin and curls his hindquarters under hers. His cloacal spurs may aid in sensing and maintaining position. Two eggs typically form a clutch. Evidence from captives suggests that more than one clutch may be laid in the course of the breeding season. Caged individuals laid eggs on the following dates: May 12, 18, 25, June 21.

Although the banded gecko is intolerant of high temperatures and extreme dryness, it is able to live in harsh desert environments by being active at night when temperatures are low and humidity is elevated. Water is obtained from its food. During the day, especially during hot, dry periods, it retreats deep below the surface. Thus its microenvironment, like that of many desert reptiles, is far less extreme than that which meets the eye of the casual visitor to the desert.

To observe geckos the best procedure is to drive dark-paved roads at night. In the Monument, on April 10, five adults were found in this way in Pinto Basin between 8:15 and 11:30 P.M. A gecko may move smoothly across a road, resembling a small piece of paper blown by a light breeze. In the headlights of a car they appear white against dark pavement and when at rest may be mistaken for a small pale stick.

IGUANID LIZARDS

Desert Iguana. *Dipsosaurus dorsalis*

Other name. Desert crested lizard.
Description. Adult 4–5½ inches in snout–vent length. A light-colored, round-bodied lizard with a rather small rounded head and long tail. Scales small and granular on sides of body and weakly keeled on back; single row of slightly enlarged scales down middle of back. Above white, pale gray, or beige, obscured on back with a network and barring of gray or brown; belly white. During breeding season adults of both sexes develop pink or orange-brown patch on each side of belly.
Range. Southern Nevada and extreme southwestern Utah south to Cape district of Baja California and to Sinaloa, and from desert slopes of the mountains of southern California east to central Arizona.
Occurrence in Monument. Widespread at low elevations; all our records from below 3000 feet. Recorded from: Little Morongo Canyon, Mar. 31; 4 mi. N Twentynine Palms, July 25 *; Twentynine Palms, July 13 *, Sept. 12 *; Blondy, Pinto Basin, May 18 *; Pinto Basin sand dunes, Apr. 6 *, 11 *, 12 *, May 8 *, July 6, 7 *, 15; Pinto Basin, Apr. 10, 30 *; Virginia Dale Mine, July 12 *.

This is a lowland species that ranges from below sea level in desert sinks to an elevation of at least 3200 feet in the Providence Mountains, California. It occupies a great variety of ground conditions—sandy flats and dunes, washes, alluvial fans, hardpan, and rocks—but it seems to be most common in sandy areas. The creosote bush is a good indicator of iguana habitat in the United States, but to the south the vegetation in its habitat is arid semitropical scrub.

Typical desert iguana country in the Monument is found in Pinto Basin. It consists of hummocks of sand topped by creosote and other bushes. There is sufficient compacted sand for maintenance of burrows, which are often concentrated about the hummocks, and there is enough vegetation to supply blossoms, leaves, seeds, and insects, upon which the lizards feed.

The iguana is the most heat-tolerant of our desert lizards, a characteristic that may relate to the fact that it lives on or near the heated ground surface and, being in large part herbivorous, perhaps must spend more time abroad gathering and assimilating food than do wholly insectivorous or carnivorous species of lizard. The mean body temperature of normal activity has been recorded as 107° F. by Norris (1953; 64 temperature records). Our figure, based on considerably fewer temperatures (23) is somewhat lower—104° F. (range 86° to 114° F.). Our temperatures, mostly of adults, were taken between 8:35 A.M. and 12:40 P.M. from April 10 to July 12. The highest temperature of normal activity ever recorded for a North American lizard appears to be one obtained in this species—115° F., reported by Norris (1953). A comparable figure, 114° F., has been obtained by us. It was recorded in an adult caught on April 25 at 11:50 A.M. outside the Monument. The ani-

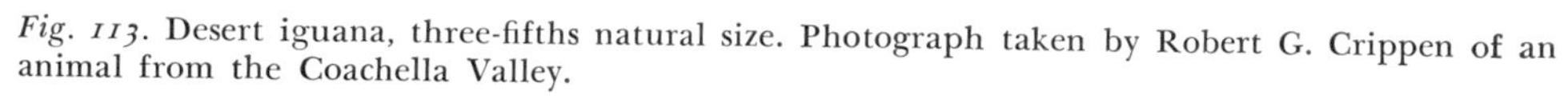
Fig. 113. Desert iguana, three-fifths natural size. Photograph taken by Robert G. Crippen of an animal from the Coachella Valley.

mal lay sprawled belly down on hard-packed sand in full sunlight, 10 feet from the nearest shade. It permitted us to come within 6 feet. When noosed, it was docile, yet bright-eyed and alert. Like the individual recorded by Norris (1953), it was shedding, which may account for its high temperature.

We have little information on tolerance of thermal extremes. One iguana carried in a collecting sack on a warm day in July was overcome by heat at a body temperature of 121° F., but it recovered fully when cooled. Emergence at low temperatures was noted in Pinto Basin on April 10. A basking adult was found on the east side of a creosote hummock at 11:20 A.M., on a cloudy, cool day. It was easily caught by hand and had a body temperature of 82° F. At noon a juvenile emerging from a burrow on the east slope of a dune had a body temperature of 80° F. Only its head and forelimbs were out. It also was caught by hand.

Perhaps low temperatures limit the northern distribution of this heat-adapted lizard. In these iguanas the thermal threshold for emergence appears to be high, since they often do not appear until most other species are abroad. However, in the heat of summer they may be out early and occasionally are found on roadways after dark. One was seen at 5:10 A.M., just after sunrise on July 15.

The iguana is a good burrower and constructs tunnels that often lead to an underground chamber in which it retreats during the heat of day, at night, and during periods of hibernation. At night it plugs the surface openings with earth, probably as protection against desiccation and snakes and other predators. It also uses burrows of other animals.

Although iguanas are somewhat gregarious, especially when attracted to a locally abundant plant food source, both field and cage studies indicate that peck orders exist and one male is dominant. Bobbing ("pushups") and chases may be seen when a subordinate or wandering male enters the territory of a dominant one. The intruder may be pursued 10 to 20 yards. Occasionally fighting occurs. The combatants approach one another, crouching and presenting their sides, often facing in opposite directions. The throat skin is extended and the sides are flattened, bringing into view the orange-colored belly patches. Suddenly one of the contestants may lash out with its tail, curling it around in front of the opponent to the opposite side, where it may strike with an audible slap. Tail slapping may be repeated until one individual gives way. A chase usually follows. According to Carpenter (1961) both sexes may fight in this way. Sometimes the combatants will stand shoulder to shoulder, facing in opposite directions, leaning and pushing as if trying to make the other give ground.

Both plants and animals are eaten. In spring iguanas feed on the blossoms of the creosote bush, but not exclusively. On April 25 outside the Monument, we found one with a fragment of a lupine pod in its mouth. In summer and fall iguanas eat the leaves of the composite *Decoria canescens*. In summer they also feed quite extensively on insects and carrion. Fecal pellets are regularly ingested. Perhaps, as in certain herbivorous mammals, reworking of materials that have passed through the digestive tract releases additional essential nutrients.

Foraging iguanas may climb high in creosote and other bushes to get buds and blossoms. On April 25 outside the Monument, a young one was found at a height of 3 feet in a creosote bush that had both flowers and seeds.

Since plants are such an important part of their diet, these lizards are seriously affected by droughts, even though they are widespread, well-established desert denizens. When there is poor growth of annuals and desiccation of the creosote bushes, emaciated iguanas may be found. A thin, perhaps desiccated individual was found in the Monument on April 6, 1951, and several in similar condition were found March 26 and 27, 1956, near Thousand Palm Junction in the Coachella Valley.

Breeding occurs in April and early May, and the sexes may be found in pairs until mid-July. A male obtained on April 25 everted its hemipenes when caught, indicative of its state of breeding readiness. A breeding male may rapidly bob its head as he approaches a female. He may nudge her flanks. In mating, he holds the skin of her neck in his jaws and curls his hindquarters under hers.

Egg laying occurs in late June and early July, 3 to 8 eggs forming a clutch. Gravid females disappear from the surface, often well in advance of oviposition; thus one seldom finds females heavily laden with eggs. They emerge again in mid-July, spent and often thin-looking. Hatchlings appear in August and remain abroad a month or so in the fall after most adults have gone into hibernation.

The geographic race that occurs in the Monument is *Dipsosaurus dorsalis dorsalis.*

Collared Lizard. *Crotaphytus collaris*

Description. Adult 3–4½ inches in snout–vent length. A stocky lizard with large head well set off from neck and with very long tail. Dorsal scales granular. A pair of conspicuous black bars across neck and shoulders; dorsal coloration varies greatly depending on locality and color phase—greenish, brown, pale gray, or yellowish often marked with light spots; usually dull yellow or rusty cross bars present on body; tail usually pale, spotted with brown. Males average larger than females and usually have blue or blue-green color on throat (in area of Monument) and bluish-slate patches on sides of belly. After copulation females develop orange or reddish spots and bars on sides of body and neck.

Range. Southeastern Oregon and southern Idaho south to central Baja California and northern Zacatecas; extends from eastern base of Cascade–Sierran system and desert slopes of mountains of southern California to eastern Missouri and central Texas.

Occurrence in Monument. Widespread in rocky areas. Recorded from: Lower Covington Flat, Apr. 9 *, Aug. 25 *; Stubby Spring, Sept. 11 *; Fortynine Palms, 3000 ft., Apr. 7 *, 9 *, May 7 *; ½ mi. S Virginia Dale Mine, July 14 *; 1 mi. NE Eagle Mountain, 3500 ft., May 14 *; north side Eagle Mountain, Oct. 21 *.

Plate VII. Collared lizard; photograph by Nathan W. Cohen.

The collared lizard is an inhabitant of open rocky places with sparse vegetation. Requirements, as pointed out by Fitch (1956) for the species in Kansas, seem to be (1) rocks that provide cover, basking places, and lookout posts from which the lizards can see their prey, (2) openness of plant growth, permitting unimpeded running, (3) a food supply of large terrestrial, diurnal insects, and (4) adequate warmth.

In the west these lizards are found on rocky slopes, rock outcrops of gullies, and on boulder-strewn alluvial fans. Rocks may vary in size from a few inches to several feet in diameter. Vegetation may consist of sagebrush, creosote bush, cactus, bunch grass, and open growth of chaparral. In the Monument, we have found this lizard ranging up into the piñon belt, and it is known to reach 5700 feet in this zone in the Providence Mountains.

In keeping with its active, predatory habits and the openness and warmth of its rock-strewn habitat, the collared lizard has a fairly high thermal optimum. We have recorded body temperatures in 19 individuals over the period from April 9 to August 25 from 9:15 A.M. to 6:25 P.M. The average temperature was 99° F. (range 92°—107° F.). This corresponds closely to the average obtained by Fitch (1956) in his study of the species in Kansas. He reports 100° F. as the "probable" optimum and 95° to 106° as the preferred range; the observed range was 69° to 110° F., based on 425 temperatures.

Collared lizards hunt by sight and usually seize their prey with a rush. They may be bipedal when running at high speed. Grasshoppers, crickets, beetles, wasps, cicadas, moths, spiders, and lizards are eaten. We have often fed them side-blotched lizards in captivity, and they are known to eat them in the wild. They have also been reported feeding on fence lizards, whiptails, horned lizards, and skinks. At Fortynine Palms we dropped two side-blotched lizards into a quart jar with a collared lizard. Later one side-blotch was missing, evidently having been eaten. A whiptail lizard confined with a collared lizard was quickly killed when the latter crushed the whiptail's head with a bite from its powerful jaws.

Enemies of the collared lizard are hawks and snakes. The coachwhip and other fast diurnal snakes no doubt catch them. Collared lizards defend themselves by biting and by inflating the body. They seek refuge under rocks, in crevices, or in mammal burrows. They are not particularly adept at digging burrows themselves.

There is marked territorial behavior in this species, and in areas of favorable habitat where there are aggregations of these lizards, a dominance order among males may exist, as noted by Fitch (1956). Females also show some aggressive behavior, but it is not as pronounced as in males. (For further details on the behavior of this species, see Fitch's monographic study.)

Two to 24 (average 7 or 8) eggs may compose a clutch. A female, 85 mm. in snout–vent length, caught in Surprise Canyon in the Panamint Mountains on June 19, contained 2 large eggs ready for laying. Fitch found that females dug tunnels in damp earth under stones and that the nests were plugged with earth.

The race in the Monument is *Crotaphytus collaris baileyi,* the widespread southwestern form of the species.

Leopard Lizard. *Crotaphytus wislizenii*

Description. Adult 3½–4½ inches in snout–vent length. A robust lizard with a spotted pattern. Head large and distinct from neck and tail long. Dorsal scales granular. When in medium to dark color phases, cross bars are evident on body and tail but spots may become obscure; ground color above light to dark gray or beige; belly whitish or yellowish; gray streaks or spots on throat. Males with large postanal scales. Females average larger than males and when gravid have bright orange markings.
Range. Great Basin and desert regions from Columbia River in northeastern Oregon south to southern Baja California and northern Sonora and Chihuahua; extends from San Joaquin Valley of California east to western Texas.
Occurrence in Monument. To be expected throughout, except on tops of higher mountains. Recorded from: Little Morongo Canyon, 2000 ft., Mar. 31 *; Black Rock Spring, Aug. 30 *; Warren's Well, May 21 *; Covington Flat at Monument boundary, May 19 *; Indian Cove, Apr. 7 *, May 1 *; Pinto Basin, Apr. 11 *, 12 *, 13 *, 30 *, May 8 *, July 11 *, 15 *, Oct. 13 *; Cottonwood Spring, July 14 *; 3½ mi. NNE Cottonwood Spring, Apr. 30 *.

The habitat requirements of this lizard seem to be open terrain firm enough for running, bushes and animal burrows for shelter, and sufficient warmth and plant growth to provide an adequate supply of lizard and arthropod food.

The leopard lizard is a lowland form probably not ranging much above 5000 feet. It prefers broad valleys and plains of arid and semiarid regions where there are stretches of hardpan or compacted sand. Patches of rocks and loose sand may or may not be present. Vegetation is sparse and usually consists of bunch grass and scattered brush such as creosote bush, sagebrush, alkali bush, burro bush, mesquite, and other low growth. Thick ground cover may impede running, as noted in areas grown to non-native grasses in the San Joaquin Valley.

In the Monument we have found these lizards on the sandy flats and dunes of Pinto Basin, among creosote bushes and galleta grass, and on up to the piñon belt at Black Rock Spring and Warren's Well, above 4000 feet. At Indian Cove on May 1, a leopard lizard was caught at the top of a massive rock 30 feet from the ground.

Leopard lizards are usually seen with great difficulty. They tend to lie in wait for prey in the broken light and shade near the outer edge of a canopy of bushes or at the edge of the cast shadow. One was seen in Pinto Basin at the edge of weak shade, 5 feet from the canopy of a creosote bush at 8:30 A.M. It may have been basking and have moved to the shadows on our approach, for its body temperature was 92° F., the lowest we have obtained for the species.

Leopard lizards feed on cicadas, grasshoppers, robber flies, bugs, beetles,

Fig. 114. Leopard lizard, approximately four-fifths natural size. Photograph of an animal taken in Little Morongo Canyon on March 31, 1951.

bees, caterpillars, and lizards such as whiptails and side-blotched lizards. We saw one whirl about and leap well over a foot in the air to catch a large robber fly. Leopard lizards have also been noted at road kills, to which they may be attracted in search of insects.

These are fast lizards, capable of catching the smaller lizards upon which they prey while on the run. The average body temperature of active leopard lizards has been recorded as 100° F. (range 92° to 106° F., based on 15 animals). These records were obtained between 8:30 A.M. and 3:15 P.M. from April 12 to October 13.

Gravid females develop orange spots and bars on the sides and an orange suffusion on the underside of the tail. These colors probably develop a few days after mating, as in the collared lizard. Their function seems to be to thwart copulation attempts by males after fertilization has occurred. Females with large ova and postbreeding colors have been seen 3 miles southwest of Warren's Well on May 21, in Pinto Basin on April 11, 13, and 30, and near Cottonwood Spring in mid-May.

These lizards may lay 2 to 7 eggs, with 4 an average-sized clutch, in late May and June. Young appear in August and September.

The race in the Monument is *Crotaphytus wislizenii wislizenii.*

Chuckwalla. *Sauromalus obesus*

Description. Adult 5½ to 8 inches in snout–vent length, the body broad, somewhat flattened with loose folds of skin on sides. Tail tapers to blunt tip. Scales small, smooth, and closely set. Males with black foreparts and limbs, variously flecked with pale gray or whitish; back usually light gray or red, the tail pale gray or straw-color. Females grayish or brownish with dark head and limbs, usually with broad cross bands on body and tail. Juveniles have conspicuous yellow and black bands on tail.

Range. Deserts from southern Nevada and southwestern Utah south to northeastern Baja California and northwestern Sonora; desert slopes of mountains of southern California east to east-central Arizona.

Occurrence in Monument. Rocky areas in creosote bush and yucca belts. Recorded from: 5 mi. W, 1 mi. N Joshua Tree, Oct. 16 *; Quail Spring, May 21 *, July 1 *; 1 mi. NW Fortynine Palms, Apr. 7 *; Fortynine Palms, Apr. 7, 8, 9 *, May 7 *, Oct. 12 *; 5 mi. SW Twentynine Palms *; Pinto Basin, Apr. 13; south side Pinto Basin (vicinity of Old Dale road junction), 2500 feet, Apr. 26 *, 30 *, May 1 *; Virginia Dale Mine, Apr. 24 *, July 14 *; north side Eagle Mountain, 4500 feet, May 16 *; Cottonwood Spring, Feb. 7, Apr. 11 *, 12 *, 29 *, 30 *, May 17 *, Sept. 13, 14; Lost Palm Canyon, Apr. 29 *, Sept. 14 *.

This lizard is strictly a rock dweller. When basking it can easily be found by simply driving slowly and keeping a watchful eye on the rocks. A chuckwalla may soon be seen poised on a rocky eminence or sprawled on the side of a boulder. On the light-colored rocks so prevalent in the Monument, it is very conspicuous. If you leave your car for a close look, the lizard will usually take refuge in a crevice.

Although these lizards seem to prefer areas of massive rock, they also occur in isolated rock piles with boulders no more than 2 feet in diameter. They frequent canyons, mountain sides, alluvial fans, borders of washes, and gulleys—wherever there are suitable rocks. Since they depend on plant food, some vegetation is always present—low-growing desert bushes and annuals, in season. One may be impressed by the scarcity of plants. It is difficult to see how such a large herbivore can exist in some of the more barren areas it frequents.

Fig. 115. An adult chuckwalla, approximately three-fifths natural size. Photograph by Nathan W. Cohen.

Fig. 116. A young chuckwalla, natural size, showing banded tail pattern. Photograph of an animal taken near Banner, San Diego County, California.

Although we have not found these lizards in the piñon habitat in the Monument, we have found them at the borders of it on Eagle Mountain, and old scats found near Pinyon Wells on October 17 were thought to be of this species. In general the creosote bush is a good indicator of suitable conditions. When this plant is found in rocky terrain, chuckwallas are almost certain to be present.

Like so many other desert reptiles, this lizard, although able to live in the hottest parts of the desert, does not have an unusually high thermal tolerance. Its rock-dwelling habits ensure its protection from thermal extremes. Of 17 cloacal temperatures obtained from basking or fully active individuals from March 25 to July 14 (7:15 A.M. to 4:00 P.M.), the average was 97° F., and the range 81° to 106° F. The highest record was from a juvenile obtained on May 7, at 4:00 P.M., at Fortynine Palms.

Chuckwallas tend to be late risers. Two factors may be responsible: their size and the insulation provided by their rocky environment. Chuckwallas have over ten times the bulk of side-blotched lizards and warm far more slowly. In summer, however, they may emerge shortly after sunrise. We found an adult female abroad at 7:15 A.M. on July 14, at Virginia Dale Mine. One might expect that they would also be slow to emerge from hibernation. However, Loye Miller received a very large individual (16½ inches long) from campers at Cottonwood Spring obtained in the second week in February. The animal was dark and sluggish. The conditions of capture are not known.

After a period of basking, these lizards leave the rocks and forage on the leaves, buds, flowers, and fruits of a variety of desert plants. The creosote bush is the staple and provides food when nearly all else fails. The incense bush (*Encelia farinosa*) is heavily used in many areas, the lizards climbing to the top of the basketwork of branches to feed on the yellow blossoms. Other plants eaten are the indigo bush, desert mallow, desert tea (*Ephedra*), and burrobush (*Franzeria*).

In years when there is a profuse bloom of desert annuals, chuckwallas find food abundant and varied; 1949 was such a year. On April 11, at Cottonwood Spring, chuckwallas were abundant, whereas on a visit in the same month several years before, none was seen. In those years it was colder and few flowers bloomed.

After feeding, the animals may sprawl out on warm rocks. One can readily identify favorite basking sites, often found at or near the top of rock piles, by the accumulations of elongate, cylindrical fecal pellets composed entirely of plant residues. Stains sometimes associated with the pellets indicate that considerable fluid may at times be voided at defecation. Since these lizards feed on moist plant material, they evidently can afford to expel considerable water.

Hawks, coyotes, and man are probably the chief enemies of chuckwallas. Their remains were found in two mammal scats (presumably coyote) found in the Yaqui Pass area, San Diego County, California. The Indians are said to have captured chuckwallas for food.

When a chuckwalla has retreated for safety into a rock pile, it may reveal its location by shifting position, making a scraping, sandpaper-like sound. In a rock crevice the animal will distend its lungs by gulping air, thus puffing up and wedging itself firmly in place. Then no amount of probing will dislodge it.

Virtually nothing is known of the breeding habits of this species. A female from Pisgah Crater, San Bernardino County, laid 10 eggs on August 1. Another from the Borrego Desert, San Diego County, contained 8 large eggs in her oviducts on August 12.

The race in the Monument is *Sauromalus obesus obesus* of the Mohave and western Colorado deserts.

Zebra-tailed Lizard. *Callisaurus draconoides*

Other name. Gridiron-tailed lizard.

Description. Adult 2½ to 3½ inches in snout–vent length. A slim gray lizard with very long legs and black and white stripes or "zebra" markings prominent on underside of tail. Dorsal scales granular, giving the skin a velvet texture. Gray or brownish ground color usually with two rows of dark blotches down middle of back and numerous dots of whitish or yellowish scattered throughout; yellow wash usual on side of body, in groin, and at base of tail; throat gray with diagonal streaks at sides. Females usually plain white below on body. Males with pair of black bars in a blue-green patch on each side of belly.

Fig. 117. Zebra-tailed lizard, approximately four-fifths natural size. Photograph of an animal taken in Pinto Basin on April 6, 1951.

Range. West-central Nevada south through deserts to Cape district of Baja California and central Sonora; extends from desert bases of Sierra Nevada and mountains of southern California to extreme southwestern Utah, central Arizona, and southwestern New Mexico.

Occurrence in Monument. Widespread and common at lower elevations, ranging up to somewhat over 4000 feet in Split Rock area on western plateau. Recorded from: Morongo Valley, 7 mi. SW Warren's Well, May 21 *; Little Morongo Canyon, Mar. 30, 31, Apr. 28 *; 4 mi. SE Joshua Tree, Sept. 8 *; Indian Cove, May 1; 3 mi. W Twentynine Palms, July 14 *; 4 mi. N Twentynine Palms, July 25; Twentynine Palms, Apr. 28, 29 *, July 13 *; Ivanpah Tank, Sept. 12 *; Pinyon Wells, 3700 ft., Apr. 10 *, 12 *; 1 mi. NW Cholla Garden, Apr. 30; Pinto Basin, Apr. 4, 6 *, 12 *, 26 *, 27 *, 30 *, May 8 *, July 6 *, 7 *, 10 *, 14 *, 15 *, Sept. 13 *, 15 *, Oct. 13, 14, 18; Virginia Dale Mine, Apr. 24 *, 25, July 12 *, 13 *; 6 mi. N, 1 mi. W Eagle Mountain, May 17 *; Cottonwood Spring, Feb. 7, Apr. 5, 12, 24 *, 26, 29 *, May 15 *, 17 *, July 7, 8 *, 9 *, 10 *, Sept. 14 *, Oct. 21, 22 *.

The zebra-tailed lizard is strictly a desert dweller that lives on open flats, washes, the gentle slopes of alluvial fans, and other areas of low relief. It may, however, enter the mountains along washes and on elevated benches where the terrain is little dissected and vegetation scant. It seems to require open areas for running.

The character of the substratum is important in its choice of habitat. Typically, there are extensive stretches of firm sand, gravel, or hardpan, with scattered patches of loose sand in which these lizards may bury themselves at night. Rocks are often present and may be used as shelter, but usually they do not predominate. On old lava flows the zebra-tail can be expected in the smoother sections where weathering has produced a pavement of small stones or where sand has accumulated.

Although this species can run rapidly on loose sand, it seems to prefer the more compacted soil around the periphery of dunes. Thus in Pinto Basin one finds large numbers on the flats surrounding the dunes but few on the dunes themselves, where the fringe-toed lizard is common.

Vegetation is usually scant in the areas occupied. Although frightened zebra-tails may run to the cover of bushes, dense growth can interfere with their running. When escaping, these lizards usually run a nearly straight course and seldom engage in the zigzag movements used often by species that live in grasslands or broken terrain with heavy plant cover.

The zebra-tail displays a number of adaptations that seem to relate to the openness of its habitat. To human eyes they are among the most effectively camouflaged lizards and are seldom seen until they move. Upon being approached their first act may be to curl and wave the banded tail. This movement usually precedes a dash for cover. The action seems to be an effort to attract attention and probably contributes to survival by diverting attack to the tail, an expendable and replaceable part. Even when running at full speed, the lizard keeps its tail curled. At the end of the run, the tail is usually lowered immediately and seems to melt into the background, along with the rest of the animal. The tail pattern and its manner of display may be classed as a type of "flash color," like the fleeting glimpse of the bright wings of certain insects. Presumably a predator, attracted by movement, may focus on the flash color and is confused when it suddenly disappears. Evidently certain predators, probably often callow young, concentrate on the color, and when it disappears lose interest and move on.

The zebra-tail is capable of running 18 miles per hour. Although at slower speeds it may run on all fours, at high speed it is bipedal and runs on the hind legs, only occasionally touching down with the front feet.

This lizard feeds on beetles, grasshoppers, crickets, flies, ants, bees, wasps, caterpillars and other insects, spiders, and occasionally other lizards, which it procures by ambush and pursuit. Its great speed no doubt aids it in catching fast-moving prey but is probably mainly used in escape.

We have obtained a total of 44 body temperatures. The mean was 101° F. and the range 81° to 108° F. Thus among desert lizards in the Monument it has higher operating temperatures than the side-blotched lizard but is exceeded by the whiptail and desert iguana.

Despite the abundance of the zebra-tailed lizard, very little is known about its reproduction. A gravid female taken near Beatty, Nevada, had a large bright orange throat patch and bright orange marks on the sides just behind the axilla. These marks evidently become accentuated in females approaching the time of egg laying. Two to 6 eggs have been recorded as composing a clutch, and laying probably occurs in June. Several young of the year were seen, but not collected, on October 21 at Cottonwood Spring.

The smallest individuals we have captured (May 15 and 17, Cottonwood Spring) were 42 mm. in snout–vent length. Males begin to get their ventral black belly bars at a snout–vent length of around 50 mm., and the marks may be fully developed, including the blue background, at a snout–vent length of 64 mm.

The race occurring in the Monument is *Callisaurus draconoides rhodostictus* of the Mohave and Colorado deserts.

Mohave Fringe-toed Lizard. *Uma scoparia*

Description. Adult 3–4½ inches in snout–vent length. A pale, flat-bodied, sand-dwelling lizard, the dorsal surfaces marked with punctations and a fine black reticulum; belly white or washed with pale greenish yellow; black crescent marks on throat; conspicuous black spot on each side of belly and black bars on ventral surface of tail, toward tip. Skin has velvet texture. Projecting, pointed scales on sides of toes.
Range. Mohave Desert from near southern end of Death Valley to Pinto Basin and Blythe, Riverside County.
Occurrence in Monument. Known only from sandy areas in Pinto Basin and vicinity of Virginia Dale Mine. Recorded from: Pinto Basin (sand dune 1¼ mi. W, 2¾ mi. S Pinto Peak), Apr. 4, 6 *, 10 *, 12 *, 25, 27 *, 30 *, July 6 *, 7 *, 10 *, 11, 14, 15, Sept. 13 *, 15 *, Oct. 13 *, 14 *; ¼ mi. W Virginia Dale Mine, Apr. 25 *.

This lizard is strictly a sand dweller, confined to dunes, hummocks, and flats of fine, aeolian sand. In some 30 years' experience with it in its native haunts, the junior author has never encountered it on any other type of substratum. The sand and gravel of washes is usually too course, and we have never seen it on firm sand compacted by rain or runoff water. It may be found in rocky habitats but only when there are suitable interspersed patches of drift sand. It does not move about among the rocks themselves.

Fig. 118. Mohave fringe-toed lizard, approximately four-fifths natural size; note the fringes showing on the right hind foot. Photograph of an animal taken in Pinto Basin on April 6, 1951.

Its distribution doubtless has become greatly fragmented with the breaking up and sorting of lake shore and river bank sand deposits, which must have been more continuous in the more humid past than now. Since the species is so completely restricted to aeolian sand, one can picture some of the populations carried about on their sand "islands" as the deposits have moved over the centuries. The lizards have probably penetrated Pinto Basin in the Monument from Palen Dry Lake, dispersing northward up Pinto Wash between the Coxcomb and Eagle mountains, thence westward.

Although vegetation is usually scant in the areas occupied by these lizards, some is required for their survival and they are seldom found far out on completely barren dunes. Some plants are eaten directly, but their chief importance seems to lie in attracting insects and other arthropod food and in providing shade. Vegetation on the Pinto Basin dune is fairly typical—scattered creosote bushes, burroweed, galleta grass (fig. 8), and annuals, in season. Hummocks of loose drift sand form at the base of bushes and are focal points of *Uma* activity.

The Pinto Basin Dune appears as a low, pale streak north of the Twentynine Palms–Cottonwood Spring highway, south of Pinto Peak. One must walk about a mile to reach it. In 1945, the dune proper was roughly $\frac{3}{4}$ mile long, perhaps 150 to 200 yards in greatest width, and 75 feet high. Its shape, its ripple marks, and the configuration of its sand hummocks indicated prevailing wind from the west. We have found *Uma* to be most abundant on the lee side at the northeast end and on the south side where there is good insolation. The eastern end is well supplied with annuals and other plants. *Uma* also occurs on the sandy flats surrounding the dune. Numerous rodent burrows, used as refuges, occur on the flats. Burrows are less common on the dune.

The fringe-toed lizard is well adapted for life in fine sand, into which it burrows or "swims" with ease, and over which it can run with great speed. In "sand swimming" it enters head first, fore limbs at its sides, propelled by rapid alternate kicks of the hind legs. Quick lateral movements of the head and neck help break the way, and vibration of the tail, after the lizard comes to rest, usually ensures its burial.

Once beneath the sand, the lizard seldom makes further progress, although we have recorded individuals moving a foot or so. These lizards can bury themselves so rapidly that the movements described may blur into a single headlong plunge. Burying is a means of escape, and if the place of entry is not witnessed, locating the quarry by sight is virtually impossible because of the slight disturbance of the sand surface.

A number of structural features aid burial—a countersunk lower jaw, valvular nostrils, ear flaps, and extensively overlapping eyelids which protect the body openings and sense organs (Stebbins, 1944). Smooth scalation reduces friction; the wedge-shaped snout facilitates the excavating action of the head; the toe fringes (fig. 118) increase the foot surface on the power stroke and fold back, minimizing resistance on the recovery stroke. These fringes probably also aid running on the surface.

The openness of the habitat has placed a high premium on background matching, and the pale, ocellated pattern blends well with the sand background, at least to human eyes. Populations of *Uma* are well known for their tendency to match the particular sand upon which they dwell.

Despite the exposed environment of this lizard in a region where daytime temperatures may be extremely high in summer, the Mohave fringe-toe does not have an unusually high temperature of normal activity. Indeed, our records place it considerably lower than the whiptail, with which it coexists. Of 16 temperatures taken between 8:20 A.M. and 7:15 P.M., April 11 to September 13, the average was 97°, range 82° to 105° F. This is several degrees lower than that obtained by Cowles and Bogert (1944) on 13 adults of *Uma notata* from the Yuma sand dunes. These workers obtained an average of 101° F. (38.3° C).

We have some evidence for early and late activity in hot weather. On July 6 a fringe-toe was seen at sunset (6:45 P.M.). On the 7th a medium-sized individual was seen at 6:40 A.M. It ran as though fully warm. On the 10th at sunset (7:15 P.M.), with sky somewhat overcast, one was seen when visibility was poor, and on the 15th some 14 animals were counted between 6:00 and 8:45 A.M. in a walk the length of the Pinto Dune.

During the heat of midday in summer, few fringe-toes are noted. Early in the year, especially on cool days, the situation is reversed. On April 6 none was encountered between 7:30 and 11:00 A.M., but five were seen between 11:00 A.M. and 12:30 P.M. The sand was damp from recent rain. On April 27 only three were seen between 10:00 A.M. and noon, but by 1:00 P.M. they were abundant.

In looking for *Uma* it is common merely to catch a glimpse of a pale form rocketing across the surface and disappearing over a hummock in a puff of sand. They are seldom seen before they move. Their vision is extremely good, and despite surface glare on bright days they may take flight at a distance of 30 yards. They often seek refuge in rodent burrows. The tunnels of the desert kangaroo rat and other mammals are used in the Pinto Dune area.

Occasionally when one of these lizards is caught out far from a burrow, it may circle the observer or even run toward him to reach safety. On April 12 a large fringe-toe ran within a few feet of us to reach its burrow. Some of these burrows seemed to be used habitually and may serve as focal points in the establishment of territories. A pair may share a burrow. On two occasions, on September 13, two adults were seen to enter the same burrow; however, their sex was not determined.

When individuals have not yet warmed fully, they are less inclined to seek refuge in burrows and it is then that one can sometimes noose them in the open. Immatures seem less inclined to use burrows to escape, and more often run to the shelter of bushes. When the lizards do not enter burrows, a common mode of escape is to dash to the opposite side of a hummock or bush and, hidden from view, to bury themselves.

Sometimes fringe-toed lizards are startled out of the sand as one walks near them. We have had them burst forth underfoot and at distances up to 20 feet. Some may have been watching with head elevated; others may have detected our footsteps. On July 15, a medium-sized individual broke cover about 10 feet in front of us. Although the sand at its hiding place was only 83° F., the animal ran rapidly.

Occasionally one can find fringe-toed lizards by tracking them, but one must search the sand early in the day before there are so many tracks that the record becomes confused. Some tracks may begin in the open. These are made by individuals which have been buried while the wind erased all marks and have emerged on a clean slate.

Fringe-toes feed on ants, beetles, caterpillars, ant lion larvae, grasshoppers, spiders, and probably other lizards. Plant fragments—buds, leaves, flowers, and seeds—are sometimes taken. In turn the lizards are preyed on by roadrunners, hawks, shrikes, burrowing owls, badgers, coyotes, and snakes.

Little is known about the reproduction of this species. Two gravid females were found on April 30 on the Pinto Dune. One of them kept in the laboratory at Berkeley laid an egg on May 22. In the related species *Uma notata* of the Algodones Dunes, Imperial County, California, Mayhew (1961) reports that laying occurred from April through August, with 1 to 4 eggs per clutch. Young of the year were seen by us, September 13 to 15, and one 36.6 mm. in snout–vent length, still with a yolk-sac scar, was found on October 14. It took refuge in a burrow about an inch in diameter, perhaps self-made.

We regard *Uma scoparia* as a distinct species from *Uma notata,* which occupies the Colorado Desert to the south and east.

Desert Spiny Lizard. *Sceloporus magister*

Description. Adult $3\frac{1}{2}$–$5\frac{1}{2}$ inches in snout–vent length. A large yellow or yellowish brown lizard with projecting, pointed, keeled scales. An incomplete row of circumorbital scales and the general yellowish color distinguish *Sceloporus magister* from *S. occidentalis.* Black wedge-shaped mark on neck in front of fore limb. Dorsal pattern in adult males in the Monument area (race *uniformis*) uniform light yellow or tan in broad band down back, grading into darker brown on sides; dusky marks or bars usually present on sides or, in young males and usually in females there may be 4 longitudinal rows of blotches; ventral surface whitish or pale yellow with throat and sides of belly blue in males, the belly patches edged with black and often joined in large individuals; in females blue markings faint or lacking. Like the western fence lizard, this species can change from dark to light color phase.

Range. Northwestern Nevada and east-central Utah south to Cape district of Baja California and southern Sonora and from inner coast range of California, west of the San Joaquin Valley, and desert bases of mountains of southern California east to central New Mexico and western Texas.

Occurrence in Monument. Widely distributed but apparently absent from barren sandy flats and some of higher western peaks. Recorded from: 2 mi. N, 4 mi. E Yucca Valley, Apr. 28 *; Little Morongo Canyon, Mar. 31; 6 mi. SE Joshua Tree P. O., Sept. 8 *; Lower Covington Flat, Aug. 23 *, 25, 29; Quail Spring, May 20 *, 21 *, July 1, 2 *, 3 *, 5 *, Sept. 8 *; Queen Valley, Apr. 10; Indian Cove, Apr. 8, 27 *, May 8, 13, 14 *, Oct. 11 *; Pinyon Wells, Apr. 10 *, Oct. 16 *, 18 *, 22 *; Twentynine Palms, July 13 *; 6 and 8 mi. S Twentynine Palms, Apr. 30, May 8; 1½ mi. SE White Tanks, Apr. 30, May 18 *; 5½ mi. SE White Tanks, Apr. 26 *, 30; 1 mi. NE Eagle Mountain, May 14 *, Oct. 21; Cottonwood Spring, Apr. 11 *, 29 *, May 1, 17 *, June 22, July 7 *, 8 *, 9 *, 10 *, 14 *, Oct. 13 *, 22.

The desert spiny lizard is the only species of *Sceloporus* present over practically all the lowlands of the extreme desert in the southwestern United States, but it does coexist with the western fence lizard on lower mountain slopes and along river courses, as in the highlands of the Monument and along the Mohave River. The two species overlap in the yucca belt, as can be observed at Indian Cove. The spiny lizard prevails in the campground and on the alluvial flats to the north, but as one proceeds up canyon the western fence lizard is found, and both species range into the highlands above. Fence lizards have been found at Split Rock, and spiny lizards at Queen Valley.

The desert spiny lizard like many animals of dry lands is a pale member of its group. It may live in regions of extreme aridity where the only surface water is that which persists briefly after storms. However, where there is water, it is attracted to it, and numbers of these lizards are usually found at oases and along river courses. In the Monument spiny lizards can be found on rocks and among trees and bushes in the vicinity of springs as at Twentynine Palms, Indian Cove, Quail and Cottonwood springs, and also on arid rocky flats far from water.

Fig. 119. Desert spiny lizard, approximately one-half natural size. Photograph taken by Robert G. Crippen of an animal from 7 miles southwest of Weldon, Kern County, California.

This species does not appear to be attracted to any particular type of vegetation. About water sources it may spend much time in trees such as cottonwood, willow, Washington palm, mesquite, and catclaw. Elsewhere it may frequent stands of Joshua tree, brushy desert, and the low, sparse growth of treeless desert plains. Vegetation provides shelter, attracts insects, and buffers the extremes of temperature and humidity.

Often one of these lizards is heard before it is seen. A rustling sound attracts attention to a Joshua tree or cottonwood, but, when one looks up, nothing may be seen. The lizard has scrambled out of sight. Stalking may be required to get the animal in view, as it repeatedly circles a tree trunk or limb in efforts to keep out of sight.

These lizards are adept climbers, ascending the sides of boulders and tree trunks with ease. One was seen to climb to a height of 20 feet in a willow tree at Twentynine Palms. Shelter is sought in tree cavities and rock crevices, beneath rocks, Joshua tree branches, and other surface litter, in wood rat nests, and in rodent burrows.

The spiny lizard is chiefly insectivorous, feeding on ants, bees, beetles, flies, grasshoppers, and caterpillars. It occasionally eats other lizards.

We have obtained a total of 10 body temperatures of individuals that were abroad on the surface, either basking or actively moving about. The average was 90° F., and the range 81° to 95° F. This average is over a degree lower than that of the fence lizard in the Monument, but more temperatures must be obtained before it can be concluded that it prefers a lower mean.

Information on reproduction is fragmentary. On the basis of individuals found in other parts of the species' range, 10 eggs may form a clutch, the number varying from 7 to 18. Females with eggs ready for laying have been found in late May and June. A gravid female with large ova was found on May 20 at Quail Spring. A captive taken near White Tanks on April 26 laid eggs between June 7 and 9 in the laboratory at Berkeley. The recently laid eggs contained embryos. One opened contained an embryo at the stage of hind-limb bud development and pigmented eyes; it measured 4.3 mm. from the cephalic to the caudal flexures.

A hatchling 35 mm. in snout–vent length was found at Quail Spring on September 8. The neck marks were joined to form a complete black collar. Robert Crippen informs us he found hatchlings in Kelso Valley in the southern Sierra Nevada on August 16.

According to the taxonomic revision of Phelan and Brattstrom (1955), the race in the Monument is *Sceloporus magister uniformis* of the Mohave and Colorado deserts.

Western Fence Lizard. *Sceloporus occidentalis*

Description. Adult $2\frac{1}{4}$–$3\frac{1}{2}$ inches in snout–vent length. Dorsal surfaces covered with projecting, pointed, keeled scales. Above black or gray with blotches or scallop marks of black sometimes elongate and run together to form longitudinal stripes; sides of belly and patch on throat blue; yellow on underside of fore limbs and back of thighs. Males have more vivid ventral blue markings than females and, when in light phase, blue-green flecks on back and tail. In dark phase, large adult males in Monument may be black over most of ventral surface. Young of both sexes resemble females in coloration. (See p. 382 for comparison with *Sceloporus magister*.)

Range. Vicinity of Seattle, Washington, south to northern Baja California and from Pacific coast to central Idaho and western Utah; absent from the desert except in a few scattered mountain areas and along the Colorado River and its tributaries.

Occurrence in Monument. Chiefly in the piñon belt; locally lower in canyons. Recorded from: Little Morongo Canyon, Mar. 31, Apr. 1; Black Rock Spring, Aug. 30 *, 31 *, Sept. 1 *, 2 *, 3 *, 4 *, Oct. 16 *; Upper Covington Flat, Aug. 24; Lower Covington Flat, Apr. 8 *, 9 *, Aug. 23 *, 24 *, 25 *, 26 *; 5 mi. W, 3 mi. S Joshua Tree, July 6; Quail Spring, May 19, 20 *, 21 *, July 3, 5 *, Oct. 15 *; Hidden Valley, May 1 *; Stubby Spring, Sept. 5 *, 6 *, 7 *, 9 *, 11 *; Key's View, July 3 *, 4 *, 15 *, 16 *; Jumbo Rocks, Oct. 15; Indian Cove, Apr. 7 *, 8, May 1 *, 8, 13 *, 14; Fortynine Palms, Apr. 7 *, May 7 *, Oct. 12 *; Pinyon Wells, Oct. 12 *, 14 *, 17 *; Split Rock, Apr. 9 *.

The fence lizard, by virtue of its abundance, widespread distribution, habit of perching in conspicuous locations, and ability to live with man, is probably the most familiar western lizard. Known often as the "blue-belly," it is a favorite of children who like to keep reptiles as pets.

It is an inhabitant of rock-strewn environments of grassland, woodland, and open forest, from sea level to over 9000 feet in the Sierra Nevada of California. It is intolerant of extreme aridity. The west end of the Monument is one of several desert areas in which it is able to live in "peninsulas" or "islands" of somewhat cooler, less arid conditions. The Mohave River and the Providence, Panamint, and Charleston mountains are other examples. The montane "island" populations are probably relicts of the last pluvial period when more humid conditions must have permitted fence lizards to live in the intermountain areas throughout most of the desert.

In the Monument fence lizards are found in rocky areas of open tree growth such as piñon, juniper, scrub oak, and Joshua tree. During the breeding season males tend to select elevated positions on the tops of boulders where they can be seen for considerable distances. They are dark-bodied and stand out conspicuously against the light-colored rocks. They are wary and often dodge out of sight when the observer is still 50 to 75 feet away. They seek shelter in rock crevices, beneath bark, under logs, rocks, and other surface objects, and in rodent burrows and loose soil. Like the side-blotched lizard, these lizards can completely bury themselves.

The fence lizard is a good climber and may be seen perched high in Joshua trees, on the trunks of piñons, or in the branches of catclaw and other bushes. A frightened individual may scramble to get out of sight but in a moment, seemingly unable to withhold its curiosity, it may peek furtively from behind its shelter.

Food consists of insects such as beetles, ants, wasps, bees, caterpillars, grasshoppers, and aphids, and other arthropods such as spiders. On October 15 an immature taken near the Quail Spring tank had a honeybee in its mouth.

Fig. 120. Male western fence lizard, approximately two-thirds natural size. Photograph taken by Robert G. Crippen of an animal from Berkeley, California, in the range of a coastal race of the species.

The lizard had come to the drinking trough, where large numbers of bees gathered to drink.

The body temperatures of 31 individuals collected in the Monument ranged from 77° to 99° F. (average 92° F.).

During the breeding season males display their blue colors from their lookout posts if other adult males enter their territory. Display consists of flattening the sides, expanding the throat and bobbing ("pushups"). Fights sometimes occur in which the contestants stand side by side facing in opposite direction with sides flattened. They may attempt to bite or strike one another by sudden movements of the body. The vanquished individual is usually pursued until he moves out of the defended area. For persons with a bent for experimentation, a territorial defense usually can be witnessed by tethering a male on a string tied around his body just in front of his hind legs and tossing him into position near a resident male.

Territorial behavior, common among lizards and many other vertebrates, helps to ensure spacing of individuals in the habitat, thereby offsetting the effects of overcrowding. It also increases the chances that females in the area will mate with the more vigorous, aggressive males. One can readily note such spacing of large fence lizard males in the Monument. Two will rarely be found together. One or more females and immatures, however, may be found within the domain of an adult male.

In mating, the male approaches the female with a rapid head-bobbing movement. Copulation occurs much as described for the side-blotched lizard. Nonreceptive females arch the back, flatten the sides, elevate the tail and turn sideways to the male, usually moving off in a stiff-legged, hopping gait. This behavior seems to deter mating. We witnessed such behavior on April 7 at Indian Cove.

Laying probably occurs in May and June. A female with large eggs nearly ready for laying was found on May 13 at Indian Cove. Hatching probably occurs in late July, August, and September. The smallest young obtained was 35 mm. in snout-vent length, and was taken at Fortynine Palms on October 12.

According to the revision of Bell (1954), the race occurring in the Monument is *Sceloporus occidentalis longipes*.

Animals from the Monument are notable for the high frequency of striped individuals. Striping results from elongation and variable fusion of the dorsal blotches. These blotches typically are arranged in four longitudinal rows —a row on each side of the midline and on each side of the body. The two sexes display striping in equal frequency, but in large adult males in the dark phase stripes may be obscure. The frequency of striping in adults in the Monument is shown in the table.

	Striped	*Unstriped*	*Total*
Males	9(19.1%)	38	47
Females	7(19.4%)	29	36
Juveniles	3	4	7
	19(21.1%)	71	90

Striping is not confined to the Monument, for a few striped individuals have been found elsewhere—along the Mohave River and near Redlands, San Bernardino County, California.

Side-blotched Lizard. *Uta stansburiana*

Description. Adult $1\frac{3}{4}$ to $2\frac{1}{8}$ inches in snout–vent length. A slim little lizard with scales small and keeled on back, larger, smooth and overlapping on belly, and granular on sides; well-developed gular fold. Tail slender and about $1\frac{1}{4}$ to $1\frac{1}{2}$ times head–body length. Black side spot present just behind fore limb, faint in females and young; dark blotches on back that often become obscured in males when in light phase, as a result of appearance of pale blue flecks; females striped, blotched, or patterned with mixed striping and blotching and no blue flecking.

Range. Central Washington south through Great Basin to southern Baja California and Durango and from Pacific coast in central California east to Rocky Mountains and western Texas.

Occurrence in Monument. An abundant species throughout. Recorded from: 5 mi. S Morongo Valley, Apr. 25 *; Morongo Pass, Apr. 28; Little Morongo Canyon, Mar. 31, Apr. 28; 4 mi. N Warren's Well, Jan. 2 *, Dec. 31; Black Rock Spring, July 6, Aug. 30 *, Sept. 2 *, 3 *, 4 *, Oct. 16 *; 5 mi. W, 3 mi. S Joshua Tree, July 6; 2 mi. SW Joshua Tree, May 19 *; Lower Covington Flat, Apr. 8 *, 9 *, Aug. 22 *, 23 *, 24 *, 25 *, 26 *, 27 *, 29 *; Quail Spring, Apr. 22 *, May 20 *, July 1, 2 *, 3 *, 5 *, Sept. 8 *, Oct. 15 *; Hidden Valley, May 1 *; Stubby Spring, Sept. 6 *, 7 *, 8 *, 11 *, Oct. 15 *; Lost Horse Valley, Feb. 10; Key's

View, July 4 *, 15, 16; Indian Cove, Apr. 8, 27 *, May 8 *, 13 *, 14, Oct. 12 *; 1 mi. NW Fortynine Palms, Apr. 7 *, 8 *; Fortynine Palms, Apr. 7; Twentynine Palms, Apr. 29 *, July 13 *; Pinyon Wells, Apr. 10 *, Oct. 9 *, 10 *, 12 *, 13 *, 14, 16, 17 *; Pinto Basin, Apr. 27 *, Oct. 18; 1¼ mi. W, 2¾ mi. S Pinto Peak, Apr. 27, July 11; Virginia Dale Mine, Apr. 24 *, 25 *, July 12, 14; 1 mi. S Virginia Dale Mine, July 14; N side Eagle Mountain, Oct. 19, 20 *, 21 *, 22, 24; 7.6 mi. N Cottonwood Spring, Apr. 26 *; Cottonwood Spring, Feb. 7, 14, Apr. 5, 11 *, 12, 26 *, 29 *, 30, May 14 *, July 7 *, 8 *, 9 *, 10 *, Sept. 13, Oct. 22 *, 23; Lost Palm Canyon, Sept. 14 *.

This is a ground-dwelling lizard that ranges from below sea level in desert sinks to over 7000 feet in the mountains. It inhabits surfaces of all sorts—alluvium, sand, rocks, and hardpan—but in most habitats some rocks and loose soil are present. Grassland, bushland, and woodland are frequented. Although it is attracted to river banks and oases and its activity is stimulated by rainfall, it is able to live in some of the most barren parts of the desert. In the Monument it ranges from the dry, windswept sandy plains to the tops of the highest mountains in the piñon belt.

Side-blotched lizards seek shelter in rodent burrows and rock crevices, beneath rocks, under plant debris, and in loose soil or sand, into which they burrow head first and usually go completely out of sight.

The species is active in the daytime and is seen throughout the year. Being of small size and capable of responding quickly to brief warm spells, it may be abroad through the winter in the Monument when other reptiles are hibernating. Loye Miller found side-blotched lizards at midday in the periods from October 29 to November 2 and December 4 to 6, "when the sun lay on certain slopes and gnats, bees, and butterflies became active." He also noted the lizards abroad on February 7 and 14, when individuals of all sizes were active. C. P. North observed quite a few active during the day 4 miles north of Warren's Well on December 31.

Fig. 121. Male side-blotched lizard in light color phase, natural size. Photograph of an animal taken in the Coachella Valley in April, 1960.

Fig. 122. Gravid female side-blotched lizard, natural size. Lower Covington Flat, April 8, 1960.

Although a diurnal lizard, like many desert species it may be active after dark in hot weather, as noted on July 14 near Virginia Dale Mine. One was seen to run rapidly under a bush at 7:15 P.M., after sundown.

The side-blotched lizard seems to be almost exclusively a ground dweller; it has seldom been noted in bushes, although it is an excellent climber and is able to ascend a vertical rock surface with ease unless the rock is polished by stream action. One was found floating dead in 12 inches of water in a smooth pothole pool on April 8 at Indian Cove. It is often seen basking on tops of boulders in the morning, and it is usually the first lizard abroad. In warm weather individuals may appear shortly after sunrise. On October 16 near Pinyon Wells a small one was seen at 7:15 A.M.

Thirty records of body temperature taken between 7:00 A.M. and 5:00 P.M. from April to October ranged from 70° to 101° F. and averaged 95° F.

The dramatic color change in males may be noted in a torpid captive individual when it is warmed up or in individuals that have just emerged to bask in the field. At the start they are dusky and virtually without pattern. Soon, however, the blue flecks begin to appear, and in a few minutes the animal may be decidedly speckled. Females may lighten in color but display no such marked pattern change.

This species feeds on a great variety of insects such as beetles, flies, mosquitoes, ants, bees, wasps, grasshopper, caterpillar, and ant lion larvae, and on spiders, mites, ticks, sowbugs, and scorpions. On September 3, at Black Rock Spring, after a heavy rain that wet the ground to a depth of 6 inches, we noted insects in abundance, and utas were very much in evidence, making catches every few seconds.

All females noted at Lower Covington Flat on April 8 appeared to be gravid (fig. 122). They looked flattened and chunky from a distance, almost like small horned lizards. They were usually seen on the ground or lower down on the rocks than the males and were less wary. The pattern of stripes or blotches tended to harmonize with the background. The males were more often up on top of the rocks and their markings were conspicuous, the head orange and the pale speckling of the back and tail giving a bluish green cast. Perching in elevated locations is doubtless a form of territorial display, which is enhanced in such exposed locations by the bright color pattern.

A gravid female collected on April 17 near Inyokern, Kern County, to the north of the Monument, laid 4 eggs after being brought to Berkeley.

On the north side of Eagle Mountain, in the fall of 1945 (October 21, 24), we found a disproportion of males and females among adults. Of 31 individuals captured, most of them taken before noon, 24 were males, and on October 22, when a special effort was made to detect females, of 13 individuals seen between 8:05 and 11:00 A.M. that day, 12 were males. A similar disproportion seemed to be present the following May near Victorville, north of the Monument. Are the females more secretive, are there differences in seasonal activity, or is there truly a difference in the sex ratio? Laboratory experiments with the striped plateau lizard (*Sceloporus virgatus*), of similar thermal preferences and diurnal habit, suggest the possibility that inherent sexual differences in seasonal behavior might be responsible.

On April 29, at 5:00 P.M. near Cottonwood Spring, copulation was observed. The male straddled the back of the female, apparently holding her by the skin between her shoulders. She moved her head from side to side in short jerks. He swung his hindquarters to one side of her tail and after several forward thrusts achieved union, which lasted for about one minute. Both held their tails high after parting.

Hatchlings have been noted over the period from July 3 to 16.

Two races of the side-blotched lizard meet in the Monument, *Uta stansburiana hesperis* of central and coastal California and *U. s. stejnegeri* of the desert. Intergradation apparently occurs along the desert slopes and passes of the mountain ranges of southern California. An unmistakable *hesperis* influence is seen in populations in the piñon–juniper areas at Black Rock Spring, Lower Covington Flat, and Stubby Spring in the western part of the Monument. Indeed, on the basis of color pattern, one is hard put to find any significant differences between some of these populations and typical *hesperis* to the west. Typical *hesperis* pattern consists of conspicuous blotching and an irregular distribution of pale blue flecks in males. The number of dorsal scale rows averages higher, and there is a tendency toward two rows, rather than one row, of postrostral scales. These scales lie between the rostral and internasals. Although we have not undertaken an extensive study of variation in scalation, a brief check supports the evidence derived from pattern.

We have compared samples of typical *hesperis* from Los Angeles County and *stejnegeri* from San Bernardino County with our sample from Stubby Spring. Dorsal scale rows were counted from the rear margin of the interparietal scale to a line connecting the rear bases of the thighs. Ten adults were examined from each region. Ranges and means are as follows: *stejnegeri* 84—94 (88.6), *hesperis* 97—120 (107.5), and Stubby Spring 90—104 (94.2). As can be seen, the animals from Stubby Spring are intermediate. As to number of postrostral scale rows, western populations in the Monument show a tendency toward *hesperis* as can be noted in the tabulations.

Region	*Number of animals*	*Number with One and Two Rows of Postrostrals* 1	2
Mohave Desert (San Bernardino County)	26	24(92.3%)	2(7.7%)
Virginia Dale Mine, Twentynine Palms, Pinyon Wells, Cottonwood Springs, Eagle Mountain	70	65(92.9%)	5(7.1%)
Black Rock Spring, Lower Covington Flat, Quail Spring, Stubby Spring	96	74(77.1%)	22(22.9%)
San Joaquin Valley and coastal southern California	28	7(25.0%)	21(75.0%)

Other populations from which we have good samples are assigned as follows: Animals from the Fortynine Palms–Indian Cove region appear to be mainly *stejnegeri,* but they seem to have more dorsal spotting than typical *stejnegeri.* A sample from Virginia Dale Mine appears to be good *stejnegeri.* Animals from the Eagle Mountain–Cottonwood Spring–Lost Palm Canyon area also seem to be primarily *stejnegeri,* but some individuals do show a tendency toward dorsal blotching.

A sample from Pinyon Wells favors *stejnegeri* in respect to postrostrals, only one of 19 animals having two rows, but many males have a dorsal pattern close to *hesperis.*

We should caution concerning the use of dorsal blotching in males as a diagnostic character in recognizing the race *hesperis.* Blotching becomes less evident with increasing size in *stejnegeri.* Immatures may be considerably blotched. There is also local variation within the range of *stejnegeri.* A number of animals from Clark Mountain and the Providence Mountains have *hesperis*-like markings. Perhaps in these mountain areas ecological conditions approach more nearly those encountered by *hesperis.* However, blotched males with subdued light flecking are found on the flats in old lava flow areas (Amboy Crater, for example). On such dark backgrounds, the utas are dark-colored. Although males may be conspicuous when poised on top of rocks, much of their time is spent on more broken backgrounds where to the human eye the pattern appears to have a concealing function. One might expect, therefore, the occurrence of local pattern types depending on the character of the predominant background. The speckled *stejnegeri* pattern seems especially effective on sand or alluvial deposits, where leaf litter is scant.

Long-tailed Brush Lizard. *Urosaurus graciosus*

Description. Adult $1\frac{7}{8}$ to $2\frac{1}{4}$ inches in snout–vent length. A slender gray or beige lizard found in bushes and trees; arboreal habits and long, slender tail (twice as long as body) usually serve to identify it. Broad band of enlarged keeled scales down middle of back surrounded by small, smooth scales. Side blotch absent; dark cross bars or blotches on back and often a whitish line on side of face, but pattern may be faint in individuals in light phase; below white with some grayish flecking; spot of reddish orange or yellow usual on throat; males with pale blue or blue-green patch on each side of belly.
Range. Mohave and Colorado deserts, from extreme southern Nevada south to northeastern Baja California and northwestern Sonora, extending east to central Arizona.
Occurrence in Monument. A lowland form; all stations of record at or below 2500 feet. Recorded from: Twentynine Palms, July 13 *; Pinto Basin (vicinity of dunes), Apr. 12 *, 27 *, Sept. 13 *; south side Pinto Basin, Apr. 30* ; Virginia Dale Mine, July 12 *.

This lizard is not likely to be seen without special effort. Even old-timers on the desert may be unaware of its presence. It spends most of its time in bushes and trees, where it rests with its body aligned with a branch or with bark irregularities with which its color usually harmonizes. It can change color rapidly, within 5 minutes or less. In the pale phase, the dorsal pattern fades out, the throat patch becomes faint or disappears, and, in males, the belly patches become obscure.

The brush lizard is strictly a desert dweller with a range corresponding closely with that of the creosote bush, a plant in which it commonly lives. The species appears to be limited to the Lower Sonoran Life-zone and so far as we know has not been found above 4000 feet anywhere in its range.

Habitat requirements appear to be loose wind-blown sand, in which the animal can bury itself; bushes or trees; aridity; and warmth. Typical habitat consists of plains where sand has accumulated to form hummocks at the bases of creosote bushes. Although compacted soil or rocks may be present, somewhere nearby there will usually be some fine loose sand. In some areas, as in Pinto Basin, the species frequents dunes where there is sparse growth.

Although the plant on which these lizards most depend seems to be the creosote bush, we have found them in atriplex, galleta grass (*Hilaria rigida*), catclaw, mesquite, and Washington palm. In Palm Springs, California, they may be seen high up near the dead fronds of Washington palms, and they are to be expected in these trees in the Monument. In desert towns they have taken to cultivated plants. In areas dominated by creosote bush, these lizards seem to prefer the larger plants. Near Thousand Palms in the Coachella Valley, we found them most often in creosote bushes in which the roots had been excavated by the wind. Perhaps the root tangle provides refuge from the attack of shrikes or other birds. Individuals have been seen to seek shelter among the roots.

In Pinto Basin, they are often found in clumps of galleta grass in which

Fig. 123. Long-tailed brush lizard, approximately four-fifths natural size. Photograph taken by Nathan W. Cohen of an animal from near Borrego State Park in the Colorado Desert.

they commonly align themselves with the leaves and stems, clinging head up or down. We have caught them by carefully slipping our hands into the grass on each side of the lizard and then clapping them together.

Brush lizards usually are encountered one or two to a bush. When two are found they are usually of opposite sex. One or more immatures may be tolerated in the same bush with an adult. They are evidently solitary animals, except when breeding. It seems that a single pair commands the use of a number of bushes, and in typical creosote bush country a hundred feet or more may separate adults.

A typical resting pose is with the body and head flat against the branch. The tail may hang in a graceful curve to one side and may look like a twig. Coloration and markings usually closely resemble the background. The body is aligned with the branch, and on branches set at an angle may be directed either up or down. The head-down position makes it easy for the lizard to drop to the ground or to another branch. Upon reaching the new location, the animal may "freeze" and a person may momentarily lose sight of it. Often a frightened brush lizard will shift to the opposite side of a branch, but it may soon give away its position by peeking at the observer. Occasional males may lower the throat skin and flatten the sides when a stick or noose is brought near them.

On warm nights these lizards can sometimes be located by flashlight in creosote bushes, where they may sleep appliquéd to the slender branches.

We have seldom encountered brush lizards on the ground, and thus the following observations seem worthy of note. One was seen to dash across the road on the south side of Pinto Basin on the morning of April 30. It ran to the shelter of a creosote bush. When caught it was in the extreme light phase and still panting. Perhaps it had been chased by a predator. On the morning of July 13 one was seen basking near Twentynine Palms oasis. It ran to an atriplex bush and climbed among the branches. An adult male was taken at 11:15 A.M. on a cool day in April as it crawled over the sand at the base of a large creosote bush at the crest of the Pinto Basin dune. Perhaps the lizard was attracted to the warm sand.

On May 2, near Garnet in the Coachella Valley, during a sandstorm, five brush lizards were found on the ground. The branches of the creosote bushes were thrashing wildly. Thus brush lizards go to the ground to hibernate, escape storms, cold weather, and predators, to bask, and to travel to other bushes.

This lizard feeds on beetles, ants, bees, leaf hoppers, shield bugs, termites, and other insects and on spiders, presumably caught mostly among the bushes where they dwell. Their arboreal habit tends to reduce their competition with the ground-dwelling side-blotched lizard of similar size and diet. Our records suggest that the body temperature may average slightly higher than that of the latter: 97° F. (range 79° to 106° F., 23 records) compared with 95° F. for the side-blotch.

Coast Horned Lizard. *Phrynosoma coronatum*

Description. Adult 3–4 inches in snout–vent length. Similar to desert horned lizard (fig. 124) but head and body spines larger and snout less blunt; two rows of fringe scales on each side of body and several longitudinal rows of prominent pointed scales on throat; dorsal and ventral coloration usually more yellowish and pattern more contrasting.
Range. California west of the Sierra Nevadan crest and south into Baja California; at north extends to vicinity of Mount Shasta.
Occurrence in Monument. Known only from western section, where locally fairly common. Recorded from: Black Rock Spring, Sept. 3 *; Covington Flat road 4 mi. SW Joshua Tree, May 19 *; Stubby Spring, Sept. 7 *, 8 *.

The coast horned lizard occupies a variety of habitats in semiarid parts of California from sea level to about 6000 feet in the mountains. It frequents grassland, brushland, woodland, and coniferous forest, but is evidently not present in the moist woods of the north coast nor in the desert except along its western fringe. The piñon–juniper woodland of the western end of the Monument provides habitat that permits it to exist in desert surroundings and to penetrate the range of the desert horned lizard. Probably there is interdigitation of the ranges of the two species elsewhere along the desert slopes of the mountains. The coastal species is evidently less tolerant of heat and dryness, however, and thus is not known to inhabit the harsh environments of the desert proper. We have as yet no evidence that the two species coexist but, in view of the occurrence of the desert form as high as Hidden Valley east of Quail Spring, it seems likely that they may come into contact.

Like the desert horned lizard, this species is usually found in the vicinity of washes and other sand deposits, into which it may bury itself. Even though it is sometimes present on steep, rocky mountain slopes, sand or fine loose soil will usually be found nearby.

We found four of these lizards on September 3 at Black Rock Spring. One, 36 mm. in snout–vent length, was probably 1½ to 2 months old. An adult male was found under a manzanita clump. The hatchling and two larger immatures were found in a patch of manzanita and other bushes. All were on bare gravel and ran under the plant cover when approached.

Typical habitat seems to consist of bare patches of sand or gravel, fairly dense bushes, and patches of loose sand. There are usually rodent burrows and ant nests present. Often the lizards lie in broken light and shade at the edges of bushes, where they watch for prey and seemingly rely on their camouflage for protection.

A captive reported by Alan Ziegler spurted blood from its eyes when it was picked up, a reaction recorded in other species of horned lizards. The lizard had been in captivity for three months and was standing in its cage with its body well off the sand and its eyes closed. When it was picked up at 10:30 A.M., blood spurted from the anterior eye corners toward the side and rear of the cage and the lizard was dropped. Then 6 or 7 large drops of blood oozed from the eyes and fell to the sand. The lizard lay with its back arched downward and its head and tail elevated. Its ground color had changed from mottled ashy white and reddish to yellow. When poked at, it tilted its body and head spines toward the disturbance and hissed with open mouth. It had resumed its former color by 11:20 A.M.

Blood spurting by horned lizards is not often seen, and the conditions required to trigger it are not known. It is presumed to be a defense reaction. The enemy may be startled into dropping the prey, as happened on the occasion reported. Perhaps the sudden and unexpected appearance of blood may have a repelling effect on certain enemies as does "reflex" bleeding in certain beetles when attacked by ants. In the lizards, the blood comes from a ruptured sinus at the base of the third eyelid (nictitating membrane).

Food consists of ants, termites, wasps, beetles, grasshoppers, flies, and caterpillars. Individuals collected in September, 1950, at Black Rock Spring and near Stubby Spring had their stomachs crammed with winged termites. Many termites were in flight at the time. Soldier termites and ants had also been eaten. The soldiers may have been caught when they gathered at openings of termitaria through which the winged forms were issuing.

Six to 16 (average 11) large ovarian eggs have been recorded in females, selected from localities widely spaced along the borders of the species' range.

A large individual from Covington Flat contained 10 shelled eggs ready for laying on May 19. Outside the Monument we have found gravid females in late April, May, and June.

The race occurring in the Monument is *Phrynosoma coronatum blainvillei* of southern California.

Desert Horned Lizard. *Phrynosoma platyrhinos*

Description. Adult $2\frac{1}{2}$—$3\frac{3}{4}$ inches in snout–vent length. A flat-bodied lizard with prominent spike-like scales projecting from back of head; body broad and rounded and tail short, flat, and wide at base. Single row of fringe scales on each side of body. Pair of large dark brown or blackish blotches on neck and brown scallop-marks on back, edged with whitish or pale yellow; general tone of background color gray, brown, reddish, or beige, depending on predominant soil color; belly whitish, often flecked with black. Juveniles have very short head spines. For characters distinguishing the closely related coast horned lizard, see figure 124.
Range. Great Basin and Mohave and Colorado deserts, from east-central Oregon and central Idaho south to shores of head of Gulf of California; extends from eastern base of Sierra Nevada to central Utah and central Arizona.
Occurrence in Monument. Most common at low elevations, but ranges to somewhat over 4000 feet. Recorded from: 6 mi. SW Yucca Valley, Apr. 30, May 1 *; Quail Spring, July 3 *; 5 mi. E Quail Spring, July 4 *; 1 mi. NW Fortynine Palms, Apr. 7 *, May 7 *; Split Rock area, May 8; Pinto Basin, (6 mi. S Pinto Peak and sand dunes), Apr. 6 *, 12 *, 13, 30 *, July 10 *, 11 *; Virginia Dale Mine, Apr. 24 *; 6 mi. N, 1 mi. W Eagle Mountain, May 19 *; $2\frac{1}{2}$ mi. N Cottonwood Spring, July 10 *; Cottonwood Spring, Apr. 12 *, May 17 *, 18 *, June 22 *.

This is a lowland species that lives in sandy habitats, but it also occurs in rocky areas and on hardpan. In most places, however, a brief search will usually reveal patches of loose sand nearby in which the lizards may bury themselves at night. Washes and dunes seem to be especially favored. In driving through the Monument, park visitors should watch for horned lizards on road shoulders in sandy areas. In the morning the lizards bask on east-facing embankments and on the tops of small boulders.

Desert horned lizards are found in association with creosote bush, burro bush, cactus, ocotillo, smoke tree, desert willow, and other desert plants. Farther north, in the Great Basin, they frequent sagebrush flats.

Horned lizards are probably more abundant than they seem. They are easily overlooked because they tend to be solitary, are well camouflaged, and usually avoid movement unless closely approached. One may nearly step on them before they will move, and sometimes individuals can be caught by hand, without a chase. A method of locating them is to search near ant nests, where they may linger for food, and to walk outward in a spiral from the nests, a procedure which ensures thorough coverage of the ground.

It appears probable to us that this lizard has been derived from a stock not greatly different from that now represented by the coastal form of horned lizard (*Phrynosoma coronatum*). With the increase of aridity, resulting in more open habitat and extensive development of sand deposits, selection

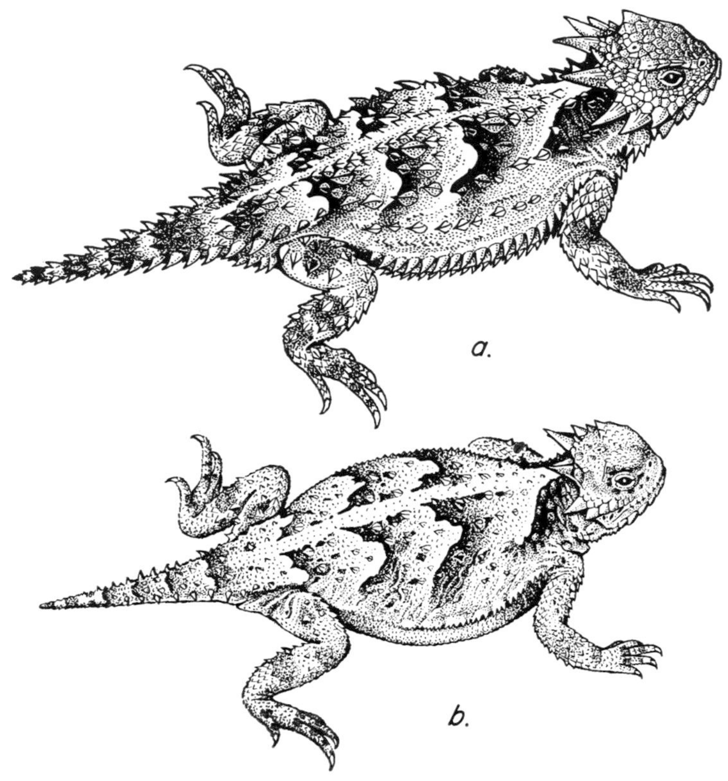

Fig. 124. Coast horned lizard (*a*) and desert horned lizard (*b*), showing differences in fringe scales and pattern. Illustration from R. C. Stebbins, Amphibians and Reptiles of Western North America, 1954, pl. 39, p. 258, copyright by the McGraw-Hill Book Company.

appears to have favored a color and form suited to concealment on a predominantly fine-grained surface. The spiny contours have been reduced and the lateral fringe scales and head and back spines are smaller than in the coastal species. Color pattern too is subdued. However, thermal characteristics, as has been noted in other desert reptiles, seem to have remained conservative, and the average temperature of normal activity does not suggest an elevated thermal optimum. The average body temperature in 19 individuals was 94° F., with a range from 84° to 102° F. These temperatures were recorded between 10:06 A.M. and 6:50 P.M. from April 6 to June 21.

The shape and behavior of a horned lizard facilitates its warming by basking. The animals flatten and tip the body to receive maximal radiation and by sprawling flat effectively absorb heat from the ground. One was found lying flat on the black-top road near Cottonwood Spring at 9:15 P.M., evidently attracted to the warm pavement. On an overheated surface they stand high, with the body well off the ground and, in bright sunlight on a warm day, they may face the sun with their sides drawn in, thereby reducing the body surface exposed to the sun.

Food consists of ants, beetles, bugs, caterpillars, spiders, and other arthropods. Banta (1961) found them feeding on the fleshy fruits of *Lycium andersonii.*

A gravid female in captivity on April 29 moved toward other lizards in the cage, presenting her hindquarters and lifting and wagging her tail. She held her body well off the ground and bobbed her head. Occasionally she would lie flat, limbs extended, and tail wagging, with her fore limbs executing crawling movements. No copulation was observed.

On April 27 a mating pair was found at 5:15 P.M., about a half-hour before sunset, near Garnet in the Coachella Valley. The male was astride the back of the female with his head against her shoulders, but he did not appear to have grasped her skin. His right hemipenis was partly everted. Tracks indicated the pair had crawled forward together for a distance of 9 feet. Copulation may have just ended.

An average of 7 or 8 (range 6—13) eggs have been found in gravid females. Laying may occur in June and July. A captive from near Blythe, California, laid 10 eggs on June 13.

The race in the Monument is *Phrynosoma platyrhinos calidiarum* of the Mohave and Colorado deserts.

NIGHT LIZARDS

Desert Night Lizard. *Xantusia vigilis*

Other name. Yucca night lizard.
Description. Adult $1\frac{1}{2}$—$1\frac{3}{4}$ inches in snout–vent length. A small, active lizard with velvety, pliable skin covered with fine granular scales over dorsal surfaces and enlarged square-shaped scales in rows on belly. Eye without movable lids and with vertically elliptical pupil. Body and tail rounded. Above yellowish, gray, olive, or dark brown with black speckling; below pale gray to greenish yellow with some dark speckling laterally and on the underside of the tail.
Range. Desert areas from Panamint Mountains of California, southern Nevada, and southern Utah south to Cape district of Baja California and northern coast of Sonora; scattered occurrences in inner coast range of California north to Panoche Pass.
Occurrence in Monument. Present wherever there are plants of the genus *Yucca.* Recorded from: 5 mi. W, 3 mi. S Joshua Tree, July 6; Lower Covington Flat, Aug. 22, 23 *; 4 mi. E Quail Spring, Apr. 23 *; Quail Spring, Jan. 20 *, Apr. 22 *, 23 *, May 20 *, 21 *, 22, June 30, July 2 *, 3; Hidden Valley, May 1 *; 1 mi. N Stubby Spring, Apr. 10 *; Stubby Spring, Apr. 9, 10, Sept. 7 *, 9 *; $\frac{3}{4}$ mi. E, 3 mi. N Key's View, July 4; Key's View, July 3 *, 4 *, 15 *; Pinyon Wells, Oct. 13 *, 14, 15; Cottonwood Spring, Apr. 11 *, 26 *, July 8 *, 9 *, 10 *.

The night lizard is one of few reptiles closely associated with a specific plant type—plants of the genus *Yucca.* It lives under the fallen branches of Joshua trees (fig. 126) and the dead rosettes of Mohave yucca and Spanish bayonet. The spiny branches provide it with hiding places, insulation against thermal extremes, protection against desiccation, and a supply of insects upon which it feeds. Litter from other desert vegetation lacks the bulk, insulation, and cover properties of the yuccas. Although yucca debris is entered by other

Fig. 125. Desert night lizard, approximately one and one-third natural size. Photograph taken by Nathan W. Cohen.

species such as the side-blotched lizard and desert spiny lizard, they use it chiefly as cover and probably do not compete with the night lizard. So dependent is the night lizard on yuccas that it occupies the more arid parts of the desert with great difficulty, if at all, in the absence of these plants. In less arid parts of its range, however, other cover may serve. At higher elevations in the desert, as on Telescope Peak in the Panamint Mountains, the night lizard lives under clumps of sagebrush (*Artemisia tridentata*). In the inner coast range, at Pinnacles National Monument, it is found in and beneath rotting digger pine logs. Along the Mohave River it may be found under rocks and boards among creosote bushes.

In the Monument, as over most of the desert range of the species, it is found most commonly under yucca litter, but at Pinyon Wells on October 13 we found two adults and a juvenile under some juniper poles and cactus stems. Yuccas were scarce.

Night lizards are active all the year in the lower, warmer parts of their range but are rarely seen unless their cover is removed. Only a few specimens were known in collections before John Van Denburgh, a western herpetologist, discovered them in 1893 under piles of Joshua tree branches. In warm weather they are most likely to be found in the larger, partly shaded heaps, but in winter they tend more to occupy the well-illuminated parts and are presumed to spend considerable time basking among the upper spines.

The common name of this lizard may be a misnomer, since recent field and laboratory observations suggest that there is considerable daytime activity, especially in winter, when night temperatures under the litter go below optimum levels for activity. Robert Glaser (MS) has reported the average body temperature as 86° F. and the range from 79° to 90° F. Substrate temperatures averaged 71° F. and ranged from 61 to 88° F. We found an individual evidently active under a Joshua tree branch at a soil temperature

Fig. 126. Prostrate, dead Joshua tree limbs, the microhabitat of desert night lizards; Lower Covington Flat, April 9, 1960.

of 55° F. during a cold rain near the town of Mojave. Thermal behavior thus appears to be comparable to that found in diurnal species that forage in the open, but the night lizard seems to have greater capability for activity at low temperature than do most sun-basking lizards.

Like the banded gecko, this small, soft-skinned animal is very susceptible to desiccation. It is able to live in the desert because of the moisture-holding capacity of its microenvironment, its avoidance of exposure, and the use of moist insect food such as flies and ants. We have often noted its predilection for dampness, and in yucca litter it is frequently found in the damper portions. Nevertheless, it is often encountered under litter that is dry.

Night lizards feed on termites, crickets, beetles, aphids, springtails, moths, caterpillars, flies, ants, ticks, and spiders. Ants, beetles, and flies are common in their diet. All these arthropods are found on or beneath the plant debris frequented.

Although in winter aggregations of over a dozen may be found under a single Joshua tree log, spacing of adults occurs in spring and suggests the existence of territorial behavior. At that time it is uncommon to find more than one adult male and female together beneath the same log.

According to M. R. Miller (1951), mating occurs in the Antelope Valley in late May and June. We can add no further observations. Night lizards are viviparous, and a single placenta is formed. The developing young increase greatly in weight by absorption of nutrients from the mother. The usual number is two, but often there is only one and rarely three. Births may occur from mid-September to mid-October. We found a recently born young in the Monument on September 9. Evidently little growth may occur over winter, for small young were noted in mid-April. Gravid females with yellow ova evident through the abdominal wall were found in early July.

Although the night lizard leads a secretive, seminocturnal, almost troglodytic existence, its eye structure (Walls, 1942:206), temperature characteristics, and tendency to bask suggest that it has relatively recently adopted this mode of life and that its predecessors lived more fully exposed to the sun. Increasing aridity, as desert conditions developed, has probably brought about this change.

The race in the Monument is *Xantusia vigilis vigilis*.

WHIPTAILS

Western Whiptail. *Cnemidophorus tigris*

Description. Adult 2¾ to 4 inches in snout–vent length. A slim lizard with slender pointed snout and long whiplike tail, characteristically crawling in jerky fashion, with much flicking out of tongue. Ground color (of race in Monument) grayish brown on head and shoulders grading to tan or rust on back and dusky toward tip of tail; dark flecks on dorsal surfaces, the flecks tending to form longitudinal stripes and cross-barred effect; below yellowish or pale gray with varied amounts of black spotting. Juveniles have alternating black and yellow longitudinal stripes and a powder-blue tail.
Range. Northeastern California, eastern Oregon, and west-central Idaho south to Cape district of Baja California and Durango; extends from California coast south of San Francisco Bay to west slope of Rocky Mountains and, in the south, across Continental Divide to western Texas.
Occurrence in Monument. Present throughout. Recorded from: Little Morongo Canyon, Mar. 30, Apr. 1, 28 *; 5 mi. W, 3 mi. S Joshua Tree, July 6; Black Rock Spring, July 6, Aug. 30 *, 31 *, Sept. 3 *, 4 *; Upper Covington Flat, Aug. 27; Lower Covington Flat, Apr. 9 *, May 19 *, Aug. 22, 23, 24 *, 25 *, 27 *, 28 *; Quail Spring, May 20 *, July 1 *, 2 *, 3 *, 4 *, 5 *; Stubby Spring, Sept. 6 *, 7 *, 8 *, 11 *; Key's View, July 3, 4, 5, 15 *; Indian Cove, May 13 *, July 6 *, Sept. 5 *; Fortynine Palms, Apr. 7, 9, May 7; 4 mi. N Twentynine Palms, July 25 *; Twentynine Palms, July 13 *, 21 *; Split Rock Area, May 8; Pinyon Wells, Apr. 10 *, Oct. 12; Pinto Basin, Apr. 6, 13, 26 *, 27 *, May 8 *, July 6, 7 *, 11 *, 15 *, Sept. 13 *, 15 *, Oct. 13; Virginia Dale Mine, Apr. 24 *, 25 *, 26, July 11 *, 12 *, 14; 10 mi. N Cottonwood Spring, Apr. 26 *; Cottonwood Spring, Apr. 5, 11 *, 12, 26, 29 *, 30, May 15 *, 16 *, 17 *, 18 *, July 7 *, 8 *, 9 *, 10 *, Oct. 22 *, 23 *.

This is one of the most widely distributed and abundant lizards in the arid and semiarid southwest. Its geographic and ecological range are almost identical to that of the abundant side-blotched lizard. It occurs from below sea level in desert sinks to about 5000 feet in the mountains. In the Monument it ranges from sandy flats in the Pinto Basin and at Old Dale to the piñon–juniper areas at Upper Covington Flat and Key's View.

Although whiptails seem to prefer habitats where the soil is suitable for burrowing, they are often found on the loose sand of aeolian deposits, along

Fig. 127. Western whiptail, approximately four-fifths natural size. Photograph taken by Nathan W. Cohen of an animal from the San Joaquin Experimental Range, Madera County, in the range of the race of central coastal California.

washes, in areas of hardpan, and among rocks. The tracks of this lizard are among those most commonly seen in sandy habitats in the Monument.

The species frequents grassland, brushland, and woodland. Although most of its range is in areas where there is no surface water except after storms, when water is present it is attracted to it, and good populations are to be found about springs and along stream courses. These lizards are especially abundant about oases in the Monument.

Usually the only time a whiptail can be approached closely is when it is basking. Once warm it will usually dart away with surprising speed and may run 150 feet before stopping. There is much variation in the reaction of these lizards to the observer, however. At Lower Covington Flat, where the population seems to be sparse, the animals were extremely wary, whereas at lower elevations they were less fearful. A startled individual at Lower Covington took off at full speed, ran up the side of a boulder, launched itself from a height of 6 feet into the air, landed on its feet 12 feet beyond, and continued running until out of sight. More often, however, they run ahead of the observer in short spurts, taking cover in bushes and finally, when persistently pursued, entering a burrow. Whether this variation represents differences in populations, changes in temperament with season, or other factors is not known.

The whiptail is basically a ground dweller and is seldom seen perched on rocks or branches as are the spiny lizards. We have, however, occasionally seen it climb. On the morning of July 15 in Pinto Basin one was seen 3 feet off the ground on the branch of a creosote bush. Another, also in a creosote bush, was seen near Kingman, Arizona, in midafternoon of a hot day in early July, on a branch 15 inches above the ground. The soil surface under the bush was very warm, and the lizard had probably ascended to escape the heat.

The whiptail is an excellent digger and constructs burrows of its own; it also uses those made by other animals. We have watched burrow construction by captives. Soil is scraped out with the front feet, and as the tunnel deepens, the lizard enters, turns about, and pushes dirt out the opening. The diameter of the tunnel constructed by an adult is 1 to 1½ inches. At night the opening may be plugged with earth. At Virginia Dale Mine we witnessed emergence of an individual from its burrow. It had pushed through the sand plug and, when first seen at 8:30 A.M., stood with half its tail still out of sight.

Anyone who has watched the behavior of this lizard cannot escape the conclusion that it is "high strung." Its almost incessant movements, rapid respiration, alert countenance, flicking tongue, and high body temperature combine to give this impression. We have taken the body temperatures of 58 individuals. The mean was 104° F., and the range 97° to 110° F. Aside from other members of its genus, among our records it is exceeded in body temperature only by the desert iguana among western lizards. Its high thermal tolerance and burrowing propensities may help to explain its success in hot, arid regions.

Whiptails feed on grasshoppers, beetles, ants, termites, caterpillars, ant lions, and the larvae of other insects, and on spiders, scorpions, reptile eggs, and carrion. At a camp outside the Monument an individual fed on discarded cooked meat scraps in our garbage pit. Confirmation of such feeding was obtained by stomach analysis. Its ability to locate and excavate buried prey and to identify and feed upon nonmoving objects suggests a well-developed sense of smell.

In May and June breeding males may actively pursue other individuals and may be seen to rub their hindquarters against the ground. A courting male crawls astraddle the back of the female, finally seizes her by the skin of her back, arches his body in a semicircle, and curls his hindquarters under her tail to achieve union.

Clutches are small, apparently usually only 2 to 4 eggs, but there is some evidence that there may be more than one clutch a year. Hatching occurs in July and August. A young individual only 42 mm. in snout–vent length was found on July 21 at Twentynine Palms.

The race in the Monument is *Cnemidophorus tigris tigris* of the Mohave and Colorado deserts.

SKINKS

Gilbert Skink. *Eumeces gilberti*

Description. Adult 2½–4½ inches in snout–vent length. A smooth-scaled, round-bodied lizard with long tail and short limbs; head little wider than neck. Scales over dorsal and ventral surfaces of body cycloid, highly glossy. Large adults plain olive or olive-brown above, the head and tail often reddish orange. Juveniles with whitish dorsolateral stripe on each side separated by olive or brown; broad dark brown stripe present on each side adjacent to the dorsolateral stripes; tail pink or blue, depending upon locality.

Range. Chiefly west slope of Sierra Nevada from Yuba River south to Tehachapi Mountains; also inner coast range of California from San Francisco Bay south to mountains of southern California; isolated mountain populations in southwestern Nevada, southeastern California, west-central Arizona, and northern Baja California.
Occurrence in Monument. Rare in western uplands. Recorded from: Lower Covington Flat, 4500 feet, May 17 *; Quail Spring, July 4. In addition Lloyd M. Smith informed us of a "red-tailed skink" found at Black Rock Spring.

This lizard ranges from the foothills that surround the San Joaquin Valley of California up to 8000 feet in the Sierra Nevada. It occurs in grassland, woodland, and open coniferous forests, usually where there are rocks and fairly dense plant growth. In the more arid parts of its range it is attracted to springs and stream courses. The specimen taken from the Monument was found in a water tunnel at Lower Covington Flat, and at Quail Spring three were seen.

Although diurnal, it is secretive and slips about among rocks and low growth, usually keeping out of sight.

This skink feeds on caterpillars, moths, beetles, Jerusalem crickets, grasshoppers, and spiders. In Surprise Canyon in the Panamint Mountains at midday we watched one foraging beneath willows near a stream. It crawled along a horizontal branch, a foot above the ground, thence 5 feet up the main trunk of a willow tree which grew at a 45° angle, whereupon it disappeared, perhaps into a crevice. Another, with a body temperature of 88° F., pursued a zigzag course over the willow leaf litter, poking its nose into cavities in the leaf mat.

The race in the Monument is presumed to be *Eumeces gilberti rubricaudatus,* the form found in the mountains of southern California. The specimen obtained was an adult and thus without the diagnostic color pattern necessary for racial comparisons.

BLIND SNAKES AND BOAS

Western Blind Snake. *Leptotyphlops humilis*

Description. Adult 9 to 12 inches in total length. An extremely slender wormlike snake, the head and body the same width, and with smooth, glossy, cycloid scales of nearly uniform size on both dorsal and ventral surfaces. Eyes vestigial, appearing as dark spots beneath the ocular scales. Above brown to pinkish, becoming lighter ventrally.
Range. Extreme southwestern Utah and southern Nevada south to Cape district of Baja California and Colima, México; extends from coast of southern California to western Texas.
Occurrence in Monument. Recorded from: Twentynine Palms (Loomis MS).

This is a snake of deserts and arid brushlands. It occupies both stony and sandy ground but usually only where there is sufficient loose soil to permit burrowing and where there are ant nests, which provide an important source of food. It prefers localities with some moisture and thus is likely to be found along the lower slopes of canyons or near streams and washes.

The subterranean habits of this snake make possible its existence in arid regions. Food consists of ants, ant eggs and larvae, termites, and probably other soft-bodied and moist arthropods.

These snakes may be found exposed on the surface of the ground at night and are best sought by driving slowly along dark paved roadways at night. The period of surface activity in southern California seems to be from April through August.

Both the subspecies *Leptotyphlops humilis humilis* of coastal southern California and the northern Mohave Desert and *L. h. cahuilae* of the lower Colorado River basin are to be expected in the Monument, the former at higher elevations. The example from Twentynine Palms belongs to *humilis* (Loomis MS).

Rosy Boa. *Lichanura trivirgata*

Description. Adult 2 to 3 feet in total length. Moderately heavy-bodied snake with small head and smooth, glossy scalation; scales small, those on top of head not enlarged. Small spur (vestige of hind limb) usually present on each side of vent. Eye with vertical pupil. Three broad longitudinal dorsal stripes of reddish brown on a ground color of light purplish gray.

Range. Panamint Mountains west of Death Valley, California, south to northwestern Baja California and from coast of southern California east to south-central and southwestern Arizona.

Occurrence in Monument. Recorded from: 3.2 mi. SW Morongo, Apr. 10 *, 21; Little Morongo Canyon, Apr. 2 *; 1 mi. S Joshua Tree P. O., May 5 *; 1 mi. NW Fortynine Palms, June 3 *; 6 mi. N, 1 mi. W Eagle Mountain, May 14 *. See Loomis and Stephens (1962) for additional records in the northwestern section and in vicinity of Cottonwood Spring.

This scarce snake is evidently less tolerant of aridity and high temperature than are most snakes in the desert. It is found in brushy areas, in the vicinity of streams, oases, and wash bottoms, and in the narrower canyons. It seems to favor rocky areas, where it finds thermal insulation and shelter from enemies. It is an excellent climber and is able to make its way among rocks with ease.

Most activity is at dusk and at night, but occasionally these snakes are abroad in the daytime. We found one stretched full length in the sun within five feet of the creek in Little Morongo Canyon at 11:00 A.M. Our few specimens were found from early April to early June.

Rosy boas probably have a low temperature of normal activity, in keeping with their nocturnal habits and sluggish behavior. We have only one record of body temperature, namely 74° F., recorded on April 10 at 9:00 P.M.; the air temperature near the ground was 75° F.

The rosy boa is a gentle, slow-moving reptile, an excellent subject to start with in overcoming fear of handling snakes. When disturbed or injured,

Fig. 128. Rosy boa, in Little Morongo Canyon, April 2, 1951.

rather than defend itself by biting, it may roll into a ball, with its head toward the center. It is live-bearing, and as many as ten young may compose a brood.

This species of snake is presently of scattered occurrence. It is unlikely that there is now a continuous connection between populations in the Monument and those of the coastal subspecies *Lichanura trivirgata roseofusca* to the west, in the San Gabriel–San Bernardino Mountain system. However, perhaps as a residual effect from a former connection, specimens from the western and northern side of the Monument have dorsal stripes with irregular edges, suggestive of the coastal type. Thus in pigmentation these individuals may be considered intergrades. Unfortunately, because of the small number of specimens available and the great overlap in scale characteristics, analysis of scutellation is of little help. We found specimens with irregular stripe borders in Morongo Pass, Little Morongo Canyon (ground color pale blue-gray), and at Fortynine Palms. Klauber (1933) had previously noted that a specimen from Twentynine Palms might be an intergrade. Loomis and Stephens (1962) observe that boas from the northwestern part of the Monument and adjacent areas have more irregular edges to the stripes than those from near Cottonwood Spring. Thus we may recognize a zone of intergradation between *L. t. roseofusca* and *gracia,* the desert race, along the elevated southern edge of the Mohave Desert, extending from the Morongo area at least as far east as Twentynine Palms. A living individual from the northern base of Mount San Jacinto east of Cabazon, seen by the junior author, suggests that intergradation may extend still farther west, into the region of San Gorgonio Pass.

COLUBRID SNAKES

Coachwhip. *Masticophis flagellum*

Other name. Red racer.
Description. Adult 3 to 4 feet, occasionally to over 6 feet in total length. Scales smooth, in 17 rows at midbody; anal plate divided; lower preocular scale wedged between upper labials, a characteristic common to racers. Adults often reddish or pinkish, with cross bands on the body becoming indefinite posteriorly and disappearing on tail; tail may be yellowish tan to grayish brown; dusky to black bands on the neck that may be more or less united; below whitish to yellowish, usually becoming pink or reddish posteriorly. Young blotched or cross-barred with dark brown, usually with no reddish color.

The common name of this snake comes from its slim shape and the appearance of its tail, which suggests a braided whip. Red color is the basis for the name "red racer," applied to the form in the western United States.
Range. Widespread in southern United States from coast to coast; extends from the Sacramento Valley in California, southern Utah, and southern Iowa south to Cape district of Baja California and to Oaxaca.
Occurrence in Monument. Ranges throughout. Recorded from: Morongo Valley, 1 mi. N county line, May 14 *; 0.6 mi. by road NE Morongo Lodge, May 10; Morongo Valley, 7 mi. SW Warren's Well, May 21 *; Warren's Well, May 21 *; Lower Covington Flat, Aug. 24 *; Quail Spring, Apr. 25; Indian Cove, May 12 *, 13–14, Sept. 11 *; 1 mi. NW Fortynine Palms, June 3 *; 6 mi. W Twentynine Palms, May 12; 7 mi. S Twentynine Palms, Apr. 10, July 11 *; Pinyon Wells, Oct. 10 *, 16 *; 3 mi. N Cottonwood Spring, July 10 *; 3½ mi. NNE Cottonwood Spring, Apr. 30 *; Cottonwood Spring, Apr. 29 *, May 15 *, 17 *, July 8 *, 9 *.

This is one of the more common snakes in the Monument, occurring in all habitats, on ground surfaces of sand, gravel, hardpan, or rock. It appears to be the fastest and most heat-tolerant snake in the area, moving with great speed, and staying abroad on bright, warm days.

When alarmed or pursuing prey, coachwhips are capable of traveling approximately 8 miles per hour, but their slender form may give an illusion of greater speed. When at bay, they fight with spirit and sometimes crawl aggressively toward an enemy. If taken in hand, they often bite, and the wound may bleed freely.

When frightened, these snakes may enter rodent burrows or climb. One found near the north entrance to the Monument ascended a goat-nut bush (*Simmondsia*) rather than entering one of many rodent burrows nearby. They may also enter bushes to bask. On April 10 at 7:55 A.M. we found one 4 feet off the ground fully exposed to the morning sun in the upper part of a bladder pod bush.

Body temperature for normal activity appears to be high. We have the following records: 86°, 99°, 99°, and 101° F. (average 96° F.).

Coachwhips hunt by day and probably detect much of their food by sight.

Fig. 129. Coachwhip. Photograph taken by Robert G. Crippen of an animal from the Coachella Valley.

Birds and their eggs, lizards (including horned lizards), other snakes (including rattlers), small rodents, and insects are eaten, the latter especially by the young. That olfaction may be used in detecting food is suggested by their habit of poking their heads into rodent burrows. A report of one feeding on a decomposed nighthawk (Cowles, 1946) is further evidence.

On April 30 at 10:50 A.M. a 3-foot coachwhip caught near Cottonwood Spring contained 4 recently swallowed nestling house finches. Circumstantial evidence for feeding on the Mohave rattlesnake (*Crotalus scutulatus*) was obtained on May 7 about 10 miles north of Yucca Valley, San Bernardino County. Rattler and coachwhip, both recently killed, lay 2 feet apart at the roadside, evidently clubbed to death. They measured, respectively, 3 and 4½ feet.

Coachwhips are attracted to the vicinity of oases, where they come to hunt and drink. At Quail Spring on April 25 at 11:00 A.M., one was found at the edge of vegetation within a few feet of the cement tank, and at Cottonwood Spring on July 9 at 9:30 A.M., a large individual (5 feet 6 inches in total length) came down to water out of thick grass. At small drinking holes where there is cover in which the snakes can hide near water, birds may be ambushed as they come to drink.

At Pinyon Wells (October 16) an emaciated coachwhip was found in water at the bottom of a well. It had evidently been unable to ascend the 20-foot vertical walls of rough granite. Perhaps it had fallen into the well in attempts to reach water.

All specimens collected in the Monument appear to be within the range of variation of the race *Masticophis flagellum piceus*. Adults tend generally to have some pink in the dorsal coloration, and the dark neck bands are usually distinct. There seems to be no tendency toward obscuring of the neck bands of adults or loss of pink color, as is characteristic of the race *M. f. ruddocki* of the Great Valley of California (Brattstrom and Warren, 1953). The race *piceus* ranges throughout southwestern United States and western Mexico, west of the Continental Divide.

Striped Racer. *Masticophis lateralis*

Description. Adult $2\frac{1}{2}$ to 4 feet in total length, occasionally over 5 feet. Proportions and scalation as described for coachwhip (p. 407). Above dusky brown or black, lighter on tail; cream to whitish stripe on each side; below usually whitish to yellow, becoming coral pink on posterior $\frac{1}{4}$ to $\frac{1}{3}$ of body, including tail; throat and neck spotted with black or dusky.
Range. Throughout most of California west of Sierra Nevadan crest and the deserts, but absent from cultivated areas, open grasslands, and heavily forested regions; extends south to central Baja California.
Occurrence in Monument. Widespread but nowhere common in piñon belt of western section. Recorded from: Black Rock Spring, Oct. 16 *; Lower Covington Flat, Apr. 9 *, Aug. 24 *; Quail Spring, Apr. 25 *; Stubby Spring, Sept. 7 *; Key's View, July 4 *. Loomis and Stephens (1962) record this species additionally from Hidden Valley, Apr. 9.

The easternmost records for the Monument are in the vicinity of Key's View and Hidden Valley, but this snake is to be expected as far east as Pinyon

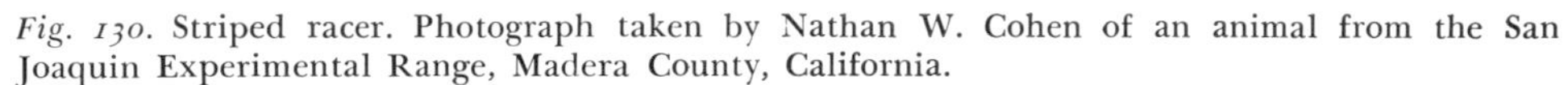

Fig. 130. Striped racer. Photograph taken by Nathan W. Cohen of an animal from the San Joaquin Experimental Range, Madera County, California.

Wells. With the exception of one caught at Quail Spring, 3900 feet, all specimens have been taken above 4000 feet.

This is a snake of brushland and open woodland; its distribution along the coast follows closely that of the California chaparral. Thus it is not surprising to find it in the piñon–scrub oak habitat in the Monument, since this habitat continues almost unbrokenly from the desert slope of the San Bernardino Mountains to the west. In its marginal habitat on the desert fringe of its range, it is especially likely to be found along stream courses and in the vicinity of springs, as evidenced by the locality records cited. It appears to require a higher humidity than the coachwhip.

Body temperature records (81° and 83° F.) of individuals abroad in mid morning suggest that the temperature of normal activity is lower than that of the coachwhip.

The striped racer feeds on frogs, lizards, snakes (including rattlers), small mammals, and birds. One seen in Fish Canyon, Los Angeles County, crawled away at high speed with the hind legs of a frog protruding from its mouth. Here may be a basis for folklore about "horned" snakes.

The subspecies of striped racer occurring in the Monument is *Masticophis lateralis lateralis.*

Western Patch-nosed Snake. *Salvadora hexalepis*

Description. Adult 2–4 feet; most individuals about 2½ feet in total length. Typically a slim beige or brown snake with conspicuous dark eyes that stand out against the pale head color. A broad beige or yellowish stripe down middle of back, bordered with brown stripes on each side; sometimes striped pattern obscured by cross bars; below whitish or dull yellow, unmarked, sometimes suffused with dull orange posteriorly; throat pale. Possesses enlarged rostral scale with free edges, which is basis for the name "patch-nose." Scales smooth, in fewer than 19 rows.

Range. Northwestern Nevada and southern Utah south to the tip of Baja California and northern Sinaloa; extends from coast of southern California east to Grand Canyon and central Arizona.

Occurrence in Monument. Recorded from: 3 mi. SW Warren's Well, May 21 *; 3 mi. SE Joshua Tree, Apr. 12; Quail Spring, Apr. 10 *; 3 mi. E Quail Spring, Mar. 30 *, Apr. 6; Queen Valley, Apr. 11; Indian Cove, Apr. 30 *; 1 mi. NW Fortynine Palms, June 3 *; Twentynine Palms, Apr. 9; Belle Camp area (near White Tanks), Apr. 11, May 1 *, June 30 *; Pinto Basin dunes, July 7 *; 1¾ mi. S Cottonwood Spring, Apr. 29. An additional noteworthy record (Loomis and Stephens, 1962) is Lower Covington Flat, 4700 feet.

This snake frequents plains and lower mountain slopes in arid and semiarid regions, generally below 6000 feet. It occurs in both rocky and sandy habitats in areas of chaparral and desert brushland.

Like the coachwhip, it is active in the daytime and feeds on lizards and small mammals. It is thought that it uses its enlarged rostral scale in digging up reptile eggs. One was found with leathery-shelled eggs in its stomach. Since it is a fast-moving, diurnal, lizard-eating snake, one wonders how it is able to coexist with the much larger snake-eating coachwhip, of similar habits. Smaller size and perhaps greater tolerance of low temperatures may permit

Fig. 131. Western patch-nosed snake. Photograph taken by Robert G. Crippen of an animal from the Borrego area in eastern San Diego County, California.

it to be active in the early morning or on cool days when coachwhips are seldom out. Use of smaller prey and specialization on reptile eggs may also help to reduce competition. Time of capture and the few body temperatures available seem to support the first point. Our specimens have been taken in the morning at 7:20, 8:00, 9:00, 9:30, and 10:50, and in the afternoon at 4:15, 5:40, 6:00, and 6:20. Body temperatures recorded were 83°, 84°, 85°, and 90° F. (average 86° F.). A low normal activity temperature would make it possible for these snakes to be active when lizards are sluggish. The whiptail (*Cnemidophorus tigris*), for example, upon which this snake is known to feed, has a normal activity temperature of 104° F. (range 97° to 110° based on 58 individuals), some 18° F. higher than the patch-nose. Lizards are probably often caught when they are basking and less likely to take flight.

The patch-nose is oviparous. Four to 10 eggs have been reported as composing a clutch. A large adult female found near Warren's Well on May 21 had eggs nearly ready for laying.

The Monument lies in an area close to the junction of three races of this snake—*Salvadora hexalepsis mojavensis, S. h. hexalepsis,* and *S. h. virgultea.* The first occupies the Great Basin and Mohave Desert, the second the lower Colorado River Basin, and the third the California coast. The coastal form is partly separated from the others by high mountains, making contact with them chiefly through mountain passes. The two desert forms are probably partly separated by the elevated southern edge of the Mohave Plateau.

The zone of intergradation between the desert races *mojavensis* and *hexalepis* has generally been regarded as extending from the Joshua Tree Monument area to a region south of the southern tip of Nevada. Intergradation between coastal *virgultea* and the desert forms evidently occurs along the western desert fringe of the Mohave and Colorado deserts. Thus one might expect to find attributes of all three races in the snakes of the Monument.

Specimens (11) from the Monument and the area adjacent to the north have been examined as to the nature of striping and the condition of the scales in the subocular region (see Bogert, 1945:14, for key characters separating the races). On the basis of these characteristics and an over-all impression of coloration, we classify most of the specimens in the Monument as *hexalepis*. Exceptions are as follows: The specimen from near Warren's Well, in style of pattern, seems close to *mojavensis*, but one supralabial on each side of the head reaches the eye, as is generally true in *hexalepis*. One specimen from Quail Spring has a supralabial reaching the eye on one side but not the other. The dorsal stripe is wide and the lateral stripes are distinct. In the suborbital character it combines the typical condition found in *mojavensis* and *hexalepis*. Another specimen from near Quail Spring has a slight narrowing of the dorsal stripe, suggestive of *virgultea*, and has the supralabial contact with the eye as in *virgultea* and *hexalepis*. Likewise, an individual from Indian Cove seems to show a slight tendency toward *virgultea* in pattern, but another individual from this locality can be classified as *mojavensis*.

A connection probably exists near the Monument between *virgultea* and the two desert races, via San Gorgonio Pass, which connects the southern California coastal plain with the desert. This connection may be visualized as Y-shaped, with the base of the Y situated in the Pass, one arm extending east into the range of *hexalepis* and the other to the northeast, via Morongo Pass, to the Mohave Plateau occupied by *mojavensis*.

Spotted Leaf-nosed Snake. *Phyllorhynchus decurtatus*

Description. Adult 12–20 inches in total length. A blotched snake with much enlarged rostral scale that appears appliquéd to the snout, hence the common name. Head only slightly wider than neck, and dorsal scales smooth or weakly keeled in some males; anal scale single. Large blotches of back brown, edged with black or dark brown, on a cream, pinkish, to light brown ground color; belly white, unmarked. Pupil vertically elliptical.
Range. Southern Nevada to Cape district of Baja California and to Sonora; desert slope of mountains in southern California east to central Arizona.
Occurrence in Monument. Recorded by Loomis and Stephens (1962) from: 4 mi. S Twentynine Palms, Apr. 13, June 3; White Tanks, May 29, June 15, 26; Cholla Garden, Apr. 2, 15, June 3, 16; Pinto Basin (Old Dale Junction), Apr. 13, May 14, 15, 21, 29; Cottonwood Spring, May 13, 14, 21, 28, June 3.

This is one of the more common snakes at elevations below 3000 feet, but it ranges to at least 3800 feet in the White Tanks area. Despite its abundance we have somehow missed it, perhaps because we have seldom been in the

Monument in late May and June, when these snakes are evidently most active.

The leaf-nose seems to prefer arid and semiarid habitats of mixed sand and rock. It is rare or does not occur in areas of continuous sand. At localities of occurrence the vegetation usually consists of sparse desert brush, such as creosote bush and cactus.

Fig. 132. Spotted leaf-nosed snake. Photograph taken by Nathan W. Cohen.

For many years this secretive, strictly nocturnal little snake was very poorly known, but more recently, with searching by driving slowly along roads, it has been proved to be one of the most abundant desert snakes locally. It retreats by day into rodent burrows and occasionally beneath surface litter or into alluvium or sand. It burrows by means of sidewise movements of its head, but it requires more time to submerge than the shovel-nosed snake (p. 422) and seems to be less well adapted for burrowing.

The chief function of the enlarged rostral scale may be to protect the snout in digging and to aid in removal of soil in excavating reptile eggs upon which it feeds. Banded gecko eggs are among those eaten. In addition it eats adult lizards and perhaps insects. Captive leaf-nosed snakes have been seen to seize the tail of a gecko and to swallow the cast tail, allowing the lizard to escape. Attacks of leaf-nosed snakes may account for some of the tail loss noted in these lizards in the field. It is of interest here that the entire range of this snake in the United States is contained within that of the banded gecko. In turn, an important enemy of the leaf-nose is the glossy snake.

The race of the leaf-nose occurring in the Monument is *Phyllorhynchus decurtatus perkinsi.*

Glossy Snake. *Arizona elegans*

Other name. Faded snake.

Description. Adult $2\frac{1}{2}$–$4\frac{1}{2}$ feet in total length. A moderately slender snake with head only slightly wider than neck; snout rather sharp and somewhat flattened in profile, the lower jaw countersunk; scales smooth and shiny, in 25 to 31 rows at midbody; anal plate single. Coloration appears faded, hence the common name; ground color light brown, buff, pale gray, or cream; row of brown to gray dark-edged blotches down middle of back with smaller ones on sides; belly whitish or pale yellow, unmarked.

Range. Central California, southern Utah, and northeastern Colorado south into northern states of México; extends from coast of southern California to east-central Texas.

Occurrence in Monument. Inhabits chiefly the lower, sandier areas, but ranges up to 4700 feet. Recorded from: 3 mi. E Yucca Valley, May *; 1.6 mi. E Joshua Tree, May 13 *; Juniper Flat near Stubby Spring, May 1 *; 1 mi. NW Fortynine Palms, June 3 *; 2.8 mi. W Twentynine Palms, July 13 *; near Pinto Basin dunes (3 mi. SW Pinto Peak), Oct. 13 *. In addition Loomis and Stephens (1962) have found it at Lower Covington Flat (4700 ft.), 5 mi. S Twentynine Palms, Cholla Garden, and near Cottonwood Spring.

This snake is primarily a lowland species that occupies arid and semiarid environments. It prefers sandy habitats but also occurs on hardpan or in rocky areas where there is adjoining loose soil; vegetation is often sparse but in some areas may consist of dense brush. The species is a good burrower, its countersunk lower jaw, wedge-shaped snout, and smooth scalation permitting it to make its way readily through loose soil. Individuals have been plowed up from depths of a foot or more. They also seek refuge in the burrows of other animals.

Over most of its range the glossy snake is nocturnal, although in Texas it is often active in the daytime. Most specimens from California have been taken after dark, from 7:00 to 11:00 P.M. It is generally regarded as cold-tolerant and is often found abroad on cold, windy nights. One obtained at 8:50 P.M. south of the Monument near Indio was encountered in a sandstorm. Its body temperature was 70° F., the air near the ground was 69° F., and the sand at the roadside was 66° F. Another road catch at 11:00 P.M. near Ajo, Arizona, had a body temperature of 90° F. Peak activity in California seems to be in late May and June.

Although this snake is usually peaceable, a juvenile found abroad about 8:30 A.M. on a bright morning near the Pinto Basin dune struck repeatedly when we attempted to capture it.

Glossy snakes feed chiefly on lizards, seeming to prefer them to small mammals, although they do eat the latter. One glossy snake was apparently killed in an attempt to swallow a horned lizard. The lizard's head spines pierced its neck.

As in the western patch-nosed snake and gopher snake, the distribution of races of the glossy snake in California suggest the possibility of three-way intergradation in the Monument area. *Arizona elegans occidentalis* of the coast is reported to intergrade with *A. e. eburnata* of the Colorado and Mohave deserts in the vicinity of Cabazon in San Gorgonio Pass (Klauber, 1946), 15 to 20 miles southwest of the western end of the Monument. *A. e. candida* of the western Mohave Desert is regarded as intergrading to the east with *eburnata*. Klauber (1946) states that specimens from Apple and Lucerne valleys are evidently *eburnata,* but that two specimens from Morongo Valley and west of Twentynine Palms have the paired preoculars typical of *candida.*

Fig. 133. Glossy snake. Photograph taken by Robert G. Crippen of an animal from Palm Springs, Riverside County, California.

These he regards as intergrades. The expected variation in the Monument seems to be borne out in our specimens. One from 3 miles east of Yucca Valley has a sprinkling of dark marks on the lowermost dorsal scale row on the rear third of the body, thus displaying a feature of *occidentalis,* and yet it has paired preoculars and dorsal blotches spanning 9 or more dorsal scale rows, in this respect resembling *candida.* A specimen from near Joshua Tree only a few miles away has paired preoculars on one side and *candida*-like cross bands; no *occidentalis* influence is evident, however. A specimen from Juniper Flat has single preoculars like *eburnata,* a few dark flecks on the lower sides suggestive of *occidentalis,* and a dorsal pattern that is a mixture of the type of blotches of *candida* and *eburnata.* Specimens from near Twenty-nine Palms and Fortynine Palms are *eburnata,* although midbody blotching of the latter is suggestive of *candida.* A Pinto Basin specimen seems clearly to be *eburnata.*

Gopher Snake. *Pituophis melanoleucus*

Description. Adult $2\frac{1}{2}$ to 6 feet long, rarely to 8 feet in total length. A blotched snake with pattern suggesting that of western rattlesnake. Scales on back keeled, becoming smooth on lower sides. Ground color usually cream to pale yellow, the blotches brown or black; blotches squarish, hexagonal, or oval down middle of back, with smaller ones on the sides; below white, cream, or yellow, often with dusky or black spots on belly scutes; throat pale; dark stripe usually present behind eye.
Range. Southern Canada south to Cape district of Baja California and central México; extends from Pacific coast to the Atlantic; absent from high elevations and from most of northeastern United States and the Mississippi Valley.

Occurrence in Monument. Recorded from: 4 mi. W Yucca Valley, May 13 *; 1 mi. W Yucca Valley, Apr. 4 *, June 2 *; 23 mi. W Twentynine Palms, Apr. 4 *; 10½ mi. E Joshua Tree, Apr. 27 *; 3 mi. SE Joshua Tree, May 19 *; Queen Valley, Apr. 10 *; Indian Cove, May 8, 14; Twentynine Palms, Apr. 9 *, May 14; 3 mi. N White Tanks, Apr. 30; 1½ mi. W White Tanks, Oct. 18 *; 1½ mi. W, 2½ mi. S Pinto Peak (Pinto Basin), May 8 *.

The gopher snake is to be expected throughout the Monument, but all records, including those of Loomis and Stephens (1962), are from its northern edge and the Queen Valley-White Tanks area, except the one from Pinto Basin.

It is surprising that so few gopher snakes have been found. Judged by acquaintance with this snake elsewhere, it is a common, adaptable species that frequents many habitats from sea level up to about 9000 feet. It is able to live in grassland, chaparral, woodland, forest, and desert, and it is one of the most abundant snakes in cultivated areas in the west.

The gopher snake displays no obvious major adaptations for desert life. However, in the desert, it has a paler ground color and more contrasting pattern than in other areas and, in hot weather, it is likely to be crepuscular and nocturnal. Over most of its range elsewhere it is primarily diurnal or crepuscular.

Temperature records available from individuals collected both within and outside the Monument suggest that it is thermally conservative, avoiding low and high extremes, as snakes generally do. The mean and range of 11 records of body temperature are 79° F. (66°–92° F.).

Temperament varies. Some individuals are peaceable and can be picked

Fig. 134. Gopher snake. Photograph of a young animal taken in Queen Valley, April 10, 1951.

up easily and carried about. Others stand ground, hiss, broaden the head and vibrate the tail, putting on a show of ferocity. Such an individual can usually be counted on to bite if picked up.

The generic name *"Pituophis"* means "phlegm serpent" and is based on the loud hissing sound made by these snakes. It exceeds in volume that made by most other species. The glottal region of the trachea is modified for sound production. Two enlarged portions, separated by a constriction, lie just behind the glottis and dilate, balloonlike, when the snake hisses. In front of the glottal opening is a vertical fin interposed in the air stream. However, we once removed this structure from a hissing snake and noted little change in the quality of the sound.

Gopher snakes feed on small mammals, birds and their eggs, and lizards. Lizards seem to be eaten mainly by the young snakes. Rarely, they are cannibalistic. They kill their prey by constriction. A Merriam kangaroo rat (*Dipodomys merriami*) was removed from the stomach of one caught on October 18 near White Tanks, and a wood rat from another obtained on April 30 outside the Monument, but in desert habitat, in the Panamint Mountains.

These snakes are good diggers and are capable of removing the bulkhead of earth used by some rodents to close their burrows. The earth is loosened with the head, caught in a bend in the neck, and hooked backward. Although mainly terrestrial, gopher snakes climb in search of the eggs and young of birds.

Eggs are laid in midsummer and are hatched between mid-September and early October. Three to 18 may compose a clutch.

Three races of gopher snakes occur in or near the Monument—*Pituophis melanoleucus deserticola* of the Mohave Desert, *P. m. affinis* of the Colorado Desert, and *P. m. annectens* of coastal southern California. According to Klauber (1947), *deserticola* integrades with *affinis* across the desert approximately along the line separating San Bernardino and Riverside counties, or slightly to the south. *Annectens* is known from Cabezon and Snow Creek in the San Gorgonio Pass area within 15 miles air line of the west end of the Monument. Because of our scant material and the variation in scale characteristics, we have judged the racial affinities on the basis of pigmentation.

Most of our specimens seem to be closest to *deserticola*, having the black neck markings of the latter and lacking the gray or brownish suffusion of the dorsolateral areas characteristic of *annectens* and the brown neck blotches of *affinis*. A single adult from Covington Road near the Monument boundary and one from the White Tanks region have some brownish suffusion present on the sides and may thus bear some *annectens* influence. An adult from Pinto Basin, although looking most like *deserticola*, has dark-bordered blotches along the middle of the back that resemble the back markings of *affinis* and may be regarded as an intergrade of *deserticola* and *affinis*. Specimens elsewhere from along the central section and north side of the Monument all appear to be closest to *deserticola*.

Common Kingsnake. *Lampropeltis getulus*

Description. Adult 3 to 4 feet in total length. Scales smooth and shiny; anal scale single. Distinctly ringed with black or dark brown and white or cream, the dark bands broader than the light ones. Juveniles tend to have a more contrasting pattern than adults.
Range. Southwestern Oregon, southern Iowa, and southern New Jersey south into northern states of México; extends from Pacific coast to Atlantic.
Occurrence in Monument. To be expected throughout. Recorded from: 5 mi. W, 3 mi. S Joshua Tree, July 5, 7 *; 2 mi. E Joshua Tree, June 2 *; Quail Spring, July 3 *, 15; Lost Horse Valley, May 20 *, 21; Key's View, July 15 *; Indian Cove, Apr. 30 *, May 1; 1 mi. NW Fortynine Palms, June 3 *. In addition Loomis and Stephens (1962) have obtained it 3.1 mi. N Cottonwood Spring, June 15.

This is a highly adaptable species that occurs in many habitats from sea level up to about 5000 feet. It frequents swampland, grassland, brushland, woodland, and forest, ranging from moist river bottoms to prairie and desert. In the Monument it occurs from the creosote bush desert up to the piñon belt and has been found in rocky areas and on alluvium.

Over much of its range this snake is diurnal and crepuscular, but in warmer areas, as in the arid southwest and in the Monument, it is nocturnal, particularly in hot weather. We have found individuals at dusk, at night, and in the morning.

Like the long-nosed snake, this species displays no obvious adaptations for desert life and its numbers seem to be greater in the watered portions, along

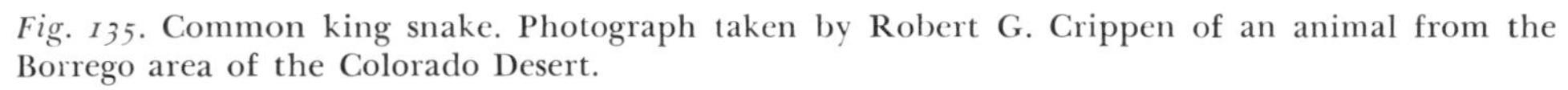

Fig. 135. Common king snake. Photograph taken by Robert G. Crippen of an animal from the Borrego area of the Colorado Desert.

streams and washes, and in the elevated, less arid sections. We have only three body temperature records: 73°, 80°, 82° F.

The diet is varied. Kingsnakes eat other snakes (rattlers, whipsnakes, and gopher snakes), lizards, the eggs of reptiles and birds (quail and others), small mammals (mice, pocket gophers), and amphibians.

The race in the Monument is *Lampropeltis getulus californiae,* which is widespread in the west, occurring in California, Baja California, Nevada, and northwestern Arizona. Kingsnakes from the desert, including those from the Monument, tend to have more contrast between the dark and light bands than those from the coastal side of the mountains. Even large individuals may have very dark brown or black bands separated by unmarked white bands.

Long-nosed Snake. *Rhinocheilus lecontei*

Description. Adult 2 to 3 feet in total length. Scales smooth; anal plate single; all or most of caudals single, an unusual characteristic. In the area of the Monument this is a distinctly cross-banded snake marked with broad black saddles separated by narrow white interspaces; white spots present laterally in saddles and sometimes pinkish color in interspaces; below cream or white, unmarked or occasionally with a few dark marks at the sides. The white spotting of the black saddles and single caudals will distinguish this snake from the similar-appearing common kingsnake.

Range. Sacramento Valley of California, southern Idaho, and southeastern Colorado south into northwestern and northeastern México; extends from coast of southern California and northern Baja California east to central Oklahoma and Texas.

Occurrence in Monument. Recorded from: 3 mi. NE Morongo Lodge, May 10; Indian Cove, Apr. 11 *, Apr. 30 *, May 14 *; 8 mi. W Twentynine Palms, July 5, 6; 4 and 7 mi. W Twentynine Palms, Apr. 7; 1 mi. W Twentynine Palms, June 2 *; 2 mi. E Twentynine Palms, July 13 *; 8½ mi. S, 1 mi. W Twentynine Palms, 4500 feet, Oct. 11; Pinto Basin, Apr. 12. In addition Loomis and Stephens (1962) record it from 6 mi. S Twentynine Palms and 2 mi. N Cottonwood Spring.

This snake seems to prefer lowland grown to brush or grass, but it is also to be found sparingly in forested districts. One was obtained at Julian, San Diego County (Klauber, 1941), in mixed grassland and coniferous forest and one at Little Rock, Los Angeles County, in piñon–juniper woodland. It is to be expected in the latter zone in the Monument. Although it is frequently found on alluvium, it is seldom encountered in extensive areas of sand.

In the southwestern part of the range of this species nocturnal habits are the rule, but it seems to be less strictly nocturnal than are the sidewinder and glossy snake. A long-nosed snake found at Indian Cove at 6:00 P.M. was abroad just before sundown, and one 5 miles north of Yucca Valley outside the Monument was crossing the road in broad daylight at 9:50 A.M. Based on Klauber's (1941) data most seasonal activity seems to be from April to September, with a peak in May. Most of our specimens have been found in April and May.

Like the glossy snake, the long-nose is adept at burrowing. Movement through loose soil is facilitated by the flattened, pointed snout and smooth scales. Captives often burrow but may prefer to take refuge in crevices if rocks are placed in their cages. In general the long-nose is a gentle snake when

Fig. 136. Long-nosed snake. Photograph of an animal taken at Indian Cove, April 11, 1960.

adult, but juveniles may strike spiritedly, and both young and adults may vibrate the tail when annoyed.

There seems to be considerable tolerance of low temperatures. One long-nose at Indian Cove at twilight on April 11 had a body temperature of 59° F., the lowest record of which we are aware for any active snake. Air temperature near the ground was similar. Other records of body temperature, with air temperatures in parentheses, are 68, 85 (77), 89 (80), 90 (86). Body temperatures often exceed those of the air because of residual heat in the substratum upon which the snakes rest. Klauber (1941) reports a long-nose blown across the smooth surface of the highway by a strong cold wind near Whitewater in San Gorgonio Pass.

Long-nosed snakes eat lizards (whiptails and others), reptile eggs, small mammals, and insects. A Merriam kangaroo rat was found in the stomach of one reported by Klauber (1941).

A red-tailed hawk on October 11 on the plateau 8½ miles south and 1 mile west of Twentynine Palms near Queen Valley had captured a long-nosed snake early in the morning (see p. 55).

This species of snake displays no very marked adaptations for desert life. Nocturnality and burrowing habits are probably to their advantage in arid regions, but these are not traits peculiar to desert snakes. That the long-nose lives in desert areas under some duress is suggested by its prevalence in irrigated sections. Indeed, over a broad expanse of low desert fanning out from the Monument to the east, records of these snakes are extremely rare.

All our specimens from within or adjacent to the Monument seem clearly to have the contrasting type of pattern normal for the race *Rhinocheilus lecontei clarus* of the Colorado Desert, although several individuals have faint red color in the pale interspaces.

Western Shovel-nosed Snake. *Chionactis occipitalis*

Description. Adult 10–16 inches in total length. Scalation smooth and glossy; dorsal scales usually in 15 rows. Snout flattened, sharp-edged, and rounded in dorsal aspect, hence the name "shovel-nosed snake." Coloration white, cream, or yellow with distinct cross bands of black or brown, the bands usually partly or completely encircling the body posteriorly; much variation in details of pattern; dorsal interspaces between bands may be suffused with pink or red or flecked with black or brown.

Range. Southern Nevada to shores of northern end of Gulf of California; extends from desert base of mountains of southern California to south-central Arizona.

Occurrence in Monument. Present in sandy areas at low elevations. Recorded from: Morongo Valley, 4 mi. N Riverside County line, June 2 *; 5 mi. S Morongo Valley, Mar. 25 *; Pinto Basin (1 mi. W, 3 mi. S Pinto Mountain), Sept. 15 *; 4½ mi. S Cottonwood Spring, June 3 *. In addition Loomis and Stephens (1962) report it from Twentynine Palms.

This snake is strictly a desert dweller, frequenting areas of dry sand. It inhabits washes, sandy flats, dunes, and sand hummocks. Hummocks, crowned with mesquite or other desert shrubs, seem to be a favorite refuge. The shovel-nose inhabits some of the most barren sections of the desert but is scarce or absent where vegetation is lacking.

For those seeking to observe this rare snake, Klauber's (1951) data on time of collection and air temperatures may be of help. He reports captures mostly from 7:00 to 10:00 P.M. at air temperatures between 70° and 90° F. Most specimens have been taken from March to August.

Shovel-nosed snakes are almost exclusively nocturnal, but there are a few records of daytime activity. The individual collected 5 miles south of Morongo Valley was crawling in the open on a rocky hillside at 2:25 P.M. on an overcast day. It sought refuge in a crevice. A sandy wash lay at the bottom of the slope 30 feet away. Air temperature near the ground was 79° F.; the surface of the soil was 83° F. This snake had probably hatched the previous fall. It had light orange color in the pale interspaces on the back and a bright orange spot in the first interspace at the back of the head.

Shovel-nosed snakes are best found by night driving. When sought by lantern or flashlight beneath bushes, they are very difficult to see. Their pale color blends with the background, and the dark bands may be mistaken for the shadows of twigs.

These are excellent burrowers, capable of gliding beneath loose sand with ease. The countersunk lower jaw (which prevents sand from entering the mouth), nasal valves, and smooth scalation are structural characteristics that aid burrowing. These snakes also take refuge in the burrows of other animals. One was dug out of a tunnel that apparently had been made by a scorpion. The snake's track led to this burrow.

Food consists of scorpions, centipedes, spiders, and insects.

The race occurring in the Monument is *Chionactis occipitalis occipitalis*, which ranges south to the southern boundary of Riverside County. Adults of this subspecies generally have brown dorsal bands and a smaller proportion of the anterior bands carried to the ventral surface than does *C. o. annulata* of the Colorado Desert.

Night Snake. *Hypsiglena torquata*

Other name. Spotted night snake.
Description. Adult 12—18 inches in total length. A moderately slender serpent with somewhat flattened head and spotted pattern. Scales smooth; anal plate divided. Pupil vertical. Ground color above pale gray, beige, tan, or dusky; row of large yellow-brown spots down middle of back, with smaller ones on sides. A large dark brown blotch usually present on each side of neck, sometimes connected at midline; below white or pale yellow, unmarked or with fine gray stippling laterally.
Range. Sacramento Valley of California and eastern Washington, south into México; extends from coast of southern California east to central Oklahoma and Texas.
Occurrence in Monument. To be expected in or near rocky areas throughout. Recorded from: 6 mi. W Yucca Valley, May 13 *; Pinto Basin, Apr. 10 *. Loomis and Stephens (1962) report it from Lower Covington Flat Camp, June 16, 23; 1 mi. N White Tanks Campground, May 29; Pinto Basin (near Old Dale Road junction), May 14, 21.

This snake occupies a great variety of habitats in arid and semiarid regions from sea level to at least 7000 feet, frequenting brush, grassland, and woodland. In many areas it seems to prefer rocks, but it has also been found beneath bark and inside rotten logs. Its flattened head may reflect crevice-dwelling habits. Our specimen from Pinto Basin was found at 10:30 P.M. at the southern end of the basin at 1750 feet elevation. Rock outcrops were present nearby. Creosote bush, Mohave yucca, and Joshua trees were dominant plants in the area. The snake had been recently killed on the road. Two specimens reported by Loomis and Stephens (1962) were caught in a can trap at Lower Covington Flat, 4700 feet, in the piñon belt.

As the common name implies, this is a nocturnal and crepuscular snake. It feeds on lizards, amphibians, and probably insects. Since remains of active diurnal lizards have been found in the stomach contents, it must catch them when they take refuge. Geckos (*Coleonyx*), which are active at night, are also eaten. Captives feed readily on the desert night lizard and side-blotched lizard, but we have no records of such prey being taken in the field. Nevertheless, by virtue of their size and accessibility they may be a staple in the diet in some areas. It is of interest here that the range of the night snake in North America is virtually identical with that of the side-blotched lizard. Prey

Fig. 137. Night snake. Photograph taken in Cedar Canyon, Providence Mountains, Mohave Desert, May 27, 1938.

is subdued by injection of toxic saliva. Enlarged teeth toward the back of the mouth help to ensure its penetration. Night snakes typically retain their hold until the struggling of the prey ceases.

Loomis and Stephens (1962) allocate their specimens to the race *Hypsiglena torquata deserticola*. However, one from near White Tanks seemed to possess some characteristics of the coastal race *H. t. klauberi*. Our specimen from Pinto Basin is clearly of the desert form, but the one from near Yucca Valley seems to have some traits of *klauberi*. This opinion is based on the size of the dorsal spots, configuration of the neck markings, and ground color as compared with other specimens from desert areas in San Bernardino and Riverside counties.

California Lyre Snake. *Trimorphodon vandenburghi*

Description. Adult 2–3 feet in total length, occasionally nearly $3\frac{1}{2}$ feet. A slim snake with broad head and slender neck; dorsal scales smooth and in 21 to 24 rows. Pupil vertical. The common name refers to the lyre-shaped pattern on the head—a large V-shaped mark with the point between the eyes. Dorsal surface with row of hexagonal brown median blotches, each blotch usually split by a light-colored cross bar; smaller blotches on the sides; ground color above cream, buff, light brown, or gray; belly whitish or yellowish, often with a sprinkling of brown dots.
Range. Coastal California east to desert borders, from near Little Lake, Inyo County, south into northwestern Baja California.
Occurrence in Monument. Recorded from: 2.1 mi. S Morongo Lodge, May 1 *. Loomis and Stephens (1962) report it from 1.9 mi. NW Cholla Garden, Sept. 5. John Todd (in correspondence) informs us of two specimens, one noted at the campground in Indian Cove and another near Cottonwood Spring, both road kills.

This snake is an inhabitant of rocky foothills, mesas, and inland valleys, on both the desert and coastal sides of the mountains of southern California. It seeks refuge in deep crevices. This and its nocturnal habit make it difficult to find. Most specimens have been found on roadways at night in the vicinity of rock outcrops and by the technique of rock flaking. The lyre snakes seem to prefer refuges beneath massive rocks to thin exfoliating shells. There appears to be abundant suitable habitat for this species along the lower slopes of the Little San Bernardino Mountains, and the scarcity of the records of it in the Monument probably results from its secretive habits and inadequate searching effort.

The California lyre snake is a back-fanged serpent with venom capable of subduing or killing the lizards and small mammals on which it feeds. Although large lyre snakes should be approached with care, this snake is probably not dangerous to man. The gape is not large and the grooved fangs are situated well back in the mouth. One would probably have to allow this snake to chew in order to have it effectively inject its venom. Although the venom itself, evidently of hemorrhagic type, does not appear to be potent in man, it should be made clear that this opinion is based on meager evidence.

Since this is a crevice-dwelling snake, it is not surprising to find that bats, which rest by day in rock fissures, have been recorded in the diet, as well as crevice-dwelling lizards.

Twelve eggs were laid in September by a captive female from the Little San Bernardino Mountains (Klauber, 1940).

Western Black-headed Snake. *Tantilla eiseni*

Description. Adult 10–15 inches in total length. A slender tan snake with a black, flattened head. Scales smooth and in 15 dorsal rows; loreal scale absent. Black cap set off from neck by narrow white collar, which is sometimes faint; below orange or coral-red in a broad band down middle of belly, whitish laterally and on throat.
Range. California from Alameda County near San Francisco Bay and the southern Sierran foothills to San Quintín, northern Baja California; also eastward to southwestern Utah.
Occurrence in Monument. Known from a single specimen found at Lower Covington Flat, Apr. 8 *.

Grassland and open woodland or brushland in arid and semiarid regions appear to be favored by this snake, but its haunts are not well known.

The individual found at Lower Covington Flat was 20 yards from the crest of a gentle southeast-facing slope under a dead branch of a juniper which lay at the base of a partly dead juniper bush. There were clumps of bunch grass and numerous large boulders scattered about, and at the crest of the slope was a jumble of huge boulders. The locality was in the piñon belt, some of the trees reaching a height of over 20 feet. The soil temperature at the collection site was 68° F. at 7:45 A.M. The area had been in the sun for some time.

This is a secretive snake, spending much time under objects, but it is occasionally found crawling in the open at night. Surface activity seems to be stimulated by rainfall. In the daytime it is usually found under logs, rocks, and boards. Although usually found singly or in pairs, as many as six individuals have been found together.

The flattened head and the tendency to burrow into duff suggest that these snakes spend much time crawling about through loose soil, leaf litter, and rocky crevices. They feed on insects, centipedes, and probably spiders, often ingesting particles of soil with their food. Captives have been fed earthworms.

Two female black-headed snakes from the Kingston Mountains of eastern California each contained a single large egg.

The Covington Flat specimen is difficult to classify racially in terms of current taxonomy. In ventral scute count it resembles the coastal *Tantilla eiseni eiseni,* but its pigmentation is more like *T. e. transmontana* of desert slopes (Klauber, 1943). The specimen is a male with 163 ventrals. However, it differs from both races in failing to have the black of the head cap carried back and below the angle of the mouth. On the basis of its low ventral count and head color, it might be classified as the form *utahensis,* a rare snake known from the west slope of the southern Sierra and isolated localities in the desert to the east. However, two specimens collected in 1946 near McFarland, in the San Joaquin Valley, within the range of *T. e. eiseni,* have only vague mottlings of black behind the mouth angle, thereby closely approaching the condition in the Covington Flat animal. They reduce the differences between *eiseni* and *utahensis* and lead us to propose that *utahensis* and *eiseni* be regarded as the same species. Since the nature of variation in *T. eiseni* is so poorly known, there is little point in indicating racial classification for our specimen.

The only generalization one might hazard at this time is that in drier areas these snakes tend to have a lighter ground color, less vivid white neckband, and less intense black cap. The number of ventrals could be a function of temperatures prevailing at egg sites, the higher counts occurring in warmer areas, in line with the observations of Fox (1948) on garter snakes. Our specimen has the pale coloration of the arid lands type, but it was found in an elevated, relatively cool area. This might account for its low ventral count. Only 18 miles south-southwest, along the Whitewater Wash, 2500 feet lower, three specimens have been taken which are typical *transmontana*. Their ventral counts are ♂ 179, ♂ 190, ♀ 193.

Another desert slope specimen, found in 1956 in Tehachapi Pass 5 miles northwest of Mojave, Kern County, California, is nearly identical in all diagnostic external characteristics, including ventral count (♂ 163) and pigmentation, with our animal from Covington Flat.

RATTLESNAKES

Western Rattlesnake. *Crotalus viridis*

Description. Adult $1\frac{1}{4}$ to 5 feet in total length. A rattlesnake with much variation in ground color—cream, gray, pink, greenish, brown, to black—the color tending to match background; usually dark in area of Monument. Usually more than 2 internasal scales present, a characteristic peculiar to this species among western rattlesnakes. Row of large brown or black blotches down middle of back, with smaller ones on the sides; dark stripe extends behind eye and is usually bordered above and below with pale color; dark and light rings on tail merge with pattern on the body, there being no abrupt change between body and tail markings.
Range. Southern Canada to central Baja California, northern Sonora, Chihuahua, and Coahuila; extends from Pacific coast to western Great Plains; absent from high elevations in the Rocky Mountains, from extreme desert, and Pacific northwest.
Occurrence in Monument. Loomis and Stephens (1962) record the species from: Lower Covington Flat, Apr. 11, 30, May 23; Jumbo Rocks, 11 mi. S Twentynine Palms, Aug. 5.

On October 12, 1953, Samuel King, then Park Superintendent, told us he had seen "two large black rattlesnakes" near Jumbo Rocks about three weeks earlier. They had crawled into a clump of Mohave yucca, in the center of which was a wood rat's nest. We visited the site on October 15 and excavated the nest but found no snakes. Vegetation consisted of Joshua trees, Mohave yucca, catclaw, scrub oaks, junipers, piñons, and creosote bush, among massive domes of granite. Mr. King reported that "black rattlesnakes" had also been seen at Belle Camp ground. Vegetation there is similar to that at Jumbo Rocks.

The western rattlesnake is a highly adaptable species that lives from sea level to over 10,000 feet in a great variety of habitats—grassland, brushland, woodland, and coniferous forest. Although apparently it is a scarce snake in the Monument, its occurrence there is not surprising in view of its ecological tolerance. The habitat in the Monument resembles that occupied on the desert slopes of the mountains of southern California and is no more arid than many parts of the Great Basin inhabitated by these snakes.

Western rattlers eat ground squirrels, cottontail rabbits, kangaroo rats, pocket gophers, meadow mice, white-footed mice, pocket mice, birds, lizards, and amphibians. There is some evidence that young snakes are more likely to feed on lizards than are the adults.

According to Loomis and Stephens (1962), their specimens are of the race *Crotalus viridis helleri,* a dark-colored form that ranges in coastal California from near Point Concepcion into Baja California. The populations in the Monument probably maintain some contact with those of the San Bernardino Mountains to the west, since suitable habitat is more or less continuous.

Mohave Rattlesnake. *Crotalus scutulatus*

Description. Adult 2–4 feet in total length. A greenish or olive-colored rattlesnake, usually with distinct diamond- or hexagon-shaped marks on back; diamonds edged with plain whitish scales, usually unmarked with dark color; tail ringed with black and whitish, the black rings narrower than the white ones and darkest near the rattle; a light stripe extends posteriorly from eye to above angle of mouth. Enlarged scales present between the supraoculars.

Range. Southern Nevada, extreme southwestern Utah, and the Mohave Desert of California southeastward through Arizona, south of the central plateau, into southwestern New Mexico, western Texas, and south in México to Veracruz. Apparently absent from deserts of California south of the Monument.

Occurrence in Monument. Loomis reports (oral communication) specimens from Black Rock Spring and near west entrance to Monument southeast of Joshua Tree.

This rattlesnake ranges from near sea level to nearly 8000 feet. It occurs chiefly in arid and semiarid environments of bunch grass, creosote bush, mesquite, and other xeric growth.

Although chiefly nocturnal, it is occasionally seen abroad in the daytime. It feeds on rodents (*Dipodomys*) and probably on lizards.

Fig. 138. Mohave rattlesnake. Photograph taken by Robert G. Crippen of an animal from the Mohave Desert.

Western Diamondback Rattlesnake. *Crotalus atrox*

Description. Adult $2\frac{1}{2}$–$5\frac{1}{2}$ feet in total length, rarely to over 7 feet. Above gray, grayish brown, buff, or pinkish; diamond- or hexagon-shaped brownish dorsal blotches with dark punctations concentrated at their edges, giving them a more or less continuous dark border; blotches bordered with whitish or beige, but scales not entirely light-colored as in Mohave rattlesnake; tail marked with black and white rings of about equal width, hence the common name "coon-tail rattler;" diagonal whitish stripe extends from eye to border of mouth. Supraocular scales separated by small scales.

Range. Extreme southern Nevada, northern New Mexico, and extreme southern Kansas south to Veracruz; extends from desert slope of mountains of southern California to central Arkansas and Texas.
Occurrence in Monument. Loomis (oral communication) obtained specimens from Cottonwood Pass and Pinto Wash Well, approximately 2000 feet elevation.

In general this snake inhabits prairies, desert flats, and foothills, ranging from sea level to 7000 feet in elevation. In the southwestern part of its range it seems to prefer brushy desert. It has been found in sandy areas grown to mesquite, as well as on rocky hillsides, and it is especially attracted to the rank growth of river bottoms. It has been found only in the lower eastern parts of the Monument.

The western diamondback rattlesnake is chiefly nocturnal but is occasionally found abroad in the daytime. It tends to hold its ground when disturbed, and its boldness, its large size, the quantity and potency of its venom, and its widespread distribution make it the most dangerous western rattler.

This snake feeds on rabbits, rats, mice, squirrels, birds, and lizards.

Speckled Rattlesnake. *Crotalus mitchelli*

Description. Adult 2–4 feet in total length. A blotched rattlesnake, usually with a speckled appearance. Dorsal blotches may be hexagonal, rectangular, hourglass, or diamond-shaped, their outlines often irregular; dark brown or black-tipped scales scattered over dorsal surface; ground color salmon, buff, pale gray, or brown; 3 to 9 rings on the tail, darker than body markings. Prenasal scales, in local race, usually separated from the rostral by small scales or granules; upper preoculars often divided horizontally or vertically.
Range. Southern Nevada and extreme southwestern Utah south to Cape district of Baja California; coastal southern California and desert base of the Sierra Nevada east to central Arizona.
Occurrence in Monument. Present in rocky habitats throughout. Recorded from: Black Rock Spring, Aug. 29 *; Lower Covington Flat, Aug. 23 *; Quail Spring, July 5 *; Stubby Spring, Sept. 7 *; Lost Horse Valley, May 18 *; Key's View, July 7 *, 16; Fortynine Palms Canyon, May 7 *; $\frac{1}{4}$ mi. N Pinyon Wells, Oct. 9 *; 1 mi. NE Eagle Mountain, May 14 *; Cottonwood Spring, Apr. 29, May 7 *, 9; $2\frac{1}{2}$ mi. ESE Cottonwood Spring, Apr. 29. In addition Loomis and Stephens (1962) report it from $3\frac{1}{2}$ mi. S Twentynine Palms, Cholla Garden, and Pinto Basin.

This rattlesnake is primarily a rock dweller. It frequents rocky slopes, canyons, outcrops, and old lava flows, and seldom ventures out onto the plains. Although it is associated geographically with several other rattlers, some of which occasionally enter rocky habitats, its close association with rocks tends to reduce its competition with these other species. In the Monument, for example, it is not likely to be found in the sandy washes and dunes inhabited by the sidewinder (*Crotalus cerastes*) or on the plains to the east where the western diamondback rattler occurs.

Although over most of its range this snake lives in desert mountains, in southern California it enters the chaparral and oak woodland of the coastal mountains. In parts of the Monument it lives in vegetation of comparable

Fig. 139. Speckled rattlesnake. Photograph taken by Robert G. Crippen of an animal from the Santa Ana Mountains, Orange County, California.

growth form, characterized by scrub oak, manzanita, piñon, and juniper. In the Panamint Mountains near Death Valley, it ranges to 7000 feet in the piñon–juniper belt.

Speckled rattlesnakes are commonly nocturnal, particularly in summer, but are not exclusively so. In the Monument we have found them abroad by day on a number of occasions. At 4:00 P.M. near Stubby Spring one was stretched out full length on the sand of the canyon bottom. The snake's body temperature was 79° F., the air 74° F. At other places we have found it in the vicinity of springs in bright sunlight and basking among rocks in the morning and late afternoon.

This species is a nervous snake, usually quick to defend itself or to seek refuge. Its rough scales and speckled pattern usually harmonize with its surroundings, and it can easily be overlooked if it does not rattle. We have often been impressed by the remarkable camouflage of individuals found on decomposed granite and among the granite mounds in the Monument.

Speckled rattlesnakes feed on kangaroo rats (*Dipodomys*), mice (*Peromyscus, Perognathus*), ground squirrels, lizards (*Uta, Cnemidophorus, Eumeces*), and birds. That they may occasionally scavenge is suggested by the fact that one was found curled around a dead *Peromyscus* caught in a snap trap near Quail Spring.

Several observations make it clear that this rattlesnake may ambush birds when they come to the ground to drink. Near a shallow pool in Smithwater Canyon one of these rattlers held a Traill flycatcher in its jaws (see p. 122).

At Cottonwood Spring Loye Miller stooped to pick up a freshly killed Swainson thrush at the spring. A slight movement in the weeds nearby caught his eye, and he drew back quickly. A speckled rattler, which had just killed the bird, lay within a short distance of the water. The snake retreated without rattling.

Klauber (1936) reports that eight birds (presumably goldfinches) were removed from the stomach of a speckled rattler killed near an aqueduct construction camp where a lawn, garden, and birdbath had been installed on the desert, 10 miles north of Desert Center, Riverside County. Birds had been attracted from great distances, and evidently the snake had lain in wait for them.

Three to 8 young may compose a litter, as determined from counts of large ova found in gravid females.

The race in the Monument is *Crotalus mitchelli pyrrhus.* Intergradation with *C. m. stephensi* evidently occurs north of the Monument, somewhat north of a line formed by the Mojave–Barstow–Needles road that runs approximately east and west across the central portion of the Mohave Desert.

Sidewinder. *Crotalus cerastes*

Other name. Horned rattlesnake.
Description. Adult $1\frac{1}{2}$ to 2 feet in total length. A pale gray to tan rattlesnake that crawls sideways. Pointed supraocular scales projecting as small "horns." Row of brown blotches along middle of back and smaller blotches on sides; dark postocular stripe; tail marked with conspicuous dark and light bands; ventral surface white.
Range. Southern Nevada and southwestern Utah south to northeastern Baja California and northwestern Sonora; extends from desert slope of mountains of California to south-central Arizona.
Occurrence in Monument. Widespread below 4000 feet; probably rare or absent in piñon belt. Recorded from: Indian Cove, May 12 *, 14; Twentynine Palms, July 13 *; 7 mi. E Twentynine Palms, Apr. 13; Pinto Basin, Apr. 10 *, 13 *, July 10 *; Virginia Dale Mine, Apr. 23 *, July 12, 14; 3 mi. N Cottonwood Spring, July 10 *. In addition, reported in vicinity of Belle Camp (3800 feet) and Cholla Garden by Loomis and Stephens (1962).

Sidewinders are most frequently found in areas of fine sand deposited by wind or water. In the Monument, they frequent chiefly aeolian deposits such as those south of Dale Dry Lake and in Pinto Basin, and washes at the bases of the mountains. Although seldom found far from loose sand, they do occasionally occur on hardpan or among scattered rocks. In sandy areas there is usually sufficient compacted soil to make possible maintenance of burrows by kangaroo rats, pocket mice, and other rodents upon which they feed.

Some of the plants found in areas frequented by sidewinders in the Monument are galleta grass, creosote bush, smoke tree, desert willow, Mohave yucca, and Joshua tree.

This snake is strictly a desert inhabitant and displays a number of structural and behavioral attributes related to life in dry sand. In adapting to the sand niche, they have reduced competition with other snakes but their

Fig. 140. Sidewinder or horned rattlesnake. Photograph of an animal found in Pinto Basin on April 11, 1960.

problems of locomotion and concealment have been intensified. Sidewinding appears to be an extension of a tendency already present in most snakes. Many species will sidewind briefly when placed on a slippery or unstable surface. This form of locomotion minimizes slippage on loose material, as is evident from the tracks of sidewinders, which are a series of J-shaped marks often containing impressions of the belly scutes. The hook of the J, made by the head and neck, indicates direction of movement.

In the daytime, when sidewinders are generally inactive, concealment is achieved by color resemblance to the background, retreat into rodent burrows, or partial burial in sand (fig. 141). In burying, these snakes typically dig a shallow pit with outward movements of loops of the body, coming to rest in a watch-spring–like coil, with the dorsal surface flush with the ground. The resting site is often in the shelter of a bush, protected from direct sunlight and where prevailing winds will not excavate the snake or bury it in sand. A brief search in sandy areas, particularly in the vicinity of rodent burrows, will usually reveal the presence of abandoned resting pits. They are saucer-shaped depressions about an inch deep.

Most foraging occurs at night, probably chiefly in the early evening. Active sidewinders have been recorded at body temperatures of 90°, 75°, and 67° F. and at air temperatures as low as 61° F. In addition to the rodents mentioned, these snakes also feed on lizards and, rarely, on birds. The frequency of sidewinder resting pits near openings of rodent burrows suggests that these snakes may deliberately bed down near burrow openings in order to ambush emerging rodents at night.

Fig. 141. A sidewinder emerging from its hiding place in the sand in the shelter of a bush in Pinto Basin, April 11, 1960.

Mating may occur shortly after emergence from hibernation and, as is true of most snakes, males apparently find females by their odor. In mid-April at Indian Wells Canyon, Kern County, California, Dr. Ray Smith found many sidewinder tracks converging on his campsite, where two female sidewinders had been found. One of the females was placed in a gallon jar. Later two males, found in the same area, were put separately into the jar with her. Her receptivity was shown by the fact that she copulated with both of them, within the close quarters of the jar.

On another occasion in mid-April, at Hopkins Well, Riverside County, California, Dr. John McSwain had forced a sidewinder, presumably a female, onto a ground cloth to secure a firm neck hold in capturing her. Later the cloth was put out on the ground with a Coleman lantern in the center to attract insects. Despite the bright light, about 11:00 P.M., a sidewinder was found on the cloth, evidently attracted by the odor. This was the first hot spell after prolonged cool weather.

Klauber (1944:102) indicates that the Riverside–San Bernardino line is the approximate boundary in California between the sidewinder races *Crotalus cerastes cerastes* of the Mohave Plateau and *C. c. laterorepens* of the low Colorado Desert basin to the south. The highlands of the southern part of the Monument mark the edge of the plateau. They include the Little San Bernardino, Cottonwood, and Eagle mountains, which project eastward, peninsula-like, from the massive San Bernardino Mountain system. This chain of mountains forms a barrier to the dispersal of sidewinders but is broken at several points by passes. Intergradation may occur in these passes and adjacent areas. Morongo and Cottonwood passes, and Pinto Wash, in the extreme eastern part, may be such dispersal routes. Farther east the two races are evidently broadly in contact, since favorable habitat is nearly continuous.

In attempting to classify specimens taken in the Monument, we encounter difficulties because of the overlapping characteristics used in separating races and the scarcity of material. Using the characteristics given in Klauber's key (1944:119), however, we can tentatively allocate our specimens as follows: Two adult females from the Virginia Dale area and 10 miles west of Yucca Valley, San Bernardino County, have a brown rattle base like *cerastes,* and *cerastes*-like scalation. Scale characters of a young individual from Twentynine Palms, however, resemble those of *laterorepens.* An adult female from Indian Cove has the brown rattle base of *cerastes* and scale characters close to *laterorepens.* One of two adult males from Pinto Basin, which tend toward *cerastes* in scalation, has a black rattle base like *laterorepens.* A juvenile from 3 miles north of Cottonwood Springs seems to share characteristics of both subspecies. Thus it appears probable that with larger samples, the Joshua Tree Monument area would be found to lie in an area of intergradation between these two races of sidewinders. It is expected that the zone of intergradation will be narrow in the vicinity of the Monument highlands but widen to the east, where the Mojave and Colorado deserts merge gradually.

LITERATURE CITED

American Ornithologists' Union
1957 Check-list of North American birds. 5th ed. Baltimore, Md., published by the Union, xiv + 691 pp.
Anderson, A. H., and A. Anderson
1957 Life history of the cactus wren. Part I. Winter and pre-nesting behavior. Condor, 59:274–296.
1959 Life history of the cactus wren. Part II. The beginning of nesting. Condor, 61: 186–205.
Banks, R. C.
1964 Geographic variation in the white-crowned sparrow, *Zonotrichia leucophrys*. Univ. Calif. Publ. Zool., 70:1–123.
Banta, B. H.
1961 Herbivorous feeding of *Phrynosoma platyrhinos* in southern Nevada. Herpetologica, 17:136–137.
Bartholomew, G. A., and T. J. Cade
1956 Water consumption of house finches. Condor, 58:406–412.
1958 Effects of sodium chloride on the water consumption of house finches. Physiol. Zool., 31:304–310.
Bartholomew, G. A., and W. R. Dawson
1954 Body temperature and water requirements in the mourning dove, *Zenaidura macroura marginella*. Ecology, 35:181–187.
Bartholomew, G. A., T. R. Howell, and T. J. Cade
1957 Torpidity in the white-throated swift, Anna hummingbird, and poor-will. Condor, 59:145–155.
Bartholomew, G. A., and J. W. Hudson
1959 Effects of sodium chloride on weight and drinking in the antelope ground squirrel. Jour. Mamm., 40:354–360.
Behle, W. H.
1948 Systematic comment on some geographically variable birds occurring in Utah. Condor, 50:71–80.
1950 Clines in the yellow-throats of western North America. Condor, 52:193–219.
1956 A systematic review of the mountain chickadee. Condor, 58:51–70.
Bell, E. L.
1954 A preliminary report on the subspecies of the western fence lizard, *Sceloporus occidentalis*, and its relationships to the eastern fence lizard, *Sceloporus undulatus*. Herpetologica, 10:31–36.

Benson, S. B.
1954 Records of the spotted bat (*Euderma maculata*) from California and Utah. Jour. Mamm., 35:117.
Bogert, C. M.
1945 Two additional races of the patch-nosed snake, *Salvadora hexalepis*. Amer. Mus. Novit., No. 1285:1–14.
Brattstrom, B. H., and J. W. Warren
1953 A new subspecies of the racer, *Masticophis flagellum*, from the San Joaquin Valley of California. Herpetologica, 9:177–179.
Brauner, J.
1952 Reactions of poor-wills to light and temperature. Condor, 54:152–159.
Brodkorb, P.
1936 Geographical variation in the piñon jay. Occas. Papers Mus. Zool., Univ. Mich., No. 332:1–3.
Bryant, M. D.
1945 Phylogeny of nearctic Sciuridae. Amer. Midl. Nat., 33:257–390.
Buechner, H. K.
1960 The bighorn sheep in the United States, its past, present, and future. Wildlife Soc. Monogr. No. 4, 174 pp.
Buxton, P. A.
1923 Animal life in deserts, a study of the fauna in relation to the environment. London, Edward Arnold, xvi + 176 pp.
Carpenter, C. C.
1961 Patterns of social behavior in the desert iguana, *Dipsosaurus dorsalis*. Copeia: 396–405.
Carter, F.
1937 Bird life at Twentynine Palms. Condor, 39:210–219.
Chattin, J. E.
1941 The distribution of pocket gophers in southeastern California. Trans. San Diego Soc. Nat. Hist., 9:265–284.
Chew, R. M.
1961 Water metabolism of desert-inhabiting vertebrates. Biol. Rev., 36:1–31.
Chew, R. M., and B. B. Butterworth
1959 Growth and development of Merriam's kangaroo rat, *Dipodomys merriami*. Growth, 23:75–95.
Cowan, I. McT.
1936 Distribution and variation in deer (genus *Odocoileus*) of the Pacific coastal region of North America. Calif. Fish and Game, 22:155–246.
Cowles, R. B.
1936 The relation of birds to seed dispersal of the desert mistletoe. Madroño, 3:352–356.
1946 Carrion eating by a snake. Herpetologica, 3:121–122.
Cowles, R. B., and C. M. Bogert
1944 A preliminary study of the thermal requirements of desert reptiles. Bull. Amer. Mus. Nat. Hist., 83:265–296.
Davis, J.
1957 Comparative foraging behavior of the spotted and brown towhees. Auk, 74:129–166.
Davis, J., and L. Williams
1957 Irruptions of the Clark nutcracker in California. Condor, 59:297–307.
Dawson, W. R.
1954 Temperature regulation and water requirements of the brown and Abert towhees, *Pipilo fuscus* and *Pipilo aberti*. Univ. Calif. Publ. Zool., 59:81–124.
1955 The relation of oxygen consumption to temperature in desert rodents. Jour. Mamm., 36:543–553.

Dice, L. R.
1945 Minimum intensities of illumination under which owls can find dead prey by sight. Amer. Nat., 79:385–416.

Fish, W. A.
1950 Nesting record of the vermilion flycatcher in the northern Mohave Desert. Condor, 52:137.

Fitch, H. S.
1956 An ecological study of the collared lizard (*Crotaphytus collaris*). Univ. Kans. Publ. Mus. Nat. Hist., 8:213–274.

Fox, W.
1948 Effect of temperature on development of scutellation in the garter snake, *Thamnophis elegans atratus*. Copeia:252–262.

Griffin, D. H.
1941 The sensory basis of obstacle avoidance by flying bats. Jour. Exper. Zool., 86: 481–506.

Grinnell, J.
1914 An account of the mammals and birds of the lower Colorado Valley with especial reference to the distributional problems presented. Univ. Calif. Publ. Zool., 12: 51–294.
1933 Review of the recent mammal fauna of California. Univ. Calif. Publ. Zool., 40: 71–234.
1936 Up-hill planters. Condor, 38:80–82.

Grinnell, J., J. Dixon, and J. M. Linsdale
1937 Fur-bearing mammals of California. Vol. 2. Berkeley, University of California Press, pp. i–xiv + 377–777.

Grinnell, J., and A. H. Miller
1944 The distribution of the birds of California. Pac. Coast Avif. No. 27, 608 pp.

Grinnell, J., and H. S. Swarth
1913 An account of the birds and mammals of the San Jacinto area of southern California. Univ. Calif. Publ. Zool., 10:197–406.

Gullion, G. W.
1956 Let's go desert quail hunting. Nevada Fish and Game Comm., Biol. Bull. No. 2, 76 pp.
1960 The ecology of Gambel's quail in Nevada and the arid southwest. Ecology, 41: 518–536.

Hachisuka, M.
1928 Variations among birds (chiefly game birds), heterochrosis, gynandromorphs, aberration, mutation, atavism, and hybrids. Ornith. Soc. Japan, Supp. Publ. No. XII, x + 108 pp.

Hall, E. R.
1946 Mammals of Nevada. Berkeley and Los Angeles, University of California Press, xii + 710 pp.

Hall, E. R., and F. H. Dale
1939 Geographic races of the kangaroo rat, *Dipodomys microps*. Occas. Papers Mus. Zool. Louisiana State Univ., No. 4:47–63.

Hall, E. R., and K. R. Kelson
1959 The mammals of North America. 2 vols. New York, Ronald Press, pp. i–xxx + 1–546 + index; i–vii + 547–1083 + index.

Heller, E.
1901 Notes on some little-known birds of southern California. Condor, 3:100.

Hoffmeister, D. F.
1951 A taxonomic and evolutionary study of the piñon mouse, *Peromyscus truei*. Illinois Biol. Monogr., Vol. 16, No. 4, x + 104 pp.

Hooper, E. T.
1938 Geographical variation in wood rats of the species *Neotoma fuscipes*. Univ. Calif. Publ. Zool., 42:213–246.

Howard, H., and A. H. Miller
1933 Bird remains from cave deposits in New Mexico. Condor, 35:15–18.
Howell, A. B., and I. Gersh
1935 Conservation of water by the rodent *Dipodomys*. Jour. Mamm., 16:1–9.
Howell, T. R., and G. A. Bartholomew
1959 Further experiments on torpidity in the poor-will. Condor, 61:180–185.
Hudson, J. W.
1962 The role of water in the biology of the antelope ground squirrel *Citellus leucurus*. Univ. Calif. Publ. Zool., 64:1–56.
Huey, L. M.
1938 A new form of *Perognathus formosus* from the Mohave Desert region of California. Trans. San Diego Soc. Nat. Hist., 9:35–36.
Jaeger, E. C.
1947 The vermilion flycatcher in the central Mohave Desert. Condor, 49:213.
1948 Does the poor-will hibernate? Condor, 50:45–46.
1949 Further observations on the hibernation of the poor-will. Condor, 51:105–109.
Johnson, D. H.
1943 Systematic review of the chipmunks (genus *Eutamias*) of California. Univ. Calif. Publ. Zool., 48:63–148.
Johnson, D. H., M. D. Bryant, and A. H. Miller
1948 Vertebrate animals of the Providence Mountains area of California. Univ. Calif. Publ. Zool., 48:221–376.
Johnson, N. K.
1963 Biosystematics of sibling species of flycatchers in the *Empidonax hammondii-oberholseri-wrightii* complex. Univ. Calif. Publ. Zool., 66:79–238.
Klauber, L. M.
1933 Notes on *Lichanura*. Copeia:214–215.
1936 *Crotalus mitchellii,* the speckled rattlesnake. Trans. San Diego Soc. Nat. Hist., 8:149–184.
1940 The lyre snakes (genus *Trimorphodon*) of the United States. Trans. San Diego Soc. Nat. Hist., 9:163–194.
1941 The long-nosed snakes of the genus *Rhinocheilus*. Trans. San Diego Soc. Nat. Hist., 9:289–332.
1943 A desert subspecies of the snake *Tantilla eiseni*. Trans. San Diego Soc. Nat. Hist., 10:71–74.
1944 The sidewinder, *Crotalus cerastes,* with description of a new subspecies. Trans. San Diego Soc. Nat. Hist., 10:91–126.
1945 The geckos of the genus *Coleonyx* with descriptions of new subspecies. Trans. San Diego Soc. Nat. Hist., 10:133–216.
1946 The glossy snake, *Arizona,* with descriptions of new subspecies. Trans. San Diego Soc. Nat. Hist., 10:311–398.
1947 Classification and ranges of the gopher snakes of the genus *Pituophis* in the western United States. Bull. Zool. Soc. San Diego, 22:1–83.
1951 The shovel-nosed snake, *Chionactis,* with descriptions of two new subspecies. Trans. San Diego Soc. Nat. Hist., 11:141–204.
Leopold, A. S.
1939 Age determination in quail. Jour. Wildlife Manag., 3:261–265.
1961 The desert. New York, Time, Inc., 192 pp.
Lidicker, W. Z.
1960 An analysis of intraspecific variation in the kangaroo rat *Dipodomys merriami*. Univ. Calif. Publ. Zool., 67:125–218.
Lindeborg, R. G.
1952 Water requirements of certain rodents from xeric and mesic habitats. Contr. Lab. Vert. Biol. Univ. Michigan No. 58, 32 pp.

Livingston, B. E., and F. Shreve
1921 The distribution of vegetation in the United States, as related to climatic conditions. Carnegie Inst. Wash. Publ. No. 284, xvi + 590 pp.
Longhurst, W. M., A. S. Leopold, and R. F. Dasmann
1952 A survey of California deer herds, their ranges and management problems. Calif. Dept. Fish and Game, Game Bull. No. 6, 136 pp.
Loomis, R. B., and R. C. Stephens
1962 Records of snakes from Joshua Tree National Monument, California. Bull. Southern Calif. Acad. Sci., 61:29–36.
McLean, D. D.
1930 The burro deer in California. Calif. Fish and Game, 16:119–120.
Marshall, J. T., Jr.
1955 Hibernation in captive goatsuckers. Condor, 57:129–134.
Mayhew, W. W.
1961 Photoperiodic response of female fringe-toed lizards. Science, 134:2104–2105.
Michener, J. R., and H. Michener
1951 Notes on banding records and plumages of the black-headed grosbeak. Condor, 53:93–96.
Miller, A. H.
1928 The molts of the loggerhead shrike *Lanius ludovicianus* Linnaeus. Univ. Calif. Publ. Zool., 30:393–417.
1931 Systematic revision and natural history of the American shrikes (*Lanius*). Univ. Calif. Publ. Zool., 38:11–242.
1933 Postjuvenal molt and the appearance of sexual characters of plumage in *Phainopepla nitens*. Univ. Calif. Publ. Zool., 38:425–446.
1936 Tribulations of thorn-dwellers. Condor, 38:218–219.
1941 A review of centers of differentiation for birds in the western Great Basin region. Condor, 43:257–267.
1946 Endemic birds of the Little San Bernardino Mountains, California. Condor, 48: 75–79.
1947 Arizona race of acorn woodpecker vagrant in California. Condor, 49:171.
1955 Concepts and problems of avian systematics in relation to evolutionary processes. *In* A. Wolfson, ed., Recent studies in avian biology. Urbana, University of Illinois Press, pp. 1–22.
Miller, A. H., and L. Miller
1951 Geographic variation of the screech owls of the deserts of western North America. Condor, 53:161–177.
Miller, L.
1932*a* Notes on the desert tortoise (*Testudo agassizii*). Trans. San Diego Soc. Nat. Hist., 7:187–208.
1932*b* The saw-whet owl in the desert. Condor, 34:258.
1946 The elf owl moves west. Condor, 48:284–285.
Miller, M. R.
1951 Some aspects of the life history of the yucca night lizard, *Xantusia vigilis*. Copeia: 114–120.
Murie, M.
1961 Metabolic characteristics of mountain, desert, and coastal populations of *Peromyscus*. Ecology, 42:723–740.
Norris, K. S.
1953 The ecology of the desert iguana *Dipsosaurus dorsalis*. Ecology, 34:265–287.
Phelan, R. L., and B. H. Brattstrom
1955 Geographic variation in *Sceloporus magister*. Herpetologica, 11:1–14.
Phillips, A. R.
1947 The races of MacGillivray's warbler. Auk, 64:296–300.

Pierce, W. M.
1921 The Bendire thrasher nesting in California. Condor, 23:34.
Pitelka, F. A.
1945 Pterylography, molt, and age determination of American jays of the genus *Aphelocoma*. Condor, 47:229–260.
1951 Speciation and ecological distribution in American jays of the genus *Aphelocoma*. Univ. Calif. Publ. Zool., 50:195–464.
Russell, W. C.
1947 The brown thrasher in California. Condor, 49:131.
Salt, G. W.
1952 The relation of metabolism to climate and distribution in three finches of the genus *Carpodacus*. Ecol. Monogr., 22:121–152.
Sargent, G. T.
1940 Observations on the behavior of color-banded California thrashers. Condor, 42: 49–60.
Schmidt-Nielsen, K., and B. Schmidt-Nielsen
1952 Water metabolism of desert mammals. Physiol. Rev., 32:135–166.
1953 The desert rat, ... Sci. Amer., 189, July:73–78.
Schmidt-Nielsen, K., B. Schmidt-Nielsen, A. Brokaw, and H. Schneiderman
1948 Water conservation in desert rodents. Jour. Cell. Comp. Physiol., 32:331–360.
Snyder, L. L., and H. G. Lumsden
1951 Variation in *Anas cyanoptera*. Occas. Papers Roy. Ont. Mus., No. 10:1–17.
Stebbins, R. C.
1944 Some aspects of the ecology of the iguanid genus *Uma*. Ecol. Monogr., 14:311–332.
1954 Amphibians and reptiles of western North America. New York, Toronto, London, McGraw-Hill Book Company, Inc., xxii + 528 pp.
Swarth, H. S.
1912 Report on a collection of birds and mammals from Vancouver Island. Univ. Calif. Publ. Zool., 10:1–124.
Taber, R. D., and R. F. Dasmann
1958 The black-tailed deer of the chaparral, its life history and management in the north coast range of California. Calif. Dept. Fish and Game, Game Bull. No. 8, 163 pp.
Turner, F. B.
1959 Some features of the ecology of *Bufo punctatus* in Death Valley, California. Ecology, 40:175–181.
Van Rossem, A. J.
1942 Four new woodpeckers from the western United States and Mexico. Condor, 44:22–26.
1945 A distributional survey of the birds of Sonora, Mexico. Occas. Papers, Mus. Zool. Louisiana State Univ., No. 21, 242 pp.
Voge, M.
1952 *Mesogyna hepatica* N. G., N. Sp. (Cestoda: Cyclophyllida) from the kitfox, *Vulpes macrotis*. Trans. Amer. Micros. Soc., 71:350–354.
Vorhies, C. T., and W. P. Taylor
1933 The life histories and ecology of jack rabbits, *Lepus alleni* and *Lepus californicus* ssp., in relation to grazing in Arizona. Univ. Ariz. Agr. Exp. Sta. Tech. Bull. 49: 471–587.
1940 Life history and ecology of the white-throated wood rat, *Neotoma albigula albigula* Hartley, in relation to grazing in Arizona. Univ. Ariz. Agr. Exp. Sta. Tech. Bull. 86:455–529.
Walls, G. L.
1942 The vertebrate eye. Bloomfield Hills, Mich., Cranbrook, xiv + 785 pp.
Went, F. W.
1948 Ecology of desert plants. I. Observation on germination in the Joshua Tree National Monument, California. Ecology, 29:242–253.

Williams, L.
1938 Birds observed in the vicinity of Twentynine Palms, California. Condor, 40:258.
Williamson, F. S. L.
1956 The molt and testis cycles of the Anna hummingbird. Condor, 58:342–366.
Willett, G.
1933 A revised list of the birds of southwestern California. Pac. Coast Avif. No. 21, 204 pp.
Woodbury, A. M., and R. Hardy
1948 Studies of the desert tortoise, *Gopherus agassizii*. Ecol. Monogr., 18:145–200.

INDEX